P9-DVX-735

BOTTLE OF LIES

ALSO BY KATHERINE EBAN

Dangerous Doses: A True Story of Cops, Counterfeiters, and the Contamination of America's Drug Supply

BOTTLE OF LIES

The Inside Story of the Generic Drug Boom

KATHERINE EBAN

ecco

An Imprint of HarperCollins*Publishers*

 For information, address HarperCollins Publishers, 195 Broadway, New York, NY 10007.

HarperCollins books may be purchased for educational, business, or sales promotional use. For information, please email the Special Markets Department at SPsales@harpercollins.com.

FIRST EDITION

Designed by Michelle Crowe

Library of Congress Cataloging-in-Publication Data has been applied for.

ISBN 978-0-06-233878-5

19 20 21 22 23 LSC 10 9 8 7 6 5 4 3 2 1

For my mother, Elinor Fuchs, and my father, Michael Finkelstein,
the first and best writers and editors in my life

CONTENTS

PART III: A CAT-AND-MOUSE BUSINESS

PART IV: MAKING A CASE

PART V: DETECTIVES IN THE DARK

PART VI: THE WATERSHED

PART VII: RECKONINGS

AUTHOR'S NOTE

This book grew out of a puzzle I couldn't solve.

In the spring of 2008, Joe Graedon, a host of the National Public Radio program *The People's Pharmacy*, contacted me. In my decade reporting on the drug industry, I'd been a guest on his radio program a number of times. But this time he wanted my help. Patients had been calling and writing in to his program with serious complaints about generic drugs that either didn't work or caused devastating side effects. The drugs, though made by different manufacturers and for a range of conditions from depression to heart disease, were all generic—less expensive versions of brand-name drugs, made legally after the patents on those medications lapsed.

Graedon had forwarded the patients' complaints to top officials at the Food and Drug Administration (FDA), but they insisted that generics were equivalent to the brand and the patients' reactions were subjective. Their response struck Graedon as more defensive than scientific. Generic drugs had become essential to balancing budgets across America. Without them, every large-scale government health program—the Affordable Care Act, Medicare Part D, the Veterans Health Administration, charitable programs for Africa and the developing world—would be unaffordable. Graedon himself had long advocated making generic drugs more widely available. But the complaints were compelling and similar in nature. He felt that something significant was wrong with the drugs, but didn't know what. He wanted someone with "investigative firepower" to look into the patients' claims.

For years I'd reported on pharmaceuticals and public health as an investigative journalist. I had broken stories about brand-name drug companies, including efforts by opioid makers to increase sales by concealing addiction risks. In my first book, *Dangerous Doses,* I exposed how a shadowy gray market allowed pharmaceutical wholesalers to sell and resell our drugs, a practice that obscured their origin and opened the door to drug counterfeiters. To the extent that I'd thought about generic drugs, I knew they made up more than 60 percent of our drug supply (90 percent today) and were an essential counterweight to the ever-rising prices of brand-name drugs.

I began exactly where Graedon pointed me: with the patients. In June 2009, I published an article in *Self* magazine that documented how patients who had previously been stabilized on brand-name drugs relapsed when switched to certain generics. Their doctors had little data and no significant comparative studies to explain these reactions. Although the FDA reviewed data from the generic drug companies and inspected manufacturing plants, it was not systematically testing the drugs. As Dr. Nada Stotland, a psychiatrist in Chicago and then the president of the American Psychiatric Association, told me, "The FDA is satisfied that generics are okay. My question is, are we satisfied?"

Even as I worked on the article, I recognized its limits. Proving that patients were harmed might tell me that something was wrong with the drugs. But what? And even if I proved there was something wrong, why was that so? The answer to those questions almost certainly lay in the laboratories, manufacturing plants, and corporate boardrooms of the drug-making companies, many of which operated overseas. Roughly 40 percent of our generic drugs are manufactured in India. A full 80 percent of the active ingredients in all our drugs, whether brand-name or generic, are made in India and China. As one drug-ingredient importer told me, "Without products from overseas, not a single drug could be made."

Ultimately, my effort to answer a single question—what is wrong with the drugs?—launched me into a decade-long reporting odyssey on four continents as I delved into how globalization had impacted the drugs we need to survive. In India, I sought out reluctant whistle-blowers, visited manufacturing plants, and interviewed government officials. In China, as I endeavored to meet with sources, the government followed me, hacked my cell phone, and sent a photo of a security official sitting in my hotel lobby, holding up an English-language newspaper, to my phone's home screen. It was an unmistakable warning: *We are watching you.* In a bar in Mexico City, a whistleblower slipped me stacks of internal communications from a generic drug company's manufacturing plant. Doctors and scientists in Ghana met with me in hospitals and laboratories. In a manufacturing plant in Cork, Ireland, I watched as one of America's most popular drugs, Lipitor, was manufactured.

I followed the trail of certain drugs around the world, trying to connect the dots. What had patients complained of? What had FDA investigators found? What actions had regulators taken? What had companies claimed? What had CEOs decided? What had criminal investigators turned up? I mined thousands of internal company documents, law enforcement records, FDA inspection records, and internal FDA communications, stacks of which piled up in my office.

My reporting led me into a maze of global deception. In 2013, I published a 10,000-word article on *Fortune* magazine's U.S. website about fraud at India's largest generic drug company. It detailed how the company had deceived regulators around the world by submitting fraudulent data that made its drugs appear bioequivalent to brand-name drugs. That article, however, left me with unanswered questions. Was that company an outlier or the tip of the iceberg? Did its conduct reflect a one-off scandal or an industry norm?

In my reporting, some important sources helped me answer that question. A generic drug executive contacted me anonymously under the pseudonym "4 Dollar Refill." He explained that there was a gulf between what the regulations required of generic drug companies and how those companies behaved. To minimize costs and maximize profit, companies circumvented regulations and resorted to fraud: manipulating tests to achieve positive results and concealing or altering data to cover their tracks. By making the drugs cheaply without the required safeguards and then selling them into regulated and more costly Western markets, claiming that they had followed all the necessary regulations, companies could reap enormous profits.

An FDA consultant who had spent considerable time in overseas factories also contacted me. She was an expert at examining the cultural "data points," or situational forces, that drive corporate behavior. One factor is company culture—the tone set by executives, the admonitions or slogans that hang on office or manufacturing-plant walls, the training that workers receive. If a company's culture permits small lapses in safety regulations, catastrophic failure is all the more likely. As a pharmaceutical manufacturing executive put it, "When I get on a plane and there are cup stains on the tray table, you wonder if they're taking care of the engine."

But company culture is also affected by country culture, the FDA consultant explained: Is a society hierarchical or collaborative? Does it encourage dissent or demand deference to authority? These factors, though seemingly unrelated, can impact manufacturing quality and lead to variance between certain generics and brand-name drugs, as well as between generics that are supposedly interchangeable with one another, the consultant posited.

Before I embarked on this project, I had always assumed that a drug is a drug—that Lipitor, for example, or a generic version, is the same for any market in the world. And since generic drugs are required to be bioequivalent to the brand and to create a similar effect in the body, I assumed that there isn't necessarily variation between different generic versions. I was wrong. Cheaply made pharmaceuticals hold up no better than cheap clothes or cheap electronics that are made swiftly in overseas sweatshops. Drugs reach the consumer "at a very low dollar price," said the FDA consultant, "but perhaps at the expense of other principles for which it is difficult to measure a dollar value."

Consumers understand that cheddar isn't simply cheese, said the consultant. There's "artisanal cheddar, Cabot cheddar, Velveeta, or a plastic block painted to look like cheese." Unwittingly, patients face similar choices about quality every time they go to a drugstore. Yet because they have no idea such distinctions exist, they have no way to request one drug that is better than the next. Patients implicitly trust the FDA to ensure the high quality of their medicine. As a result, most patients do more due diligence when they change their cell-phone provider or buy a car, yet will walk into CVS "and not think for one second about what they put in their mouth that could kill them," a lawyer who represented a pharmaceutical whistleblower told me.

We are dependent on distant drug manufacturers, yet have little visibility into their methods. In the factories where my reporting led, the FDA's investigators are an infrequent presence and pressure for profits is intense. The result is a facade of compliance that papers over a darker reality. "It's like it was at the turn of the twentieth century," a Dutch pharmaceutical executive, who encountered a frog infestation at a manufacturing plant in China, told me. "It's like *The Jungle,*" he said, referring to the book by Upton Sinclair that exposed gruesome conditions in America's meatpacking plants.

There is no disputing the benefit of well-made generic drugs. When generics work perfectly, and many do, the results can be miraculous. "Basically, the ability of India and other countries to produce generic medicine at a fraction of the cost of the patented drugs saved the lives of millions of people in developing countries," said Emi MacLean, formerly the U.S. director of the Access to Essential Medicines Campaign at Doctors Without Borders. The plunge in prices has also made medicine affordable and treatment possible for millions of Americans who have no alternative to generics without significant price regulation of brand-name drugs.

Generic drugs are essential to our health care system, and their quality is critical to us all. Nonetheless, in my effort to answer the question that Joe Graedon posed ten years ago—what is wrong with the drugs?—I uncovered the labyrinthine story of how the world's greatest public health innovation also became one of its greatest swindles.

Katherine Eban
Brooklyn, New York
March 2019

ABOUT THE REPORTING

This book, including all the scenes, dialogue, and assertions in it, is based on extensive interviews, firsthand reporting, and documentation. I interviewed over 240 people, a number of them multiple times, including regulators, drug investigators, criminal investigators, diplomats, prosecutors, scientists, lawyers, public-health experts, doctors, patients, company executives, consultants, and whistleblowers. Primary reporting for this book took place between January 2014 and November 2018, and included on-the-ground reporting trips to India, China, Ghana, England, Ireland, and Mexico and travel throughout the United States. The book also includes material I gathered from 2008 to 2013, while reporting a series of articles about generic drugs in both *Self* and *Fortune* magazines.

In every scene with dialogue, I have reconstructed quotes from the recollections of participants as well as documentation, including meeting minutes, handwritten notes, and memoranda of interviews by criminal investigators. The quotes I use from emails and other documents are verbatim, and I have not corrected spelling errors. No names of characters have been changed.

In the course of reporting, I obtained a significant number of confidential documents. These include roughly 20,000 internal documents from the U.S. Food and Drug Administration, including emails, memorandum, meeting minutes, reports, and data; thousands of internal government records related to the investigation of the generic drug company Ranbaxy; and thousands of internal corporate records from several generic drug companies, including emails, reports, strategy documents, correspondence, and sealed court records.

Documentation also came from sixteen Freedom of Information Act requests that I filed with the FDA, as well as from a lawsuit I filed to obtain calendar and meeting records for an FDA official. I also read through years of publicly available FDA inspection records.

Wherever an individual or company has chosen to respond to questions or allegations, relevant portions of their statements can be found incorporated into the book's endnotes or main text. The endnotes are intended to guide readers to publicly available resources and documentation or to offer more detail on certain topic areas. They do not contain citations for nonpublic material, such as private emails, sealed court records, or other confidential documents.

Funding for this book came only from impartial sources with no stake in the outcome of the events described. These include an advance from HarperCollins and grants from the Carnegie Corporation, the Alfred P. Sloan Foundation, the McGraw Center for Business Journalism at Craig Newmark Graduate School of Journalism, and the George Polk Foundation.

IMPORTANT PEOPLE AND PLACES

Affiliations listed are those held at the approximate time characters appear in the book. Dates have been included for titles held by multiple characters. The names of some FDA divisions have since changed due to government reorganization.

Drug Companies

Ranbaxy

MANAGING DIRECTORS

Arun Sawhney, CEO and managing director, 2010–2015
Atul Sobti, CEO and managing director, 2009–2010
Malvinder Singh, CEO and managing director, 2006–2009
Shivinder, brother
Brian Tempest, CEO and managing director, 2004–2005
Davinder Singh "D.S." Brar, CEO and managing director, 1999–2004
Parvinder Singh, chairman and managing director, 1992–1998; joint managing director, 1976–1991, with his father, Bhai Mohan
Bhai Mohan Singh, chairman and joint managing director, 1976–1991; chairman and managing director, 1961–1975

RESEARCH AND DEVELOPMENT

Rajinder "Raj" Kumar, director, 2004–2005

Rashmi Barbhaiya, director, 2002–2004

Rajiv Malik, head of formulation development and regulatory affairs

Arun Kumar, associate director of regulatory affairs

Dinesh Thakur, director and global head of research information & portfolio management

Sonal Thakur, wife

Andrew Beato, attorney at Stein, Mitchell, Muse & Cipollone LLP

U.S. BUSINESS

Jay Deshmukh, senior vice president of global intellectual property

Abha Pant, vice president of regulatory affairs

OUTSIDE LAWYERS AND CONSULTANTS

Kate Beardsley, partner, Buc & Beardsley

Christopher Mead, partner, London & Mead

Warren Hamel, partner, Venable LLP

Agnes Varis, consultant

Cipla Limited

Yusuf "Yuku" Khwaja Hamied, chairman and managing director

Khwaja Abdul "K.A." Hamied, founder

Daiichi Sankyo Company

Tsutomu Une, head of global strategy

Mylan N.V.

EXECUTIVES

Heather Bresch, CEO
Rajiv Malik, president
Deborah Autor, senior vice president, head of global strategic quality

Indian Government

Central Drugs Standard Control Organization

Gyanendra Nath "G.N." Singh, drug controller general

Ministry of Health and Family Welfare

Harsh Vardhan, minister

U.S. Government

Congress

David Nelson, senior investigator, House Committee on Energy and Commerce

Food and Drug Administration

OFFICE OF THE COMMISSIONER

Scott Gottlieb, commissioner, 2017–present
Margaret Hamburg, commissioner, 2009–2015

OFFICE OF THE CHIEF COUNSEL

Marci Norton, senior counsel
Steven Tave, associate chief counsel for enforcement

CENTER FOR DRUG EVALUATION AND RESEARCH

Janet Woodcock, director

Robert Temple, deputy center director for clinical science

OFFICE OF COMPLIANCE

Deborah Autor, director

Thomas Cosgrove, director, Office of Manufacturing Quality

Carmelo Rosa, director, Division of International Drug Quality

Edwin Rivera-Martinez, chief, International Compliance Branch

Douglas Campbell, compliance officer

Karen Takahashi, compliance officer

OFFICE OF PHARMACEUTICAL SCIENCE

Office of Generic Drugs

Gary Buehler, director

OFFICE OF GLOBAL REGULATORY OPERATIONS & POLICY

Office of International Programs

FDA India Office

Altaf Lal, director

Atul Agrawal, supervisory consumer safety officer

Muralidhara "Mike" Gavini, senior assistant country director

Peter Baker, assistant country director

Regina Brown, international program and policy analyst for drugs

Office of Regulatory Affairs

Dedicated Drug Team

Jose Hernandez, investigator

Office of Criminal Investigations

Debbie Robertson, special agent

Department of Justice

OFFICE OF CONSUMER LITIGATION

Linda Marks, senior litigation counsel

U.S. ATTORNEY'S OFFICE, DISTRICT OF MARYLAND

Stuart Berman, assistant U.S. attorney

Doctors and Patient Advocates

Joe Graedon, cohost of the NPR program *The People's Pharmacy*

William F. Haddad, generic drug advocate

Harry Lever, director, Hypertrophic Cardiomyopathy Center, Cleveland Clinic

Randall Starling, head, Section of Heart Failure and Cardiac Transplant Medicine, Cleveland Clinic

Manufacturing Plants

Fresenius Kabi

Kalyani, Nadia District, West Bengal, eastern India

Mylan

Morgantown, West Virginia, southeastern United States

Nashik, Nashik District, Maharashtra, western India

Pfizer

Dalian, Liaodong Peninsula, Liaoning Province, northeastern China

Ringaskiddy, County Cork, southern Ireland

Zhejiang Hisun (affiliate), Taizhou, Zhejiang, eastern China

Ranbaxy

Dewas, Dewas District, Madhya Pradesh, central India

Mohali, SAS Nagar District, Punjab, northern India

Ohm Laboratories, New Brunswick, New Jersey, northeastern United States

Paonta Sahib, Sirmour District, Himachal Pradesh, northern India

Toansa, Nawanshahar District, Punjab, northern India

Wockhardt

Chikalthana, Aurangabad District, Maharashtra, western India

Waluj, Aurangabad District, Maharashtra, western India

BOTTLE OF LIES

PROLOGUE

MARCH 18, 2013
Waluj
Aurangabad, India

Peter Baker, a drug investigator for the U.S. Food and Drug Administration, traveled two hundred miles east of Mumbai, along a highway choked by truck traffic and down a road with meandering cows, to get to his assignment. Behind a metal fence lay a massive biotech park, run by the Indian generic drug company Wockhardt Ltd. Amid the dozens of buildings, Baker's job was to inspect a particular area of the plant—Plot H-14/2—to ensure that it could safely make a sterile injectable drug used by American cancer patients.

Baker, thirty-three, had arrived lightly provisioned. He had just a few items in his backpack: a camera, a gel-ink pen, a green U.S. government–issued notebook, and his FDA identification. He had a graduate degree in analytical chemistry and a command of the Code of Federal Regulations, Title 21, the rules that governed drug manufacturing. But more importantly, he had his instincts: a strong sense of what to check and where to look, after completing eighty-one inspections over four and a half years at the FDA.

At 9:00 a.m., the sun already burning, Baker and his colleague, an FDA microbiologist, showed their identification to guards at the gate and were ushered into the plant, where the vice president of manufacturing and other company officials waited anxiously to greet them. In a world of drab auditors toiling with checklists, Baker stood out. He was handsome and energetic. He wore his brownish-blond hair in a buzz cut. On one bicep, he sported the oversized tattooed initials of his motorcycle group. As the officials began their opening presentation, he interrupted with a staccato burst of questions. Was there any other manufacturing area on-site that made sterile drugs for the U.S. market, aside from Plot H-14/2? he asked repeatedly. No, the officials assured him.

Baker's job—part science and part detective work—had been transformed by the forces of globalization. From 2001 to 2008, the number of drug products imported into the United States had doubled. By 2005, the FDA had more drug plants to inspect abroad than it did within U.S. borders. Baker had been dispatched to Wockhardt, in an industrial area of Aurangabad, because of a global deal that had evolved over more than a decade. Drug makers in India and other countries gained entry to the U.S. pharmaceutical market, the world's largest and most profitable. In return, the American public got access to affordable versions of lifesaving drugs. But this boon came with a serious caveat: foreign drug manufacturers had to comply with the intensive U.S. regulations known as "current good manufacturing practices" (cGMP) and submit to regular inspections. If everything went according to plan, the result was a win-win for foreign drug makers and American consumers alike.

Though few Americans knew Wockhardt by name, many took its medicine. The company manufactured about 110 different generic drug products for the American market, including a beta blocker—

metoprolol succinate—to treat hypertension, which reached about a quarter of U.S. patients taking a generic version of the drug. Because the Aurangabad plant manufactured sterile injectable medicine, the regulations it had to follow were particularly strict.

Every detail mattered. Every digit of data had to be preserved in its original form. As one moved closer to the plant's sterile core, where vials of medicine sat exposed, the rules became even more restrictive. Employees there had to move slowly and deliberately, so as not to disturb the unidirectional air flow. Even to take notes, FDA investigators had to use sterile, lint-free paper. There was a good reason for these rules. One small slip—a failure to filter air properly, a misreading of bacterial samples, the exposed wrist of a technician—could result in a contaminated product that would kill instead of cure.

Given the high stakes—lives on one side, and profits on the other—fear governed the inspection. Baker feared that he would miss something that would endanger the lives of U.S. patients. Wockhardt officials feared that he would find something that would restrict the company's access to the U.S. market. They needed every advantage to survive the FDA's inspection. Here, Wockhardt had several things stacked in its favor. The plant was massive, roughly the size of a small city. Baker and his colleague had just one week to inspect the site. With only five working days, how much could they find?

But Wockhardt had an even bigger advantage. Company executives had known for weeks that Baker was coming to inspect their factory. In the United States, FDA investigators simply showed up unannounced and stayed as long as was needed. But for overseas inspections—due to the complex logistics of getting visas and ensuring access to the plant—the FDA had chosen a different approach: to announce its inspections in advance. As was typical, Wockhardt had "invited" the FDA to inspect and the agency had accepted. Plant officials served as hosts and Baker was their guest—albeit one whose arrival they dreaded.

Given weeks of lead time, the officials worked feverishly to prepare for Baker's arrival. They polished floors, cleaned equipment, and combed through files to rid them of anomalies. They warned their employees to remain polite but silent, and to let their supervisors answer questions. They had fixed everything in every place where investigators were likely to look, a drill they'd endured fifteen months earlier when a different inspection team had arrived from the FDA.

On that visit, the investigators had found some troubling shortcomings: live bugs in a water storage tank, flooring in disrepair, ineffective cleaning procedures. But the investigators had recommended, not demanded, that the plant make corrections. In the FDA's coding system, they had given the plant a passing grade, one known as "Voluntary Action Indicated" (VAI). This meant that the Wockhardt operation had survived the inspection with no restriction to its most lucrative franchise—the sale of drugs to the United States.

This time, though company officials had planned for an inspection, they had not planned for Peter Baker. Unlike so many other FDA investigators, he was hard to prepare for—and control. He wouldn't tolerate an opening slide show or a guided tour, which were typical ways for plant officials to run down the clock. He seemed to be everywhere at once. He studied the employees for signs of evasion as he questioned them repeatedly. Company officials quickly ascertained that his visit posed a serious threat, one that would require drastic action on their part if the plant was to emerge unscathed.

On the second day of his inspection, Baker and his colleague entered a hallway far from the sensitive areas of the facility. It was a place where he could let down his guard. But as he looked down the long, gleaming corridor, he noticed a man at the far end who was walking toward him just a little too quickly. The man, a plant employee, had a furtive demeanor. In one hand, he carried a clear garbage bag full of papers and assorted refuse, making his hurried walk seem even stranger. As the man glanced up, he noticed Baker and momentarily froze. The two men locked eyes.

Abruptly, the man pivoted and returned the way he had come. Baker followed him, quickening his pace. The employee sped up, too, until the two men were engaged in a low-speed chase beneath the fluorescent lights.

"Stop!" yelled Baker's colleague, the microbiologist. The man broke into an open run. As he bolted, the investigators gave chase until the employee flung open a side door, careened from the hallway, and hurled the bag onto a pile of garbage in a dim storage area beneath a stairwell, then scrambled up a flight of stairs and vanished into the building's concrete maze.

Baker, close behind, retrieved the bag. Inside, he discovered roughly seventy-five manufacturing records for the company's insulin products. They had been hastily torn in half, but he was able to piece some together. As he did, his concern grew. They revealed that many of the vials contained black particles, potentially deadly contaminants, and had failed visual inspection.

Under good manufacturing practices, every record created at the plant had to be made available to regulators. But these documents were marked "for internal dept. use only." Baker suspected that the records were secret for a reason. The testing results were so bad that, had they been disclosed, the plant would have had to launch a costly internal investigation and likely reject every batch it produced.

Over the next three days, Baker demanded that Wockhardt officials open their computers, as was his right, and he began scouring records. One by one, he uncovered the company's deceptions. As he suspected, the records from the garbage bag had not been logged into the company's official system. The drugs flagged in the records had been released to patients in India and the Middle East. Baker discovered that the drugs had been manufactured in a secret formulation area that the FDA had never known about or inspected. Once he arrived there, he learned that Wockhardt had used the same defective equipment, in the same secret area, to make medicine for the U.S. market—including the injectable drug adenosine, which treated irregular heartbeat.

The result was a disaster for Wockhardt. Two months after Baker's inspection, the FDA restricted the import of drugs from the Waluj plant into the United States, a potential $100 million loss in sales for the company. The next day, Wockhardt's CEO held an emergency conference call with anxious investors to assure them that the company would bring the plant into compliance "in a month or two months maximum."

At a glance, the plant appeared perfectly run, the equipment shiny and new, its procedures meticulous and compliant. But the torn records Baker had uncovered led him beneath the plant's impeccable surface and into a labyrinth of lies, where nothing was what it appeared to be. The records were false. Drugs were manufactured in a secret area. Some of them contained visible contaminants that endangered patients. Baker, who had pieced all this together over the course of five punishing days, was left to wonder: if so much inside this plant was fake, what, if anything, was real?

PART I

SHIFTING GROUND

CHAPTER I

THE MAN WHO SAW FURTHER

LATE FALL 2001

Hopewell, New Jersey

Dinesh S. Thakur was fastidious. He wore perfectly ironed khakis, a white button-down shirt, a dark sports jacket, and well-polished loafers. Stocky and of medium height, he had a round face, full head of dark hair, and deep-set eyes that gave him a doleful appearance. On this chilly afternoon, the leaves just beginning to turn gold and crimson, the thirty-three-year-old information scientist set out across the grassy slope toward the man-made lake. It was a favorite destination on the Bristol-Myers Squibb campus, where employees went to clear their heads or escape the highly regimented corporate culture, if only for their lunch hour.

But today Thakur had come at the behest of an older and more senior colleague, who'd invited him for a walk to discuss an unspecified opportunity.

Bristol-Myers Squibb's research and development center sat on a manicured campus, just beyond a network of leafy residential streets and imposing stone homes. Inside the gated guard posts, low-slung concrete buildings with dark windows dotted the hillsides. The few trees were planted at regular intervals. The lush grass around the lake had been mowed with such precision that it looked like a striped carpet. Every hundred feet stood an emergency assistance pole, to summon aid if needed. Cars were kept to fifteen miles per hour. Even the lake's turtles had a demarcated crossing lane.

The orderly grounds reflected the painstaking research that went on there. Scientists from this campus developed drugs that had entered the worldwide lexicon, from Pravachol for high cholesterol to Plavix to prevent blood clots. Decades earlier, what was then Squibb had developed an antibiotic to treat tuberculosis, for which its scientists won the prestigious Lasker Award. Bristol-Myers had forged new ground in cancer research. In 1989, the two companies merged. Nine years later, Bristol-Myers Squibb was awarded the National Medal of Technology and Innovation at a White House ceremony.

Thakur played a small but cutting-edge role in the company's endeavors. He ran a department that built robots, automated laboratory helpmates intended to make the work of drug testing more efficient and reliable. Thakur's lab buzzed with innovation. More than a dozen scientists reported to him. Pulleys, motors, bells, and levers were scattered about, and bright-eyed college students cycled through, pitching in as needed. Thakur set his own hours, which were long and sometimes involved staying overnight to watch the robots. They needed to repeat the same tasks faultlessly, with the goal of eliminating human error from the laboratory.

The results often did not turn out as desired, which was standard for a manufacturing scale-up. On these occasions, Thakur and his team were forced to scrap their work and start again. Yet they felt confident that the company viewed these failures as a normal part of the scientific process. When it came to Thakur's lab activity, the old advertising slogan from Squibb seemed to still hold sway: "The priceless ingredient in every product is the honor and integrity of its maker."

The work, with its scrupulous attention to detail, suited Thakur's temperament. He was promoted steadily with strong performance reviews, one of which noted that he was "very logical, ethical and loyal" in dealing with peers and superiors. Over six years, he had steadily ascended to his hard-won title: Director, Discovery Informatics.

Punctual as ever, he made his way to the walking path that looped around the lake, where his older colleague, Rashmi Barbhaiya, was waiting. Heavy-set, with snowy white hair and dark circles beneath his eyes, Barbhaiya had been developing drugs at BMS for twenty-one years. He had an intimidating aura and the smooth manner of a senior executive. By contrast, Thakur was reserved and somewhat awkward, with little gift for small talk. But this had not hindered him at BMS, where few people understood his robotics work or sought to discuss it.

Both men were originally from India. Two years prior, Thakur had built an automated computer program for Barbhaiya's group. More recently, as Barbhaiya oversaw BMS's purchase of a small pharmaceutical company, he'd tapped Thakur to help transfer and reconcile the data. Today Barbhaiya was about to propose an opportunity Thakur did not predict.

As they walked along the footpath, Barbhaiya disclosed that he was leaving BMS—and the United States—to become the research and development director of India's largest drug company, Ranbaxy Laboratories, which made generic medicine. Thakur was surprised. Barbhaiya had spent his whole career climbing to the upper ranks of one of the world's top pharmaceutical research companies. At BMS, he had lived and breathed the prestige of creating new molecules. He'd become a recognized expert in whittling down the long odds that any drug maker undertakes when setting out to develop a new cure.

But Barbhaiya was planning to leave it all behind. To go from the brand-name sector in the United States to the generic one in

India. By name, it was the same work—pharmaceutical research—but it was a seismic identity shift. The BMSes of the world invented. The Ranbaxys of the world duplicated. BMS did innovative science versus Ranbaxy's copycat engineering. But the more Barbhaiya explained his decision, the less skeptical Thakur became.

In India, Ranbaxy was legendary, and the family that built it, the Singhs, were hailed as corporate royalty. As one of India's oldest and most successful multinationals, Ranbaxy had reinvented the perceived capabilities of an Indian corporation. In 2001, it was on track to clock $1 billion in global sales, with its U.S. sales reaching $100 million after only three years in the American market. The FDA had already approved over a dozen of its drug applications. Ranbaxy had offices and manufacturing plants around the world, including in the United States, but was headquartered in India. Looking to the future, Ranbaxy was going to be investing heavily in innovative research. The company was aiming to develop new molecules. Barbhaiya would be building the company's research capacity, almost from the ground up. "Why don't you come with me?" he proposed. "You'll be closer to your parents and doing something for the country."

It was an offer that, on the face of it, made little sense. BMS had paid for Thakur's ongoing schooling, a master's program in computer engineering. He'd received years of in-house training on the best manufacturing and laboratory practices. But like Barbhaiya, Thakur knew that the ground was shifting beneath his feet. The generic drug business was booming around the world. Generic drugs—legally produced copies of brand-name drugs—comprised half of the U.S. drug supply, a number that was steadily growing. The patents that protected dozens of best-selling drugs, from Lipitor to Plavix, would expire within the next decade, meaning that generics companies would be able to manufacture and sell copies approved by the Food and Drug Administration (FDA). With the demand for generics growing, all their jobs would be reconfigured soon enough. One of the main drivers behind this shift was India itself, which was fast becoming a global player in the pharmaceutical industry.

As Thakur contemplated the pros and cons of Barbhaiya's offer, he had a further thought. The goal in the brand-name world was to make the best possible drugs for the highest possible price. It was the heyday of the branded drug industry, with companies reaping billions in profits on the success of big-name drugs. The largesse at BMS reflected this. Office Christmas parties included caviar and champagne. Sometimes Thakur caught an empty seat on the corporate helicopter that shuttled executives between the company's hubs in Princeton, New Jersey, and Wallingford, Connecticut, marveling at the easy commute for those at higher pay grades.

In the generic world, the culture would be different because the goal was different: to make the best cures affordable and available to all. But it would mean leaving the United States, where he'd spent decades focused on building the best possible life he could.

Thakur had first come to know America through its movies. In college, as an engineering student in Hyderabad, he had gone to see classic films such as *Citizen Kane* and *Gone with the Wind*.

In college, he took the GRE, applied to graduate programs in the United States, and got a scholarship to the University of New Hampshire, where he lived in the graduate dorm as one of only a few minority students. He had never been out of India before, had never seen snow. In his new home, he marveled at the beauty of the White Mountains, the serenity of old New England towns, each with its own church and town square. He drove to Acadia National Park whenever he could and loved biking its rocky shoreline. Otherwise, he studied almost continuously, producing a doorstop of a master's thesis, which he later published in a journal under the title "Soluble and Immobilized Catalase: Effect of Pressure and Inhibition on Kinetics and Deactivation."

Shortly after graduating, he was hired by a small biotechnology company to help automate its laboratories. There, though a picture of Thakur and his robots made it into the company's annual report, his unsupportive boss told him that he lacked the requisite talent for the job. So he moved on to BMS, where he continued the same work successfully.

As he climbed the corporate ladder, his mother grew worried that he had not yet married. Through a family connection, his parents visited the parents of a young woman named Sonal Kalchuri, who was fun loving and well educated and had long dark hair and almond eyes. Thakur met her on a trip to Mumbai, and the two began a phone relationship and correspondence over the following eight months.

In most ways, they were opposites. He was compulsively organized. She was laid-back. He was a workaholic who "never let his hair down," as she put it. She was social and loved parties. But they shared an interest in science. Sonal was just completing her undergraduate degree in engineering. And they both loved to sing. His childhood home had been filled with music. Both his parents sang in the classical Hindustan style. Over the years, Thakur had developed an excellent voice and a love of the genre's sinuous improvisations. He and Sonal would go on to perform together in classical Hindustani bands.

They married in 1995 and had a traditional days-long wedding, with both of them draped in flowers. Thakur wore the customary turban for grooms. Sonal was wreathed in gold jewelry, and her hands were hennaed with intricate patterns. She loved the event, but Thakur found the socializing taxing. Afterward, the couple made a home in Syracuse, and he returned to work. For Sonal, the transition was painful. The twenty-three-year-old had never been away from her family before. Now she was alone in a house in a foreign country.

Nonetheless, she enrolled in a computer-engineering graduate program at Syracuse University and emerged with a master's degree. She got an excellent job at the Carrier Corporation as a software engineer. Thakur moved up steadily at BMS. In 1999, he was promoted to an associate director position. This involved a move from the Syracuse office to the research institute in Hopewell, New Jersey, just a few miles from the company's Princeton offices. The couple found a spacious home with a high-ceilinged family room that appealed particularly to Sonal. They were getting ready to start their own family.

Their son, Ishan, was born a week after the 9/11 attacks. The Princeton area was devastated. Typically, the parking lot at the Princeton Junction train station filled with the cars of workers who commuted an hour to their Manhattan jobs every weekday, then emptied every night. But after 9/11, cars remained, waiting for commuters who never returned from work.

Though Ishan was born amid tragedy, he brought unalloyed joy into the Thakurs' lives. Sonal's mother came to stay for eight months. And Thakur's parents came to visit, too, for the first time since he'd left for graduate school in the United States eleven years earlier. It was during these hectic months that Barbhaiya proposed to Thakur that he return to India.

Thakur did not immediately share the offer with Sonal. He continued to think about it quietly, as his work progressed at BMS. The family moved again, to Belle Mead, New Jersey, which had better schools and was closer to Sonal's work. Thakur continued his ongoing schooling, a master's program in computer engineering, for which BMS was paying. And in-house training in the best manufacturing and laboratory practices also continued. To leave all of this for an Indian generics company seemed like a big step down.

But Thakur was getting restless at BMS and knew that he'd probably risen as far as he could, with little opportunity—at least in the short run—for further advancement. During a summer vacation in 2002, he went to India and stopped by Ranbaxy's research and development center in Gurgaon. He was impressed by the company's bustle and sense of potential. He'd have far more freedom and authority there. The offer was excellent. To his surprise, Sonal also became interested. She missed her family and wanted to return home. They resolved to give it a try.

Thakur set about recruiting several members of his BMS team. It struck his colleague Venkat Swaminathan, a software engineer, as an exciting opportunity. If Ranbaxy was really looking to develop new medications, it could be a welcome change from BMS's restrictive bureaucracy. Dinesh Kasthuril, too, was intrigued. Though he loved his current job and was halfway through Wharton business school, also on BMS's dime, he was impressed that Ranbaxy wanted to try to develop new drugs. And though they were all born in India, none of them had ever worked there before. All three wanted to contribute to their native country's emergence onto the world stage. "A lot of it was from the heart," Kasthuril recalled.

Their similar views further bolstered Sonal's confidence about the move. She felt that her young family would have friendship and support. The three colleagues thought of themselves as setting off on a momentous adventure: to help build an Indian company dedicated to research, a Pfizer for the twenty-first century. Even as Kasthuril's boss at BMS tried to convince him not to leave, he had to acknowledge that Dinesh Thakur was "able to see things further" than most people could.

Three months before their departure for India, Thakur achieved a long-awaited milestone: he became an American citizen, a fact that he noted proudly atop his curriculum vitae. But by then, he and his colleagues had set their course.

CHAPTER 2

THE GOLD RUSH

AUGUST 17, 2002

New Delhi, India

On a humid day one year before Dinesh Thakur arrived at Ranbaxy, a company executive boarded a plane at the Indira Gandhi International Airport, bound for Newark, New Jersey. He'd left the office in a "crazy rush," an employee recalled, to catch the almost sixteen-hour flight.

His mission was top-secret. In his luggage were five binders, each about three inches thick, containing reams of data. The documents comprised key portions of what would become an Abbreviated New Drug Application or ANDA, to be filed with the FDA. The application, once completed, would become known in industry parlance as a "jacket."

But this was no run-of-the-mill jacket. The executive carried the most potentially lucrative dossier ever compiled in the generic drug world: the data the company would use in its application to make the first U.S. generic version of the world's best-selling drug of all time: Lipitor. Pfizer's vaunted cholesterol fighter was "the Sultan of Statins," as Wall Street analysts called it. The molecule itself, atorvastatin calcium, was a descendant of Nobel Prize–winning science. Coupled with Pfizer's marketing might, it had become the world's first $10-billion-a-year drug.

Had the Ranbaxy executive's mission been known, a good portion of people in the United States—from patient advocates to members of Congress to the 11 million Americans who relied on Lipitor to lower their cholesterol—would have welcomed him. Everyone in America, it seemed, wanted cheap equivalent drugs. State and federal budgets were buckling under astronomical drug costs. Brand-name Lipitor, though less expensive than its competitors, cost the many uninsured Americans who depended on it roughly $800 a year. Even for some with insurance, the copays alone were a reach.

In theory, the contents of Ranbaxy's binders could solve that. The data showed that Ranbaxy's version reached roughly the same level of absorption in the bloodstream as Pfizer's and used the same active ingredient, the atorvastatin calcium molecule. If all the claims in its application were true, Ranbaxy's version of the drug would be a godsend for American patients.

At Newark airport, the sun had just risen when a waiting car whisked the man to 600 College Road East in Princeton, New Jersey, Ranbaxy's U.S. corporate headquarters. There the regulatory team—headed by Abha Pant, an intense company loyalist and the only woman to have climbed into Ranbaxy's executive ranks—immediately got to work, combining the core documents from the five binders with other necessary paperwork.

By that night, the final submission was ready. It spanned seventeen volumes and totaled more than 7,500 pages. The jacket covered four different dosage strengths that Ranbaxy planned to make and package at its plant in Paonta Sahib, in the northern Indian state of Himachal Pradesh. The application was handed off to an overnight courier and arrived at the FDA's Rockville, Maryland, campus the next morning, where it was stamped "RECEIVED: August 19, 2002."

But neither Pant nor her colleagues were satisfied, because they had no idea if they'd filed first, which was what mattered most. The first company to file its application, if approved, won the exclusive right to sell the generic for six months, before others joined in. There was a rumor that the generic drug company Teva had already filed. And word had it that the generics companies Sandoz, Mylan, and Barr had also been doing clinical tests. Days and weeks went by with ominous silence.

Inside the FDA, Ranbaxy's application—the cornerstone of the company's plans to reach $1 billion in U.S. sales by 2015—became Abbreviated New Drug Application 76-477. Meanwhile, Ranbaxy executives waited.

Jeffrey Myers, Pfizer's senior patent attorney, was in his office at the company's headquarters on East Forty-Second Street in Midtown Manhattan when he got the notice—a generic drug company had filed an application to make generic Lipitor. That application contained a full-on challenge of the Lipitor patent, known as a "paragraph IV certification." Myers sat up straighter in his chair. At that point, Lipitor had been on the market for five years, and its patents were not set to expire until 2011.

Myers learned of patent challenges all the time, but this one drew his attention. "We didn't have any forewarning," he recalled.

Of course, Myers knew this day would come. But he was expecting the challenge to come from a well-established generics company, like Mylan or Sandoz. He regularly had lunch with his counterparts from those companies. This was the first challenge he'd ever gotten from an Indian company, one he barely knew. To him, the move was about as legitimate as a pirate scaling the side of his ocean liner.

As he scrutinized the fine print of Ranbaxy's challenge, he began to see problems. The drug had to be in the same dosage form.

Lipitor was sold in tablets. Ranbaxy had filed for capsules, as though its chemists had never seen the original drug. It also proposed to make it in a different molecular form, amorphous rather than crystalline—which was a trick, as Myers well knew, since Pfizer scientists had tried for years to make an amorphous version but had failed because the drug became highly unstable.

Lipitor could not be replicated with ease. It had taken a team of scientists to formulate, the industry's best marketers to launch, and a manufacturing team that understood its intricacies and challenges. Since 1998, all of the active ingredient for the world's supply of Lipitor had been made in Cork County, Ireland, at three vast Pfizer-owned plants. Pfizer had expected production to top out with 50 metric tons of active ingredient. Just five years after the drug had launched, that number had quadrupled to 200 tons.

The Ringaskiddy manufacturing plant, which sits on a 200-acre campus and operates twenty-four hours a day, has a "quality culture," which means that it aims to operate with as close to zero defects as possible. Its employees are routinely trained to GUARD PFIZER QUALITY, as a sign on the wall of one company plant admonished them.

Lipitor is as moody as the slate-gray landscape, but Ringaskiddy has developed a failproof manufacturing system. "The drug is finicky, and we know how to make it," said Dr. Paul Duffy, Pfizer's vice president of biopharma manufacturing operations. "When you work with something for twenty years, it's like your baby, you know its moods."

In New York, Myers—a lawyer with a PhD in chemistry from Cornell University—suspected that Ranbaxy's chemists were outclassed in the face of a drug they barely understood and probably couldn't even make. Given that, he felt an inkling of excitement for the battle that lay ahead. "I live to obliterate these guys," he later reckoned of his generic drug opponents. "My job is to stop them."

How you saw Ranbaxy depended on where you sat. Myers's view

from Pfizer's Midtown Manhattan headquarters was, "once you get down to Ranbaxy, you start to swim with the bottom-feeders." But in many ways, it was an upstart's market. The triumphalism of the branded drug industry was being eroded from beneath by a surging generic drug industry that had both public and political support. Once Ranbaxy's bid to make Lipitor became public, a CNN business reporter assessed it as "a classic David versus Goliath scenario—Pfizer's revenues are about 50 times the size of its diminutive challenger."

Prior to 1984, the Ranbaxys of the world had no way to challenge the Pfizers. There was no clear pathway for a generic drug to be approved in the United States. Under FDA rules, even if a drug's patent had expired, generic drug companies were required to repeat extensive and costly clinical trials, even though the brand companies had already proven the safety and effectiveness of their drugs.

A crusading journalist at the time, William F. Haddad, who relished his role as an underdog, set out to change that. According to one of his colleagues, Haddad had an "extra gland that produces publicity instead of sweat," and he became a media-savvy advocate for generics. He had first worked as an assistant to Senator Estes Kefauver (D-TN), who, as chair of the Senate Antitrust and Monopoly Subcommittee, had fought for consumer protections and battled the pharmaceutical industry. Kefauver had told Haddad about a suspected Pfizer-led cartel to control the price of the antibiotic tetracycline in Latin America. After Kefauver's death in 1963, Haddad wrote a high-profile series about the price-fixing cartel for the *New York Herald Tribune*.

Haddad left journalism and became the head of the all but invisible Generic Pharmaceutical Industry Association. With a small cadre of sympathizers, he began lobbying Congress to create a distinct process for the FDA to approve generic drugs. Politically, the brand companies "had control of every avenue," he recalled. So Haddad and his group walked the hallways of Congress, trying to make their argument to the few who would listen.

The turning point came in the early 1980s, when Haddad got a meeting with Senator Orrin Hatch, a conservative Utah Republican. He expected the senator to be aligned with the Big Pharma cause. Unexpectedly, Hatch listened with real engagement and interest. In a two-hour meeting, Haddad explained to the senator that the patents for more than 150 drugs had expired, yet the brand-name medications faced no competition because there was no way to get a generic approved through the FDA. As a result, Americans were forced to pay too much for their medicine. "He was questioning me like a district attorney," Haddad recalled.

He was stunned when, a few days after the conversation, Hatch called him up and said, "I think you may be right." The senator joined forces with a Democratic congressman from California, Henry Waxman. Together, they pressured the Big Pharma CEOs into agreement and drafted a law that established a scientific pathway at the FDA to get generic drugs approved. It was the Abbreviated New Drug Application. No longer did generic drug companies have to prove the safety and effectiveness of their drugs from the ground up, as the branded companies did with costly long-term clinical trials. Instead, the companies could win FDA approval with more limited tests to prove their drugs were bioequivalent and performed similarly in the body.

But there was another major hurdle. During the deliberations, one generic drug executive pulled Haddad into a corner and said, "Look, what if I sue and I win. What do I get?" What incentive was strong enough to justify the up-front costs of developing a generic version of a drug, the possibility of litigating against brand-name drug companies intent on defending their patents, and perhaps failing on both counts?

The solution, called the "first-to-file" incentive, transformed the generic drug industry. It allowed the company that first filed its generic application with the FDA to reap a big reward: the right to sell its drug exclusively for six months at close to the brand-name price, before other generic competitors jumped in and the price plunged. Being first became the difference between making a fortune and making a living.

The Drug Price Competition and Patent Term Restoration Act, which became known as Hatch-Waxman, passed unanimously in the House of Representatives, 362–0, in 1984. Though a huge victory for generic drug makers, it also extended by a few years the length of patents for brand-name companies. President Ronald Reagan signed the legislation in a Rose Garden ceremony that September. Touting the benefits of lower-cost drugs, he told his audience, amid laughter, "Senior citizens require more medication than any other segment of our society. I speak with some authority on that."

The Hatch-Waxman bill "really started the generics industry," said Haddad. "It gave it its footing, it gave it its foundation, it allowed the companies to grow, it reduced the prices dramatically."

It was also clear from the outset that generic drug companies could make a huge profit. The day the bill went into effect, companies sent "tractor trailers full of ANDAs" to the FDA, recalled a former agency bureaucrat. "We got one thousand applications within the first month of the program." The volume of bids—coupled with the potential jackpot of first-to-file—underscored that a generic drug factory was, as one of the FDA's earliest generic drug chiefs, Dr. Marvin Seife, claimed, "a place where you put raw materials into a mixing vat, turned the spigot and out comes gold."

Inside generic drug companies, the first-to-file incentive ignited a frenzy. "Nothing was more important," said Jay Deshmukh,

Ranbaxy's former senior vice president for global intellectual property. At issue was not just what day the application arrived at the FDA's Rockville, Maryland, campus, the agency's headquarters for generic drugs, but in what order. "Minutes mattered," said Deshmukh.

As the competition grew, so did the waiting. In the run-up to a patent expiration date, it was not uncommon to see generic drug executives asleep in their cars in the FDA parking lot overnight in order to be first at the door when the building opened. Periodically, a tent city would sprout in the parking lot, with executives camping out for weeks at a time. Each company had a strategy for how to wait and how to be first. Some paid line-sitters to wait in the parking lot. Teva booked hotel rooms nearby and rotated staff throughout the night.

On the cold, clear night of December 23, 2002, with Christmas just two days away, the FDA parking lot was crowded. The FDA had shut its doors hours earlier. But representatives from four different generic drug companies—Ranbaxy, Teva, Mylan, and Barr—were waiting in line, stamping their feet and clapping their gloved hands to stay warm. Ranbaxy had sent two of its most reliable staffers in a stretch limousine, so they could take turns sleeping and waiting.

Everyone had just one goal: to be first through the FDA's doors when the agency opened the next morning. They had all brought applications to make generic versions of a drug called Provigil, manufactured by Cephalon, to combat daytime sleepiness—a bonanza for whichever generics company filed its application first.

As the sky began to lighten, a Ranbaxy executive fully intended to keep his place at the head of the line. But just as the doors opened, a Mylan employee, a petite young woman, pushed him out of the way and rushed through the door to get the coveted time stamp, signifying first place.

Back at Ranbaxy corporate headquarters, the director of U.S. regulatory affairs, Abha Pant, had to console herself with second place. It was not a total loss, because being first was not a guarantee of success. The FDA would consider only applications it deemed "substantially complete." This was to prevent generics companies from tossing down half-baked applications as placeholders so as to be designated first while they figured out how to actually make the drug. So Pant never gave up hope. Being second was as critical. She would be waiting for the first one to trip and fall.

The FDA struggled to put a stop to the camping problem. In July 2003, the agency amended its rules so that any generics company that delivered its application on a certain set day could potentially share six months of exclusivity. In written guidance to the industry, the FDA noted:

> Recently, there have been a number of cases in which multiple ANDA applicants or their representatives have sought to be the first to submit a patent challenge by lining up outside, and literally camping out adjacent to, an FDA building for periods ranging from 1 day to more than 3 weeks. Concerns about liability, security, and safety led the property owners to prohibit lines of applicants before the date submissions may be made.

Though shared exclusivity was somewhat less attractive, first-to-file remained the most lucrative opportunity for generic drug companies.

For Ranbaxy, the applications remained essential to its strategic plan, dubbed "Garuda Vision," after a soaring Hindu eagle. Lest any employee forget the company's goal, a framed poster headlined "2015 Strategy" hung on the walls of the New Jersey office. The first bullet point, in bold, was "significant FTF filings annually" below the headline "USA: $1 billion sustainable profitable business by 2015." As one of Ranbaxy's CEOs, Davinder Singh Brar, explained in a company-sponsored book, the billion-dollar dream was a "vision . . . etched in every employee's mind."

Inside Ranbaxy, overseeing the first-to-file applications fell to Jay Deshmukh, the lean, sardonic attorney who specialized in intellectual property. Back in 1998, as a bored young lawyer in Cincinnati, he saw a surprising ad in the *Journal of the Patent and Trademark Office Society*. Ranbaxy was looking for a patent lawyer. "I had never seen an Indian company asking for a patent lawyer," Deshmukh recalled. He applied on a whim.

Deshmukh, an Indian-born chemical engineer by training, was intrigued by the prospect of working at Ranbaxy, especially after meeting the visionary managing director Dr. Parvinder Singh, whom he found to be "extremely smart and personable." Deshmukh ended up taking the job and doubled his salary. He relocated his young family to Princeton, New Jersey. Though it seemed like an excellent career move, above all he viewed the job as a "going home kind of thing, contributing to India."

Knowing little about Indian corporate culture, he immediately found himself in a "very paternalistic" environment, where "your boss is your father—he's always right." Deshmukh immediately locked horns with his boss. Within a year of joining the company, he requested a meeting with Parvinder, in which he asked to report directly to the CEO, D. S. Brar, instead. Parvinder consented. In doing so, Deshmukh cemented his future role in the company, as Brar became Ranbaxy's managing director a year later. It was Brar who encouraged Deshmukh to aim for generic Lipitor.

Inside the company, the Lipitor quest was not just your average commercial endeavor. "The lure of the drug was irresistible, like a gorgeous naked woman who is not your wife," said Deshmukh. "It's hard for guys to say no. How could we not?"

On October 9, 2002, almost two months after Ranbaxy had filed Abbreviated New Drug Application 76-477, the FDA broke its silence, first with a phone call and then with an official letter: Ranbaxy had indeed filed first, and its application to make atorvastatin, the company's generic version of Lipitor, would be evaluated.

The news led to rejoicing inside Ranbaxy. The FDA parking lot had been empty when Ranbaxy filed its application, because the company was so far ahead of its rivals. Now there was a path to the largest generic drug jackpot in history. But huge obstacles remained. First, FDA regulators had to deem the science in the dossier to be worthy. Ranbaxy's testing data would have to demonstrate, to the agency's satisfaction, that its generic Lipitor would release an equivalent amount of the active ingredient into a patient's bloodstream. After that, Ranbaxy would have to survive attack by an army of Pfizer's patent lawyers, who'd successfully been standing sentinel around the drug for years. Ranbaxy would have to follow the careful choreography, and withstand the scrutiny, of the world's most dominant drug market.

In theory, all the companies had to follow the same rigid set of good manufacturing practices. But for companies that were inclined to emphasize profits instead of quality, there were many avenues for improvisation—and shortcuts. Deshmukh acknowledged that the incentive of first-to-file created a "Wild West" environment in which companies had to not only become first filers but protect those applications at all costs. That drive—to be first and stay first—led to a stark choice for Ranbaxy, just months before Dinesh Thakur arrived at the company.

In May 2003, Ranbaxy's top executives gathered in the conference room of a hotel in Boca Raton, Florida, for what was supposed to be a nuts-and-bolts operational meeting. The company CEO, D. S. Brar, presided in his impeccable turban. Rashmi Barbhaiya, the director of research and development who had recruited Thakur, was there. So was company president Brian Tempest. Their discussion was quickly subsumed by a topic that had dominated email chains and led to the creation of a closely guarded report, its contents restricted to those in the room.

Three months earlier, the company had launched Sotret, its version of the brand-name anti-acne drug Accutane made by Roche, onto the U.S. market. As the first low-cost version available to American patients, it resulted in instant market share and was another vital step toward the company's larger goal to achieve $5 billion in global sales within a decade.

But just a few days prior to the meeting, the Ranbaxy executives had suspended the profitable Sotret launch. They told U.S. regulators that they had seen a "downward trend" in how rapidly the 40-milligram capsule dissolved and would temporarily withdraw three lots of it from the market while they completed their probe of the cause. But that was a lie. Their random tests of Sotret had shown that the formulation was failing. Under rules set by the FDA, they had but one choice: to make a full disclosure to regulators, recall Sotret from the market, and return to the laboratory to try to reengineer the drug until it worked.

Unless there was another choice. "Go find Malik," snapped Brar, glaring at his deputies, one of whom went scrambling from the conference room in his direction. Rajiv Malik, a canny and ebullient process chemist who was Ranbaxy's head of formulation development and regulatory affairs, was viewed by his colleagues as a Houdini of the generic drug world. He had unsurpassed skills at reverse-engineering and seemed to know how to turn anything into anything. If the problem could be escaped, he'd figure out how.

On that day, the usually jovial Malik—who'd been with the company on and off for eighteen years—appeared uneasy as he entered the room. Malik had led the laboratory effort to develop Sotret. Now his colleagues wanted him to fix the situation quickly. He could feel their impatience.

"This is not a quick-fix problem," he told the group. "I don't have a magic wand."

For his visibly frustrated Ranbaxy colleagues, Malik then reviewed the excruciating history of the Sotret development he'd overseen. Despite more than five years of costly laboratory work, Ranbaxy's chemists still could not get the drug to dissolve correctly. As a soft gel product in a suspended form, it was difficult to control the particle size.

The batches they'd tested for approval from the FDA had been manufactured in a controlled environment and had finally worked well enough, mimicking the original drug. But when they'd scaled up the manufacturing to create commercial-sized batches, impurity levels spiked and the drug dissolved incorrectly. Malik explained his working hypothesis: that when the soft gelatin was exposed to oxygen, it set off a reaction that affected the dissolution. It would take time to engineer a solution. In the meantime, they would need to stop selling the drug.

"I have no idea how long it will be before Sotret can be marketed," he told the executives.

He hadn't said what everyone in the room already knew. Even if formulated perfectly, the drug was uniquely dangerous. The FDA required a "black box" warning on the label, alerting patients that the drug could cause severe birth defects or miscarriages if taken while pregnant or prompt suicidal tendencies in those who took it, many of whom were teenagers. The drug, in its brand version, had been the subject of a congressional hearing after a U.S. congressman's son had killed himself while taking it. To restrict the drug, regulators required companies to report any sales, expiration, or destruction of it. The dangers demanded caution and transparency.

Given the circumstances, the FDA's regulations required Ranbaxy executives to withdraw the drug from the market and suspend making it until the failures could be remedied. But the heated discussion kept returning to the commercial pressures. If the company fumbled, there would be a rival drug maker waiting to launch its own version right behind them. *Not* marketing the drug would gut their profits.

Malik looked around the table. He saw "irrationality," as he later put it, in the mind-set of his colleagues. The executives who ringed the conference table faced a stark choice. Stopping the launch would mean abandoning the company's financial goals. Continuing it with no further disclosure to regulators would endanger patients and violate the FDA's rules.

The push for profits won out. They chose to continue the launch and conceal the problems from regulators, even as they returned to the laboratory in search of a solution. Years later, D. S. Brar would say that he could not remember the Boca Raton meeting specifically, but remarked of his tenure at Ranbaxy, "Nowhere at any time did I hear any executive say that we shall short-circuit the process and the procedure to circumvent time to market." To the contrary, he added, "we were always very scared to do anything which would go wrong in the U.S. That was the kind of scrutiny the company had internally."

But shortly after the Sotret meeting, the executives memorialized the faulty quality of their acne drug in a document titled "Sotret—Investigation Report," which Abha Pant, the vice president of regulatory affairs, later filed away in her office at the company's New Jersey headquarters. The cover page read, in bold letters, "**Do Not Give to FDA**."

CHAPTER 3

A SLUM FOR THE RICH

AUGUST 2003

Gurgaon, India

If globalization can be said to have a headquarters, then it may well be Gurgaon, an entire city built from the outsourcing efforts of the world's Fortune 500 companies. Gurgaon lies just eighteen miles southwest of New Delhi. Two decades ago, it was a sleepy farming town ringed by forest, nestled below the beautiful Aravalli mountain range. As global multinational companies looked to move their back-office functions to India, developers sensed opportunity. Office towers rose from the fields. Roads named Cyber City and Golf Course were built. In short order, Gurgaon became known as "Millennium City."

Its skyline was branded by global capitalism: Accenture, Motorola, IBM, Hewlett-Packard, and many others tacked their logos onto newly acquired buildings. Thousands of people and cars, as well as numerous shopping malls, followed, with the encouragement of the Haryana Urban Development Authority, which seemed to have no urban plan other than to welcome developers. Ranbaxy, too, established its research headquarters here, on an elegant, well-guarded campus.

Amid Gurgaon's building frenzy, there were few restrictions and little in the way of infrastructure. A patchwork of after-the-fact water treatment plants, sewers, subway stations, and power lines could not keep up with the demand. Corporate residents and their well-heeled employees were left to scramble for dwindling water and electricity. They bought most of the latter privately, at exorbitant rates, using diesel generators that further fouled the already polluted air.

Donkeys and pigs wandered amid the chauffeured town cars on the clogged, potholed streets. Officials estimated that, owing in part to the numerous bore-wells drilled by private residences and businesses seeking water, the plummeting water table would be entirely depleted within two decades. India had intended Gurgaon to be a showcase of the nation's central role in the twenty-first-century economy. Instead, the BBC suggested it was a "slum for the rich."

Still, for corporations and their employees, it was the place to be. In the summer of 2003, the Thakurs settled into a freestanding gated house with a small guard post outside, manned overnight by a hired guard. Their very address reflected Gurgaon's dramatic development: they lived in "Phase 1," the first area of Gurgaon to be built. Their house had a planted lawn, white tiled floors, and gracious rooms for entertaining. Its own diesel generator kicked in when Gurgaon's overtaxed grid blinked out. Thakur set up a home office in the basement, right next to a play area for Ishan, where the boy watched Barney and Clifford videos as Thakur worked through the weekends.

As a foreign citizen of Indian origin, Thakur was supposed to register with the local police upon his arrival. So he went to the decrepit Gurgaon police station, where the police officers looked baffled as Thakur explained that he was there to try to meet the requirements of his visa. Determined to educate them about the policies they were supposed to enforce, Thakur went home, printed out the relevant forms, and returned to explain them in more depth. His effort to comply with the law took almost a whole day.

He left the police station with a new form, filled out by hand and signed in multiple places, which he laminated to forestall ever having to return. In India, paperwork seemed to be a hedge against chaos, as well as a contributor to it. "We've created mountains of paper to create the assurance that if something happens tomorrow, there's a file," Thakur later mused. "It's a great mechanism for justifying any action you take. It has to be in a file *someplace*."

Thakur was not naive about the challenges of living in India. But he was determined not to compromise his ethical values to do so. He would remain a stickler for rules in a country where subterranean agreements, often accompanied by cash payments, governed so many interactions. As Thakur focused on his work at Ranbaxy, he remained confident that private Indian companies operated differently than the corrupt and bloated public sector. And he believed that the efficiency of corporations could help lift India into the twenty-first century.

Motorcycles, trucks, taxis, and auto-rickshaws whizzed along Gurgaon's main thoroughfare, the Mehrauli-Gurgaon Road. Fruit carts pulled by donkeys dotted the roadsides, as did stray goats and buffalo. Encampments of people by the hundreds lived on the roadside beneath fraying tarps.

Tucked along a side street, behind a guard post and sliding gate, sat the main research and development center of Ranbaxy Laboratories. The entryway was framed by impeccable shrubs and plantings. Inside the main entrance, above the gleaming tile floors, hung a portrait of the company's former managing director Parvinder Singh, who had died four years earlier of cancer, at age fifty-six. The white-bearded Parvinder sat with hands folded beneath an opulent red drape, in a white Sikh turban with a matching handkerchief in the breast pocket of his dark suit, a serene smile on his face, as though monitoring—and blessing—the goings-on. The Indian press had dubbed Parvinder, the son of the company's founder, "the alchemist who saw tomorrow." Under his leadership, Ranbaxy had become a global company—the transformation that had helped create Thakur's new job.

Thakur's mandate, as director of research information and portfolio management, was to impose some order and transparency on the burgeoning global pipeline. He was an information architect, expected to build a scaffold for the company's data. This newly created position made him one of the few company executives with a clear grasp of all Ranbaxy's far-flung world markets. He threw himself into the work. He hired six people, trained them in portfolio management, and developed a complex Excel spreadsheet to map the progress of the drugs the company was making for each part of the world.

Thakur frequently stayed long after his colleagues left. Sometimes, in the evenings, the family driver, Vijay Kumar, brought Sonal and little Ishan to pick Thakur up at work. The toddler would scribble letters on the whiteboard in his father's office or gleefully race up and down the empty hallways before the family returned home for the night.

Though Vijay had to take the Mehrauli-Gurgaon Road every day by necessity, he tried to get off it as quickly as possible. Stretches were deteriorating. Traffic backed up. When it flowed, it was a free-for-all, with few traffic lights, rules posted but ignored, the actual lanes a mere suggestion that few heeded. At night the potholes, poor streetlights, and wandering buffalo made it downright dangerous.

Vijay had first met Thakur the previous summer, when the executive had come to interview at Ranbaxy. In his early twenties, he worked at a taxi company and had been assigned to drive Thakur around. Thakur had been impressed with his quiet and responsible demeanor, as well as his skill in navigating the awesome hazards of Gurgaon's roads. When Thakur moved back to India, he hired Vijay as the family driver, a big step up for a young man who'd come from a family of subsistence farmers.

One late evening, just months into his employment, Vijay picked up Thakur at Ranbaxy, then reentered the Mehrauli-Gurgaon Road. Motorcycles and trucks sped around them on the dark thoroughfare. Suddenly, vehicles ahead swerved around what looked like a pile of trash in the road. As Vijay drove closer, they saw it was a man's motionless body.

Like most of the other drivers, Vijay had deemed it smartest and safest to ignore the man. But as other cars swerved around the body, Thakur ordered him to pull over. Vijay begged for permission to keep going, but Thakur refused. He made Vijay pull onto the shoulder and get out with him. As traffic broke around them on the shadowy road, they reached the man, who was alive but drunk and bleeding from the head. They lugged him to safety. It was a crazy rescue, one that violated every rule of safe driving—and defensive living—that most Indians knew: namely, keep your head down and your car moving, and don't volunteer aid to strangers. No good could come of it. Such a rescue was more likely to invite any number of bad outcomes.

But for Thakur, getting the man off the road was not enough. Though he was generally a wary and reluctant observer, he had a countervailing tendency: to lean into whatever project he'd started, regardless of the possible result. He insisted to Vijay that they carry the man to a local hospital, a block and a half away.

To Vijay, lugging the man down the street was one of the stranger projects he'd ever undertaken. Why do this when they had no stake in whether the man lived or died? At the hospital, the surprised staff seemed to agree with Vijay and refused treatment unless Thakur paid in advance. He peeled off 7,000 rupees (about $140, a lordly sum and twice what Vijay earned in a week) and left his business card behind. He did not even think to help anonymously. To Vijay, his new boss seemed to have weird American ideas about coming to the rescue. And as the young driver correctly predicted, the outcome was bad.

The next day, a police officer showed up at Ranbaxy and accused Thakur of running the man over. Why else would Thakur have paid so much money to help him, unless he'd done something bad to him in the first place? Thakur called the Human Resources Department and directed them to deal with the situation. The officer ultimately went away, which to Thakur almost certainly meant he was paid to no longer press his inquiry.

Coming to the attention of "the system" in India had almost no upside. Altruism was often greeted with suspicion. Had Thakur not worked for a company with resources, one able to pay to make an allegation vanish, who knows where the confrontation with the police officer might have led? Companies were king in India, while people were more likely to be treated as dispensable. Though the incident left Thakur uneasy about his decision to return to India, his lingering doubts were soon erased by an extraordinary event at Ranbaxy.

On November 21, 2003, Ranbaxy's corporate communications director, Paresh Chaudhry, watched in amazement as U.S. Secret Service officers swept rooms and posted snipers on the rooftop of Ranbaxy's headquarters. Former U.S. president Bill Clinton was arriving that day to publicly thank Ranbaxy and two other Indian generic drug companies for agreeing to manufacture AIDS drugs that would be sold for about 38 cents a day in African and Caribbean countries. The price was roughly 75 percent less than the lowest possible cost for equivalent brand-name drugs. Although U.S. taxpayers would foot the bill, it was the William Jefferson Clinton Foundation that had closed the deal.

Chaudhry, hardworking and innovative, had never in his wildest dreams imagined that he'd be running logistics for an event involving a former U.S. president, let alone Bill Clinton, who was particularly beloved by Indians. When Clinton had visited India in late March 2000, he became the first U.S. president to do so in more than twenty-two years. On that visit, he'd emphasized the need for the two countries to cooperate in tackling diseases such as AIDS.

Clinton's passion for the country was more than pro forma. He'd returned in April 2001, three months after a devastating earthquake in Gujarat killed 20,000 people. Through the American India Foundation, which he was chairing, Clinton raised millions to resurrect devastated villages. He declared to adoring throngs, "I intend to come back to India for the rest of my life."

He was back now, the third time in four years, making good on his pledge to unite India and America in the fight against AIDS. For almost two weeks straight, Chaudhry had been immersed in preparations, his world becoming a blur of Secret Service agents, lists of names, badges for employees who would be there, what food would be served, which route Clinton would take, and who was going to receive him. The visit was a potential marketing bonanza. As Chaudhry well knew, he could work his whole career and never have a chance like this again.

For years, he had invited international journalists to visit Ranbaxy. He'd shown them the company's research facilities and modern manufacturing plants. He'd explained the robust efforts to develop new chemical entities. But the response was generally the same: "Thank you very much. We'll come back to you if we're interested in something." And then silence.

As Chaudhry knew, he was up against the widespread assumption that low cost equals low quality. Indian generic drug makers—who, by and large, invented little but rather reengineered existing brand-name drugs—were viewed around the world as fakers and copycats. Even in Africa, their drugs were held in low esteem. In Cameroon, doctors referred to their products as *pipi de chats*—cat urine. Things had begun to shift in their favor in 2001, during the U.S. anthrax scare following the attacks of September 11. Bayer had proposed to sell the U.S. government ciprofloxacin—an antibiotic that was one of the few remedies for anthrax poisoning—at almost two dollars a pill. Ranbaxy's price was one-fifth that. "Bayer and the lobbyists in the U.S. and Washington tried their best to say that this is a fake company," Chaudhry recalled. But patents blocked the U.S. government from purchasing Ranbaxy's version. Today's event, Chaudhry hoped, would finally be the game-changer.

As Clinton walked through the sliding doors in a dark suit and deep red tie, Chaudhry was among the first to extend a hand in welcome. The others in the vestibule, all in their best suits, canted forward, anxiety and excitement etched into every face as they stood ready to anticipate the former president's wishes.

Eager employees gathered in the auditorium to listen, along with many members of the media whom Chaudhry had been chasing for so long. Thakur got a front-row seat. Clinton stood shoulder-to-shoulder with Ranbaxy's CEO, D. S. Brar, who wore a black turban, dark suit, and patterned tie against a crisp white shirt. He had been appointed Ranbaxy's CEO after Parvinder Singh's death and was hailed as a professional manager, his appointment viewed as a milestone for the family-owned company.

Clinton explained that he had come to thank the Indian companies—including Ranbaxy, Cipla, and Matrix Laboratories—for agreeing to make such low-cost drugs. "It's very important to give these companies some credit here, because they have basically believed in us," he told those assembled. Their efforts, he added, ensured "that we can go out and work with other countries and convince them that treatment was a viable option and an affordable option."

Brar then had his turn. "The drugs can be priced at a lower price only if there are volumes, and for that the large AIDS-inflicted nations have to come forward and purchase in bulk. This was not happening till the Clinton Foundation stepped into the picture and brought together the nations and the producers of the drugs."

After the speech, as Clinton pressed into the crowd, Thakur got a turn to shake his hand.

The visit was everything Chaudhry had hoped for, and more. From that moment, the company's sales increased and its reputation improved. "We arrived as a company," Chaudhry recalled. Ranbaxy was poised to surge in Western markets. "We can kill these big guys in the U.S. with all our products," Chaudhry summed up the feeling. "It's good for humanity. It's good for the government. It's good for the people. Why should anyone block us, for heaven's sake?"

Clinton's visit was a boost for the entire industry. Before it, governments around the world had faced aging populations, the AIDS epidemic, and soaring drug prices. How could they afford to treat these patients? Clinton had shown them a solution. India's drug companies, it appeared, were on the side of the angels. As Ranbaxy's next managing director, Dr. Brian Tempest, later told the *Guardian:* "We don't make a lot of money out of selling AIDS treatments cheaply. . . . This is really out of social responsibility, because we are based in the developing world and have all its issues on our doorstep."

Whether the companies could profit by selling vastly discounted AIDS drugs or not, Clinton had vouched for Indian generics. His visit led to new opportunities and potential profits all over the world. On a later trip, Clinton went to Cipla's facility in Goa, where he planted a pine tree in the gardens, a tradition for important visitors started by the company's legendary chairman, Yusuf Hamied. For companies that had no fancy lobbyists in Washington, D.C., the visits proved invaluable. "Our humanitarian effort has been our public relations," Hamied later explained. "Every door is open now."

After his stop at Ranbaxy, President Clinton used the same visit to head to Agra to see the Taj Mahal, the white marble mausoleum built in the seventeenth century by a Mughal emperor for his wife. There, Clinton followed visitor protocol and took an electric bus from the outskirts of the monument to its gates, a requirement that protected the World Heritage site from air pollution. But on the return to his hotel, the electric bus stalled, and the former president had to walk the rest of the distance. In a country that prided itself on spectacular ceremony and magnificent hospitality, the lapse risked becoming an international embarrassment—suggesting that the nation's old, broken-down infrastructure still lurked beneath the gleaming facade.

CHAPTER 4

THE LANGUAGE OF QUALITY

FEBRUARY 25, 2000

New Iberia, Louisiana

Jose Hernandez sniffed the air. The forty-three-year-old FDA investigator stepped deeper into the K&K Seafoods crab-processing plant, which looked none too appetizing. His mind clicked over the regulations enshrined in the U.S. Food, Drug, and Cosmetic Act. Gazing inside the plant, he could practically see the relevant pages: the fish and fishery products "Hazard Analysis Critical Control Point" plan, Title 21, Code of Federal Regulations, part 123, 6 (b).

But his nose signaled the most trouble. What was that odor? It reminded him of his Labrador, Livy, after a bath—that sodden, damp-wool smell of a dog soaked to the skin. Not a good sign in a seafood facility that was supposed to be operating under good manufacturing practices. He doubted the plant was safe for consumers.

Hernandez, balding with a dark mustache, glasses, and a runner's physique, was the FDA's resident-in-charge of the Lafayette,

Louisiana, office, a four-man outpost. His job, as a badge-carrying FDA investigator, was to inspect the seafood manufacturing plants and small medical centers in the area. Hernandez got paid $45,000 a year, a salary on which he was supporting his wife and four children. The agency had no laptops, so Hernandez took handwritten notes while inspecting, then signed up for time to use the single desktop computer at the office to type them up. He wore coveralls and plastic boots while inspecting seafood plants.

It would not be everyone's choice of jobs, but Hernandez thrived in it, and he'd begun to earn a reputation as one of the FDA's smarter, more intuitive, and more energetic investigators. He lived in a rambling 5,000-square-foot house, which he was painstakingly renovating. He'd learned expert carpentry skills from his grandfather, who'd helped raise him in Puerto Rico. Hernandez graduated from the Inter American University in San Juan and began at the FDA as a generalist investigator in 1987. Though he didn't have fancy graduate degrees, he had a mechanical mind. He knew how things were supposed to fit together and could tell when they didn't. He could readily organize facts and remember them. He also had an uncanny sense for when something was wrong.

To relax, Hernandez worked on his house and, whenever possible, took his kids camping. But his mind was never at rest. It ranged continually over Title 21 of the Code of Federal Regulations of the Federal Food, Drug, and Cosmetic Act. He knew the regulations almost encyclopedically, but also kept them close at hand to reread. They were his scripture. "The guy preaching the Mass for thirty years always goes back [to the Book]," he said. "I never tried to answer from memory. I never tried to guess. You can never charge anybody with anything unless there's a regulation."

He mulled over patterns—the visible workflow in a plant versus the invisible machinations, the relationship between what he saw and what the regulations stipulated. During everyday activities, like drinking water from a plastic bottle, he'd recite the regulations to himself: 21 CFR165.110 was for bottled water. The container holding the water was regulated differently than the water itself (21 CFR1250.40). To him, the inspections were puzzles, and he was always trying to find the missing pieces.

Under FDA regulations, he only had to show his badge and any manufacturing facility regulated by his agency had to allow him full access to the plant and grounds. He never gave advance notice—none was required. Refusal to admit an FDA investigator could lead to a plant being shut down. He would stay as long as he felt was needed for a thorough review. That could mean one day or two weeks. He began each inspection by driving around the perimeter, taking in the broadest view. He thought of it as first looking through the wide lens of a camera. He could then zero in on the important stuff. When it came to K&K Seafoods, he knew exactly what to do. He needed to return when least expected, and presumably least wanted. Nighttime was when they cooked the crabs. "If you want to build a case," he said, "you have to be there when things are happening."

He went home for dinner. Once the kids were asleep, he returned to the plant at 9:00 p.m., and an unhappy-looking employee let him in. This time the wet-dog smell was even stronger. Hernandez followed it to the back of the plant, where he found a small kitchen and a pan on the stove with pieces of meat in it—dog meat. He headed out to the plant floor, where a man cooking the live crabs was chewing as he worked. He'd caught the employee red-handed, though the violation itself, 21 CFR110.10 (b)(8), was rather understated: no eating allowed in areas where food is being processed.

The FDA's regulations were narrow and specific. It made no difference whether the man was chewing a cracker or a dog haunch. What Hernandez thought about it made no difference either. He couldn't impose deeper sanctions for something particularly repulsive.

On the face of it, the K&K crabmeat plant was a noisome place that might have raised anyone's suspicions. But Hernandez had a gift not just for aggressively following up obvious clues but for seeing beneath the surface of even pristine-looking manufacturing plants. He'd proven this during his inspection at the Sherman Pharmaceuticals plant in Abita Springs, Louisiana, which made eye lubricants for contact lenses and prescription eye solutions. In 1994, he'd arrived, with two trainees, seven months after the plant had emerged unblemished from an inspection.

His domain was plant and grounds. So as usual, Hernandez began with the grounds, working his way from the outside in. He wandered into the woods ringing the plant. In the distance he noticed a smoldering pile, as though for a barbecue. He directed one of the trainees to find a stick and poke around in the embers. They uncovered a pile of charred medicine that the company was burning. But why? The investigators were able to spot lot numbers on the partially burned containers. The medicine, as it turned out, had not yet expired. "You're not going to destroy product that is actually sound, so what happened to the rest?" Hernandez wondered. As it turned out, the company was burning medicine that had been returned because of contamination. Instead of investigating the cause and reporting it to the FDA, as required, the company chose to try to destroy the evidence. Hernandez detailed his findings in an inspection form called a 483.

The FDA's investigators codify their findings in three ways: No Action Indicated (NAI) means that the plant passes muster; Voluntary Action Indicated (VAI) means that the plant is expected to correct deficiencies; and Official Action Indicated (OAI), the most serious designation, means that the plant has committed major violations and must take corrective actions or face penalties. Under Hernandez's watchful eye, both K&K and Sherman Pharmaceuticals received an OAI, putting them at risk of even greater sanctions.

In 1995, the FDA imposed on Sherman Pharmaceuticals its most stringent penalty, a so-called Application Integrity Policy (AIP)—one of only about a dozen such restrictions imposed by the

FDA. This placed the plant under strict monitoring and required it to prove that it was not committing fraud. Sherman Pharmaceuticals went out of business shortly afterward. Hernandez never felt any sympathy toward the company—or any other, for that matter. It was not his job to take things lightly or to look the other way.

The Food and Drug Administration serves one of the most important functions of any government agency. Its job is to safeguard public health by ensuring that our food, drugs, medical devices, pet food, and veterinary supplies are safe for consumption and use. In doing this, the FDA regulates about one-fifth of the U.S. economy—essentially, most of the products Americans are exposed to and consume. It operates from a sprawling headquarters in Silver Spring, Maryland, and has a workforce of over 17,000 employees, twenty satellite offices around the country, and seven offices overseas.

Whatever one may think of regulators—heroic public servants or pests with clipboards counting the number of times workers wash their hands—there is no doubt that in the world's estimation, the FDA is viewed as the gold standard. If you hold its regulators up against those from most other countries, it's like comparing "the latest model Boeing to an old bicycle," said a senior health specialist for the World Bank.

Part of the FDA's vaunted reputation comes from its approach. It does not just regulate with a checklist or scrutinize the final product. Instead, it employs a complex, risk-based system and scrutinizes the manufacturing process. Under FDA standards, if the process is compromised, the product is considered compromised too.

The FDA requires companies to investigate themselves under a review system called Corrective Action and Preventive Action. The drug company Merck was famous for doing this and discarding drug batches if it had the slightest concern about their quality. "You have to look to know the truth, and you have to have people who know how to look," a former FDA investigator explained. "[And] unless agencies start looking, companies don't look."

Inspector Hernandez's methods might seem simple: sniffing, looking, poking with a stick. But he was armed with concepts and regulations that had evolved over more than a century, related to both drug and food safety, which developed in sync. Today a manufacturing plant must disclose and investigate quality problems, rather than simply burn bad drugs in the woods. Workers can't eat dog meat (or anything else, for that matter) as they work on canning crab, since a manufacturing plant must control its environment against contaminants. The concepts of control, transparency, and consistency fall under current good manufacturing practices (cGMP), the elaborate architecture of regulations that govern the processing of food and the manufacturing of medicine.

Such regulations did not exist at the dawn of the twentieth century. The phrase "good manufacturing practices," now ubiquitous in facilities around the world, made its debut in a 1962 amendment to the U.S. Food, Drug, and Cosmetic Act. For today's drug manufacturers, cGMP is understood to be the minimum requirements a manufacturer must follow to ensure that each dose of a drug is identical, safe, and effective and contains what its packaging says it does. Those requirements evolved after a centuries-long debate over how to best guarantee the safety of food and drugs.

The medicine men of the Middle Ages were among the first to promote the idea that the quality of a drug depends on how it's made. In 1025, the Persian philosopher Ibn Sina penned an encyclopedia called the *Canon of Medicine* in which he laid out seven rules for testing new concoctions. He warned experimenters that changing the condition of a substance—heating honey, say, or storing your St. John's wort next to rat poison—could change the effect of a treatment.

Medieval rulers recognized the perils of inconsistency and the temptation for food and drug sellers to cheat their customers by replacing edible or healing ingredients with poor substitutes. In the mid-thirteenth century, an English law known as the Assize of Bread prohibited bakers from cutting their products with inedible fillers such as sawdust and hemp. In the sixteenth century, cities throughout Europe began publishing standardized recipes for drugs, known as pharmacopoeias. In 1820, eleven American doctors met in Washington, D.C., to write the first national pharmacopoeia, which, according to its preface, was meant to rid the country of the "evil of irregularity and uncertainty in the preparation of medicine."

The same year, a German chemist named Frederick Accum published a controversial book with a mouthful of a title: *A Treatise on Adulterations of Food, and Culinary Poisons. Exhibiting the Fraudulent Sophistications of Bread, Beer, Wine, Spirituous Liquors, Tea, Coffee, Cream, Confectionery, Vinegar, Mustard, Pepper, Cheese, Olive Oil, Pickles, and Other Articles Employed in the Domestic Economy. And Methods of Detecting Them.* Accum railed against factories' use of preservatives and other additives in packaged foods, such as olive oil laced with lead and beer spiked with opium. Widely read throughout Europe and the United States, Accum's treatise brought the issue of food safety and the need for oversight to the public's attention. In the United States, it wasn't until 1862 that a small government office, the Division of Chemistry, began investigating food adulteration, with a staff housed in the basement of the Department of Agriculture—a fledgling effort that would later morph into the FDA.

In 1883, a square-jawed, meticulous doctor from the Indiana frontier, Harvey Wiley, took over the division. At thirty-seven, Wiley was known as the "Crusading Chemist" for his single-minded pursuit of food safety. He rallied Congress, without success, to introduce a series of anti-adulteration bills in the 1880s and '90s. By 1902, his patience worn out, Wiley recruited twelve healthy young men and fed them common food preservatives such as borax, formaldehyde, and salicylic, sulfurous, and benzoic acids. The diners clutched their stomachs and retched in their chairs. The extraordinary experiment became a national sensation. Wiley called it the "hygienic table trials," while the press named it the "Poison Squad." Outrage fueled the movement for improved food quality.

Meanwhile, officials at the Laboratory of Hygiene of the Marine Hospital Service (later the National Institutes of Health) grappled with a different public health crisis. In 1901, an epidemic of diphtheria—a sometimes fatal bacterial disease—broke out in St. Louis. The disease was cured by injecting victims with an antitoxin serum, which was produced in the blood of horses. That October, a five-year-old patient who had gotten a shot of the antitoxin began to show strange symptoms: her face and throat contorted in painful spasms, and within weeks she was dead. The antitoxin intended to cure her diphtheria had actually given her tetanus. Officials traced the contamination to a retired milk-wagon horse named Jim, who had come down with tetanus some weeks before.

Though the St. Louis Health Department learned that the horse was sick in early October and consequently shot him, department officials had bled Jim twice before his death—in August and in late September. The August blood was clean, but there wasn't enough blood to fill all of the vials. The officials topped off the remaining vials with the September batch, but failed to update the labels. As a result, some bottles marked "August" contained September's tetanus-tainted blood, which killed thirteen children.

In response, Congress passed the Biologics Control Act in 1902, also known as the "Virus-Toxin Law." It required producers to follow strict labeling standards and to hire scientists to supervise their operations. The law also authorized the Laboratory of Hygiene to regulate the biologics industry through inspections.

By then, journalists had begun exposing troubling practices in the food and drug industries. In 1905, an eleven-part series in *Colliers' Weekly*, "The Great American Fraud," shocked Americans by exposing "cough remedies," "soothing syrups," and "catarrhal powders" as being worthless and deadly. In June 1906, Congress finally passed the legislation that the chemist Harvey Wiley had spent decades lobbying for. The Food and Drug Act, or "Wiley Act," banned dangerous additives in foods and prohibited manufacturers from making "false or misleading" statements and from selling misbranded and adulterated drugs. Additionally, medicines sold under a name listed in the *United States Pharmacopoeia* had to meet the published standards of strength, quality, and purity. As impressive as the law was for its time, it was marred by loopholes. It allowed harmful substances such as morphine in products, so long as they were disclosed on the label. And although the law made fraudulent claims a crime, it fell on the government to prove that a salesman intended to deceive his customers. Swindlers easily avoided prosecution by insisting that they believed in their fake remedies.

The FDA formally began in 1930. In 1933, its officials created an exhibition of hazardous food and medical products, which they displayed to Congress and at public events. The collection included an eyelash dye that blinded women, a topical hair remover with rat poison in it that caused paralysis, and a radium-based tonic called Radithor that was said to restore one's sex drive but in reality caused deadly radium poisoning. The press called the exhibit "The American Chamber of Horrors."

Several years later, Congress proposed a new food and drug act, but only after another tragedy nudged the bill into law. In 1937, 107 people, many of them children, died from taking a liquid antibiotic called Elixir Sulfanilamide. They died excruciating deaths. One grieving mother wrote to President Franklin D. Roosevelt of her daughter's painful end: "We can see her little body tossing to and fro and hear that little voice screaming with pain. It is my plea that you will take steps to prevent such sales of drugs that will take little lives and leave such suffering behind and such a bleak outlook on the future as I have tonight."

Sulfanilamide effectively treated streptococcal infections. Since its discovery in 1932, doctors had administered the drug in tablets and powders. But in 1937, a chief pharmacist at S. E. Massengill Company came up with a formula for a children's syrup that called for dissolving the drug in diethylene glycol—a sweet and, as it turned out, deadly poison that would be used decades later as an ingredient in antifreeze. When FDA agents investigated Massengill's plant, they were amazed to find that "the so-called 'control' laboratory merely checked the 'elixir' for appearance, flavor, and fragrance," but not for toxicity. As one FDA agent reported, "Apparently, they just throw drugs together, and if they don't explode they are placed on sale." Spooked by the disaster, Congress finally passed the Food, Drug, and Cosmetic Act in 1938, which authorized the secretary of agriculture to approve new drugs before they could be marketed. A company hoping to sell its concoction had to submit an application that described the drug's ingredients and production process, as well as submit safety studies to convince the secretary that its manufacturing methods, facilities, and controls were adequate.

But what could be deemed "adequate"? That question came into sharp relief between December 1940 and March 1941, when nearly three hundred people fell into comas or died from taking antibiotic sulfathiazole tablets, made by the Winthrop Chemical Company in New York. On its FDA application, Winthrop claimed to have "adequate" controls. But a batch had been contaminated with as much as triple the typical dose of Luminal, a barbiturate antiseizure drug. Patients who swallowed the tainted antibiotics unknowingly overdosed on the barbiturate. In its investigation, the FDA learned that the company assembled the antibiotics and the barbiturates in the same room and often swapped the tableting machines. The company couldn't account for what came out of its tableting machines—because it had little idea what went into them.

In the wake of the crisis, FDA officials met with an industry consultant who told them that most drug makers in the United

States lacked adequate controls, in part because there was no agreement on what a good control system should be. The FDA's drug chief wrote a memo to his division, arguing that, going forward, "the mere perfunctory statement that adequate controls are employed will not be sufficient."

But it was the specter of an averted tragedy that had the most profound effect. In 1960, the Cincinnati manufacturer William S. Merrell applied to the FDA to sell a drug called Kevadon, widely known as thalidomide. Introduced in Germany in 1956, thalidomide was being marketed to pregnant women throughout Europe, Canada, and South America as a sleeping pill and to treat morning sickness. In the United States, the Merrell company had begun to distribute samples to doctors, but the drug was not yet commercially available. Frances Kelsey, an FDA medical officer, was assigned to review the application. She could have rubber-stamped it, but the company's limited safety studies gave her pause. She questioned company officials about how the drug worked in the body, but they refused to answer. Instead, they complained to her superiors and pressured her to approve the drug. Kelsey refused.

By the winter of 1961, it was clear that she'd made the right decision. A growing number of physicians abroad were linking thalidomide to babies born with severely deformed limbs, such as shrunken legs and flipper-like arms. More than ten thousand mothers who had taken the drug gave birth to disabled children. Kelsey was hailed as a hero. Because of her refusal to capitulate, American patients were spared the worst, with only seventeen cases of birth defects linked to the samples. The near-miss once again galvanized Congress, and in 1962 it passed an update to the Food, Drug, and Cosmetic Act known as the Kefauver-Harris Amendment, which required applicants to prove that their drugs were not only safe but effective, disclose potential side effects on the package, and report adverse events to the FDA. Most significant, the amendment redefined what it meant for a drug to be adulterated. Products made in plants where the process didn't conform to "current good manufacturing practices" were deemed tainted.

That was a seismic shift. The manufacturing *process* became the key to quality, as it is today. This new definition gave the FDA the power to enforce good manufacturing standards. But the question remained: what should they be?

In late 1962, a group of FDA investigators met to hash out a first draft of these practices. The new regulations, published the following year, established new categories for standards in the "processing, packing, and holding of drugs." Each "critical step" of the manufacturing process had to be "performed by a competent and responsible individual." Workers were required to keep a detailed "batch-production record" for each drug lot, which included a copy of the master formula and documentation of each manufacturing step. As the new regulations rolled out, manufacturers struggled to comply. Drug recalls rose.

In 1966, the agency undertook a major survey of the most clinically important and popular drugs on the U.S. market. Of 4,600 samples tested, 8 percent were either more or less potent than they were supposed to be. The FDA decided that the best way to get manufacturers up to speed was through rigorous inspections. In 1968, the agency launched an intensive three-year blitz. Investigators showed up unannounced and camped out at scores of companies, sometimes for as long as a year. They badgered. They educated. They collaborated. They bullied. If a manufacturer couldn't—or wouldn't—comply with investigators' demands, they were put out of business. The effort effectively launched the FDA's modern-day inspection program.

In the decades-long journey to improve quality, the pivotal shift was from product to process. No longer could drug makers simply wait until after a drug was made to test for passing results, a hallmark of bad manufacturing. Perhaps at the end, you could test a few pills from a batch, but a million pills? That was impossible. Quality had to be built into the process, by documenting and testing each result along the way.

This practice, known as "process validation," gained currency in the late 1980s. The data from each manufacturing step became the essential road map. The acronym ALCOA stipulated that data had to be "attributable, legible, contemporaneously recorded, original or a true copy, and accurate."

As Kevin Kolar, formerly Mylan's vice president of technical support, explained, a finished drug cannot be separated from the data created in the process of making it. "One without the other is not a product. . . . If it's not documented, it didn't happen. Meticulous attention to detail. That's your business, your entire business."

As the years passed, it became evident that Jose Hernandez, in Louisiana, was destined for something more complex than dog-meat detection. By the year 2000, manufacturing began to move offshore. Over the next eight years, the number of drug products made overseas for the U.S. market doubled. By 2005, for the first time, foreign manufacturing sites regulated by the FDA exceeded those in the United States.

The FDA was growing desperate for investigators willing to travel overseas. Hernandez volunteered and began doing inspections in Japan, Austria, Germany, India, and China, in quick succession. By 2003, he had joined the overseas inspectorate, a small cadre based in the United States but dedicated to visiting foreign plants. The work was exhausting and difficult. He kept his green government-issued notebook by his bedside, jotting down observations even as they came to him in half-sleep. He had little respect for his supervisors, whom he found to be more concerned with office politics than public health. His sense of commitment to the American consumer kept him energized.

The FDA knew that the best way to keep drug plants compliant was for investigators to show up unannounced, when they were least expected and wanted. So long as a drug plant remained fearful of a surprise visit, it would be more likely to follow good manufacturing practices. But the dynamic of the inspections in the international realm was completely different. No longer could Hernandez simply walk in, show his badge, and conduct an inspection. Instead, the FDA notified foreign plants of upcoming inspections months in advance. The plants then issued a formal invitation, which the FDA's investigators used to secure travel visas. This system of advance notice was not legally required, nor was it the best way to run an inspection. But as the FDA scrambled to deal with a growing backlog of foreign inspections, advance notice became the jury-rigged solution to a host of challenges. It helped to ensure that the right employees from the plant were available during an inspection and served as a diplomatic gesture to foreign governments. Under this system, however, the foreign inspections were not a candid assessment of a plant's true condition, but more of a staged event.

The plants coordinated the investigators' trips and arranged their local travel. "The element of surprise is out the door," said Hernandez. This made him even more reliant on his instincts and on everything he'd learned over the years. As he found himself in remote countries with languages he did not understand, he kept returning to the idea of "plant and grounds." It became a kind of mantra. To him, it meant a "broad scope of thinking."

In this way, he came to believe that—despite the different languages, the different cultures, the different time zones—quality was a language all to itself. And he was certainly fluent in that. Facilities either had controls or they didn't, and he could look, sniff, or poke to figure it out. For example, he often couldn't read what was written in the records themselves. Instead, he studied the way the records looked. Were they smudge-free or were there fingerprints on the copier? Was one equivalent batch of records smaller than the other? Were records creased or frayed, and if not, why not? In this way, he discovered things that some of his colleagues missed. In one instance, an overseas drug company printed records on a heavy fiber paper. He discovered that the director of quality was having staff alter data by scraping words off the page with a sharp blade. In another instance, he inspected a Chinese factory. Before entering the sterile manufacturing area, managers there requested that he wash his hands with soap and then don a double set of gloves, which everyone there was required to do before entering. *I am watching the show,* he thought as they went inside. He then said to the manager, "If everyone has to put on double gloves, how come there are fingerprints on the doorknob inside?"

These clues were all pieces in the massive jigsaw puzzle of a manufacturing plant. But the puzzle now stretched across continents.

Globalization cast a shadow over a process that required transparency, making distance the biggest challenge to everything the FDA had learned about safety over the last 170 years. Dr. Patrick Lukulay, former vice president of global health impact programs at the United States Pharmacopeia (USP), explained: "The issue of globalization, that's the issue of countries where you are not [there]. . . . You almost have to be on your toes, doing unannounced inspections, listening to whistleblowers. Regulation," he asserted, "is a cat-and-mouse business."

CHAPTER 5

RED FLAGS

2004

Gurgaon

Haryana, India

Bristol-Myers Squibb had been a staid, legalistic environment. Every employee at every level was expected to attend workshops on topics ranging from how to maintain proper audit trails to matters of gender sensitivity.

At Ranbaxy, Thakur encountered chaos. The company was bristling with ambition and big ideas but had a seat-of-the-pants feel. The vice president of clinical research chain-smoked four packs a day. At Ranbaxy's New Jersey manufacturing plant, sensitive pharmaceutical ingredients wound up in the employee refrigerator next to the half-and-half. Disputes at executive meetings sometimes escalated into fistfights. Thakur assumed that the freewheeling environment was the result of an aggressive company expanding too fast: "There was no structure. It was completely the antithesis of everything I'd learned for ten or twelve years."

But as 2003 drew to a close, rather than be discouraged by the disorder and lack of training, Thakur took it as a sign that he was very much needed. His plan was to collect and archive all the company's data. Moving from paper-based chaos to digitized order was part of the larger shift from an insular India-centric company to an outward-looking multinational corporation with accepted norms of record-keeping.

His team began to standardize the most basic things, down to the templates and fonts used for company presentations. They worked with a sense of mission, full of insight and new ideas in the service of transforming Ranbaxy. "Coming in, you are going to change the world," said Venkat Swaminathan, who had come with Thakur from Bristol-Myers Squibb. "You are going to do things differently." He even saw an upside to the chaos: the group would not "have to worry about approvals for this, and approvals for that," as they had at BMS. They could simply move ahead with their plans.

But the chaos was impeding progress. Ranbaxy had no company-wide system for managing its drug portfolio. Different divisions did not communicate with one another. There was no way to track data. Different divisions even presented earnings variously in Euros, dollars, and rupees. Most of the company's record-keeping was paper-based. Through a survey, Thakur learned that more than half the time scientists couldn't even find the documents they'd created a year earlier. His team computerized and standardized systems, allowing scientists to access and store key documents, such as standard operating procedures and study reports.

One of Thakur's early efforts was to digitize the records from the company's clinical trials, including forms for informed consent, patient medical records, and lab results. He sent Dinesh Kasthuril to Majeedia Hospital, where Ranbaxy ran a unit that conducted clinical trials. The visit was tense. Afterward, Thakur got a call from the unit's director, explaining that the connectivity at the hospital unit was poor, so the records would be impossible to digitize. Thakur assured him that they would put in a new link between the hospital and Ranbaxy's data center. Thakur sent another member of his team back. This time they weren't even allowed to enter the facility.

To Thakur, the most logical explanation for this behavior was that he'd entered a hierarchical old-boys' network, whose long-term staff felt usurped. Not only was he new to the company, but he had come from the world of branded drugs. He figured his new colleagues might be on the lookout for a superior attitude. Thakur resolved to move slowly and politely, so he would not be accused of pushing them around—a point echoed in a management review that he'd undergone not long after he arrived at Ranbaxy.

The finished report noted Thakur's self-confidence, self-reliance, and high expectations of others, as well as his emotional control under pressure. But it also noted his "desire to make things happen and implement the result of his analysis with a degree of pace and sense of urgency." The evaluation continued: "He recognises that Ranbaxy is a different culture and that his desire to be direct and open does not always achieve the result that he is looking for. Further to this, his high expectations are not always fulfilled which can cause him to demonstrate a lack of patience with people."

His attitude was not the only problem. Thakur was put on a committee tasked with developing a corporate records retention policy. After a few meetings, the company's chief information officer notified the committee that a decision had been made to delete email records after two years. Thakur spoke up forcefully against the plan. Most research and development projects could span up to a decade, he pointed out in an email. Prematurely deleting records could cause the company to lose critical work product and fall afoul of regulators.

A few days later, he got a call from the information officer, directing him to delete his response and any records from the meetings he'd had on the topic. This was a direct instruction from the CEO's office, the executive told him. The effort to formulate a better records retention policy sputtered out with the instruction to delete all records of the debate.

For Thakur and his team, navigating the company—and Indian corporate life—had a *Through the Looking-Glass* feel as they encountered obstacles that ranged from troubling to absurd. The men wanted to buy a software program called Documentum, an electronic document management system. In their view, there was nothing comparable on the market. When they took their plan for approval to the purchasing committee, they were told, "We need three quotations."

"But there's only one such program," they tried to explain. Nope.

"We need three," came the response. The purchasing committee urged them to "just get some local guys to show up."

In another instance, Kasthuril held a meeting with the formulation team directors to discuss how to digitize lab work. The group murmured some objections until one of the directors piped up with a question: "If we do this, how do we backdate documents?" The vice president of formulation jumped in to explain that the man was asking a hypothetical question. Whether hypothetical or not, there was clearly opposition to systems that created transparency. But Thakur's team set aside their concerns and pressed ahead. They were there, after all, to improve the company and its systems.

In January 2004, turmoil inside the company broke out into the open. Ranbaxy CEO D. S. Brar announced that he was stepping down, having lost a power struggle with the company's heir apparent, the founder's son, Malvinder Singh. Dr. Brian Tempest, an Englishman and chemist with tousled graying hair and a rumpled appearance, was promoted to CEO. His ascent was viewed as keeping the seat warm for Malvinder, who at thirty-two-years old was promoted to president of pharmaceuticals. To many, it seemed like a loss for professional management and a win for a "dynastic family concern run on one or two people's fancy," as Swaminathan saw it. The news did not bode well for the vision, held by Thakur and his allies, of Ranbaxy as an Indian Pfizer of the twenty-first century.

A bigger shock was yet to come. At a celebration of the Hindu festival of Holi, held outside on the company campus, employees and their families ate from food trucks and listened to live music. Thakur was standing in the crowd when he noticed his boss, Rashmi Barbhaiya, beckoning to him. Once the two men reached a quiet spot, Barbhaiya said, "I am leaving Ranbaxy." Thakur was stunned. Barbhaiya had been with the company for less than two years. It was Barbhaiya who had created Thakur's job and defended his innovations internally. On his assurance, Thakur had left a good-paying job and a settled life in the United States. Now Rashmi was leaving? "What will I do?" Thakur asked.

"You will survive, and I am not leaving just yet," Barbhaiya said. "I will be around for a few more months and we can talk." In the ensuing months, his mentor became furiously negative about the company. The older man told Thakur that Ranbaxy wasn't the type of place where somebody like him should work. Over dinner at an elegant hotel with a group of American scientists who had come to help with training, Barbhaiya bad-mouthed the company, making those at the table uncomfortable. Thakur felt baffled by his animosity.

Afterward, Thakur took him aside and asked him to explain his fury. Barbhaiya remained elusive but suggested that he knew of enough "shenanigans" that could bring down the company.

Months later, Thakur went to lunch at Barbhaiya's home. At the table, Thakur broached the topic again.

"Dinesh, I was trying to change the tires on a car running at sixty miles an hour," his mentor replied. Thakur asked him to elaborate. Barbhaiya brought up the budget projections for 2004. Thakur recalled those quite clearly, as he had compiled the data across departments for each regional portfolio. "Did the math make any sense to you?" Barbhaiya asked.

Thakur thought back to the number of products in development—

around 150. As Barbhaiya explained, in the United States it could cost a minimum of $3 million to develop each generic drug. The cost in India could be about half of that, since labor was so much cheaper. Ranbaxy's development budget to support that should have been around $225 million, but it was closer to $100 million instead, as Thakur recalled. The company was dramatically shortchanging its own work.

Thakur filed the information away and the conversation moved on. But the whole experience of Barbhaiya's departure made Thakur uneasy. Without a high-level ally, his own future at Ranbaxy looked grim.

In July 2004, Thakur's hopes soared when he met his new boss, Dr. Rajinder Kumar. Tall and handsome with elegant manners, Kumar had an open and warm disposition, a reputation for integrity, and a thoroughly blue-chip background. He had arrived from London, where he'd served as GlaxoSmithKline's global head of psychiatry-clinical research and development.

Kumar had completed his medical training in Scotland at the University of Dundee, then specialized in psychiatry at the Royal College of Surgeons. He went on to join SmithKline Beecham and became vice president and director for clinical development and medical affairs in neurosciences. There, he helped to develop Paxil, a blockbuster antidepressant. He was compassionate, oriented toward patients, and rigorous in his approach to good manufacturing practices.

Unlike the glowering Barbhaiya, who stayed in his office with the shades pulled down, Kumar kept his office door open. He frequently strolled around to meet subordinates at the laboratories and other work areas. Thakur liked and respected Kumar immediately, as did most who met him. Both men had been trained in an environment that valued transparency, and Thakur felt an immediate allegiance.

On the evening of August 17, after just six weeks at the company, Kumar sent Thakur an urgent email, asking him to report to his office early the next morning. Always punctual, Thakur arrived so early that he passed gardeners watering impeccable shrubs and cleaners still polishing the lobby's tile floors. He passed by the large portrait of Parvinder Singh, Ranbaxy's renowned CEO, as he headed to Kumar's office.

When Thakur stepped into his new boss's office that morning, he was surprised at his appearance. Kumar looked underslept and uneasy, his eyes puffy and dark. He had returned the day before from South Africa, where Ranbaxy's new CEO, Dr. Tempest, had dispatched him to meet with government regulators. It was clear from Kumar's demeanor that the trip had not gone well. The two men strolled into the hall to order tea from white-uniformed waiters.

"We are in big trouble," Kumar said to Thakur intently as they returned and then motioned for him to be quiet. In his office, Kumar handed Thakur a report from the World Health Organization (WHO). It summarized the results of an inspection that the WHO had conducted at Vimta Labs Ltd., a company that Ranbaxy had hired to administer clinical tests of its AIDS medicine. The WHO had done the inspection on behalf of the South African government, which was buying Ranbaxy's antiretroviral (ARV) drugs to treat its AIDS-ravaged population.

The inspection, conducted by a French inspector named Olivier LeBlaye, had uncovered astonishing fraud. Many of the "patients" Vimta had enrolled in the study did not seem to exist. Much of the data purporting to measure the drugs' dissolution in the patients' blood appeared to have been fabricated. The graphs from tests on entirely different patients were identical, as though photocopied. As Thakur read, his jaw dropped. There could be no assurance that the medicine had even been given to actual patients, owing to the lack of documentation. And there was no evidence that Ranbaxy had monitored the work or audited the results, as was required. This level of fraud meant that the drugs—destined for terribly sick AIDS patients—had essentially been untested.

With the company's credibility on the line, Dr. Tempest had sent Kumar to reassure South Africa's drug regulators that the situation at Vimta was isolated. Once there, however, Kumar went further, assuring the South Africans that he would do a full review of the antiretroviral portfolio and redo patient tests if need be.

Thakur listened intently as Kumar spoke. On the plane back to India, Kumar's traveling companion, the director of bioequivalence studies for the company's entire generic portfolio, told him that the problem was not limited to Vimta or to the ARVs.

"What do you mean?" asked Thakur, barely able to grasp Kumar's point.

The problem went deeper, said Kumar. He told Thakur that he wanted him to put aside all his other responsibilities for the foreseeable future, go through the company's entire portfolio—every market, every product, every production line—and determine what was real, what was fake, and where Ranbaxy's liabilities lay. Kumar then asked him to check in by day's end, as they set this plan in motion.

Thakur left Kumar's office stunned. Were more of Ranbaxy's drugs compromised? If so, how could the company have gotten approvals from the FDA, the world's toughest drug regulator?

As directed, he returned at the end of the day, but Kumar was not in. Thakur waited. Finally, Kumar arrived, looking visibly upset. Without a word, he sat at his desk and worked intently for twenty minutes before finally looking up. "I need a drink," he said darkly. Kumar explained that he'd spent the day fighting with the corporate office over what to do with the fraudulently tested ARV drugs. Kumar had insisted that there was only one right course: to withdraw the drugs from the market immediately and conduct the biostudies properly.

Though the corporate office had initially agreed, it drafted a press release stating only that Ranbaxy would *look into* the problem. Kumar revised the draft to state that the company was pulling the drugs off the market, effective immediately. But corporate kept returning the initial, vaguely worded press release for his approval. He sent back his revision *again*. "I am a physician, and I cannot sign off on something knowing full well it will cause harm to patients," Kumar declared. "I don't care how much money or face Ranbaxy loses. Either this stuff comes off the market or I am gone." Thakur did not want to contemplate losing another boss—especially one he liked so much.

Thakur later returned home to find his three-year-old son, Ishan, playing on the front lawn. He suddenly recalled an incident from the previous year when the boy had developed a serious ear infection. The pediatrician prescribed Ranbaxy's version of Amoxyclav, a powerful antibiotic. Despite his son's taking it for three days, the boy's 102-degree fever persisted. So the pediatrician changed the prescription to the brand-name antibiotic made by GlaxoSmith-Kline. Within a day, Ishan's fever was gone. Thakur took the boy in his arms, resolving not to give his family any more Ranbaxy medicine until he knew the truth.

PART II

INDIA RISES

CHAPTER 6

FREEDOM FIGHTERS

1920
Ahmedabad, Gujarat, India

For many years, few people wanted to take Indian medicine, let alone praise the companies making it. To the brand-name pharmaceutical companies that had spent decades and millions developing drugs, the Indian companies that copied their products were no better than thieves. They deserved to be sued rather than thanked. And to patients around the world, the MADE IN INDIA label connoted flea-market quality they'd prefer to avoid.

Behind the scenes, however, one man did more than anyone to change this perception and lay the path that led to Bill Clinton's visit: Dr. Yusuf K. Hamied. For years, as the chairman of the Indian drug giant Cipla Ltd., he had made drugs that even his own government wouldn't buy and made bold public offers that most people ignored. He didn't care what anyone thought of him. He relished the ire of the brand-name companies he battled. And then one day in 2001, he made an announcement that attached a whole new set of words to Indian drug makers: iconoclasts, visionaries, saviors.

But Dr. Hamied's story—and the launch of India's modern-day

pharmaceutical industry—really began a century ago, in an ashram. Not just any ashram, but the Sabarmati Ashram founded by Mahatma Gandhi in what today is Ahmedabad, in the western state of Gujarat. From there, India's most revered activist set about trying to liberate India from British rule, through what became known as the noncooperation movement.

Around 1920, Gandhi began urging all Indians to turn their backs on anything British. Civil servants abandoned government posts, Indian students left British-run universities, and civilians stayed home during the Prince of Wales's visit in November 1921.

A charismatic young chemistry student, Khwaja Abdul Hamied, heeded Gandhi's call. Handsome with a regal carriage, K. A. Hamied was a natural leader among his peers. He left school and went to Sabarmati, where Gandhi directed him and another student, Zakir Husain, to leave the ashram and start an Indian-run university. They did so. To the students, Gandhi was a "great prophet for the freedom of the country," as Hamied later recalled. "His words were law unto us." Husain would later become India's third president. The school he helped to found, Jamia Millia Islamia in New Delhi, remains in operation today, with a mission statement to prepare Indian students to be masters of their own future.

In 1924, after Gandhi—fiercely committed to nonviolence—suspended his movement because of an outbreak of violence among his followers, Hamied left to study abroad. He received his doctorate in chemistry in Berlin. On a school outing to a lake, he met and fell in love with a Lithuanian Jewish girl with Communist leanings. They married in 1928 in Berlin's only mosque. With Hitler rising to power, they headed to India, and later sponsored a dozen Jewish families to come there, saving them from almost certain death.

Back home in 1930, Hamied found a depleted and impoverished research environment, lacking even functional laboratories. The pharmaceutical market was almost entirely dominated by multinational companies, from Boots and Burroughs Wellcome to Parke Davis. With few exceptions, the Indians acted as mere distributors.

A young man with little money, Hamied dreamed of launching a grand laboratory and was finally able to found Cipla in 1935. Four years later, at the start of World War II, Gandhi visited Hamied's manufacturing plant and wrote in the guest book, "I was delighted to visit this Indian enterprise." But Gandhi's visit was not just a social call. It had a vital purpose. Britain had promised India its independence, so long as Indians helped in the war effort. Part of that effort was to make medicine. While demand for medicine from the Indian army had surged, supplies from Europe's drug makers had collapsed. Quietly, Gandhi urged Hamied to step into this void, which he did. Cipla became the largest producer of the antimalarial drug quinine and vitamin B12 for soldiers suffering from anemia.

Although Hamied lived modestly, the vagaries of Indian rent laws enabled him to lease a palatial apartment, the 7,000-square-foot ground floor of the Jassim House, in Bombay's elegant seaside neighborhood, Cuffe Parade. There, as India's famous freedom fighters visited the Hamied household, the children learned about the need for both political and personal independence. "If you are going to prosper in life, you have to do it yourself" was the lesson Hamied's young son, Yusuf, gleaned from his father.

India gained independence in 1947 (Gandhi was tragically assassinated just one year later). By then, Hamied had been elected to the Bombay Legislative Council. In 1953, he was appointed sheriff of Bombay, a ceremonial role akin to the city's leading ambassador. As politics consumed more of his time, his son Yusuf stepped up to run Cipla.

Known as Yuku to family and friends, Yusuf had his father's keen scientific mind, striking features, intense narrow eyes, and wry smile. At eighteen, he left to study chemistry at Cambridge and earned his PhD by the time he was twenty-three.

He returned in 1960 to work at Cipla, only to run into a thicket of red tape. Indian rules stated that, because the company was publicly traded, the government had to approve the hire of anyone related to a company director and set that person's salary. Consequently, for a year, Yuku earned nothing, and then for the next three years he made the equivalent of $20 a month, until he could reapply to the government for a raise.

Like his father, Yuku became a voracious reader of scientific literature. He essentially taught himself how to manufacture tablets and injectable drugs. He also revolutionized the making of active ingredients, a drug's essential components. These are often manufactured separately and sold to other pharma companies, which then add ingredients known as excipients to make a finished drug. Under Yuku's leadership, Cipla eventually became one of India's biggest manufacturers of bulk drugs. At the time, India followed the outdated British patent laws of 1911, which made medications in India even more expensive than in Europe. Under these laws, most pharmaceuticals were blocked for development. But Yuku found possibility in one medical area, where most of the patents had expired in the 1940s.

Looking around Bombay, with its profound poverty and exploding population, Yuku set out to develop a line of birth control medication. He offered it to the government, India's main purchaser and distributor of medicines, for two rupees (then around 20 cents) a month. Birth control in the United States cost around $8 a month. The government turned him down, seemingly indifferent to its population problem.

The birth control disappointment came at the same time as one of his biggest accomplishments. In 1961, he helped form the Indian Drug Manufacturers' Association (IDMA), which set out to amend India's antiquated patent laws. Prime Minister Indira Gandhi was highly sympathetic and would later say at the World Health Organization in 1981: "My idea of a better ordered world is one in which medical discoveries would be free of patents and there would be no profiteering from life or death."

The 1970 Indian Patents Act made it legal to copy an existing molecule, but illegal to copy the process by which it was made. This gave India's chemists freedom to remake existing drugs so long as they altered the steps to formulate them. The law created fierce antagonism between India's generic drug makers and the multinational brand-name drug companies, many of which left the Indian market.

India's golden age of pharma began as the new patent law took effect. Indians became accomplished reverse-engineers, and their companies began not only supplying the Indian market but exporting to Africa, Latin America, Iran, the Middle East, and Southeast Asia.

In 1972, Yuku's larger-than-life father, Khwaja Abdul Hamied, died, and Yuku became Cipla's CEO. Even as he amassed wealth—buying homes, polo ponies, and art—he did not relish India's "boom" in the same way as his peers did. All he had to do was walk outside his art-and-light-filled corporate headquarters in Mumbai to find an unignorable ocean of suffering. Millions of its residents lived in desperate slums, with no reliable access to electricity, sanitation, or food. Thousands lived on the roadside, without even a tarp over their heads.

The future he saw was one with too many people, too many diseases, and not enough medicine. It was this perspective—the world he was unable to *not* see—that led in large part to what happened next. It would take the son of an Indian freedom fighter and a Jewish Communist, with a background imprinted by Gandhi, to radically reimagine the horizon for Indian drug makers.

By contrast, the drug company Ranbaxy grew out of a set of values diametrically opposed to Cipla's. Unlike Hamied, Bhai Mohan Singh was not the product of Gandhian inspiration. He was a veteran of the so-called License Raj, India's antiquated business regime in which the government set every quota and dispensed every permit and license. The system required high-level connections and wads of rupees, not just to get licenses but to block the licenses of competitors.

Few got better results than Bhai Mohan Singh, a distinguished financier with exquisite manners, who was born in Punjab in 1917 to a wealthy Sikh family. His father had been a construction magnate. Bhai Mohan initially refused to enter the family business, but as World War II raged, his father's company got a massive contract to build barracks for the Indian army. Bhai Mohan's father sent his son to the Kangra valley to oversee the arrival of building supplies. More contracts followed the successful project, including one for a major highway to take British troops to the border of what was then Burma.

Through infrastructure, the family became one of the wealthiest in Punjab, and Bhai Mohan secured his place in a rarefied world. By 1946, his father had retired and transferred most of his assets to Bhai Mohan, making him a magnate. With this windfall, he formed a finance company, Bhai Traders and Financiers Pvt. Ltd., and loaned money to numerous businesses. One was a small drug distribution company called Ranbaxy & Co. Ltd.

His cousins Ranjit and Gurbax, a clothier and a pharmaceutical distributor, had started the company in 1937 (and named it by combining their two first names), with the modest goal of distributing foreign drugs in India. In 1952, after Gurbax was unable to repay the initial loan, Bhai Mohan bought the company, kept Gurbax on as president (because he was knowledgeable about the pharmaceutical trade), and incorporated in 1961. Bhai Mohan had three sons, including Parvinder. In time he would refer to Ranbaxy as his fourth.

Among Bhai Mohan's many assets, Ranbaxy was hardly a prize. At the time, the reputation of most Indian drug companies rested on the prestige of the foreign drugs they distributed. Ranbaxy primarily distributed Japanese drugs, then held in low esteem, so its reputation suffered. Chemists mocked the company salesmen for the cheap drugs they sold.

Bhai Mohan, who had no experience with pharmaceuticals, might have turned his back on the company entirely. But when Gurbax tried to oust him from the board, the duel for corporate control—which Bhai Mohan won—piqued his interest in the industry.

If Bhai Mohan knew little about pharmaceuticals, he knew everything about besting his rivals through boardroom maneuvers and the deployment of his connections, which included government ministers and powerful bank executives. When his alliance with an Italian drug company no longer suited him, for example, he had his allies surreptitiously report the company's legal violations to the government. The company was asked to leave the country, enabling Bhai Mohan to take over the Italian company's shares on his own terms.

However, his mastery of the License Raj did not help him make a cutting-edge pharmaceutical company. Ranbaxy had no particular mission or vision. Nor did it have any capacity to manufacture its own products. It had a single factory that reformulated bulk drugs into tablets and capsules from raw ingredients acquired elsewhere.

Two things propelled the company into the future. In 1968, Ranbaxy had its first pharmaceutical success when it launched a generic version of Roche's Valium under the product name Calmpose and rolled it out with a phrase from the nineteenth-century poet Ghalib: "When the day of death is set, why does sleep elude me all night?" The drug became India's first "superbrand." More consequential, however, was the return of Bhai Mohan's eldest son, Parvinder. In 1967, he came back from the United States with a doctorate in pharmacy from the University of Michigan at Ann Arbor and immediately joined the company, bringing with him skills and a seriousness of purpose that Ranbaxy had previously lacked.

Intense and more austere than his father, Parvinder had left India a middling student, more committed to refining his golf game than applying himself to his studies. But in Michigan, he had spent day and night in the laboratory. The dean there had written to Bhai Mohan that a student like Parvinder came along only once in a decade. He had also returned with a more profound sense of his own spirituality. He married the daughter of a guru who led a spiritual organization that prohibited alcohol and meat.

Unlike his father, Parvinder believed that Ranbaxy had to achieve independence. The company could have no control of its destiny, or the quality of its products, until it could make its own active ingredients and do its own research. Both required huge infusions of capital. To finance it, Parvinder helped take the company public in 1973.

Though Parvinder envisioned Ranbaxy as a global company, the rest of the world still held a dim view of Indian medicine. In Thailand, Ranbaxy's business was so poor that the company held a prayer ceremony with sixteen local monks to try to turn things around. The U.S. market was the world's largest and most lucrative, but also the most difficult to enter, with the most vigilant regulators. And though Indian companies were selling active ingredients to drug companies there, finished doses were a whole other matter.

In 1987, two Ranbaxy executives traveled to the United States to gauge the possibility of exporting there. As recounted in a company-sponsored book:

> They met representatives from 20 companies and all of them were bewildered to learn that an Indian company wanted to export finished formulations to the U.S. [One executive] recalls visiting a top American distributor who

> made him wait for two hours and then, after breaking a pencil [which had the words] "Made in Israel" [on it], he said, "Israelis are already invading the market here, and now we will have pharmaceuticals and medicines coming from India! What is the world coming to?"

However, some in the United States saw the future more clearly—and Parvinder forged alliances with them. He visited a fine chemicals importer named Agnes Varis at her Manhattan office. Varis was a firebrand and iconoclast, ahead of her time. She was one of eight children of Greek immigrants. Her father sold ice cream from a pushcart until his death, when Agnes was fourteen. Her mother, who was illiterate, sewed buttons in a garment factory. Varis got a degree in chemistry at Brooklyn College and in 1970, at age forty, launched her company, AgVar Chemicals, which brokered pharmaceutical ingredients from foreign manufacturers. Her company made millions, she became a major donor to the Democratic Party, and she was on a first-name basis with the Clintons.

Varis was immediately taken with Parvinder and his colleagues. "They were brilliant," she recalled in 2010, the year before her death. "They were gorgeous. They dressed beautifully. Their English was perfect. I felt they were on a very high standard." With her pharmaceutical know-how and golden Rolodex, she would become their political benefactor in their quest to sell finished doses in the United States.

Back in India, as the company drove to expand, Bhai Mohan was looking to ensure family peace. In 1989, at age seventy-one, he split the growing family businesses among his three sons, a settlement intended to ensure an amicable division before he died. He gave his eldest, Parvinder, all of his Ranbaxy shares. Manjit, the middle son, got an agrochemical company and some luxury properties. Analjit, the youngest, inherited Max India, a fine chemicals company whose biggest customer was Ranbaxy. Because Ranbaxy was the largest of the three companies, the two younger brothers inherited additional funds.

As the businesses inherited by the younger sons foundered, Manjit and Analjit grew bitter and believed that they had been shortchanged. They came to resent Parvinder, whom they viewed as the favored son. The brothers' relationship soured to the point where Analjit believed that Ranbaxy was orchestrating secret smear campaigns against him and his company. He told his close aides that he found Parvinder cold, calculating, and emotionless. "My brother loved to push a chili up my back every day," he recalled of his childhood.

For his part, Parvinder thrived as Ranbaxy's managing director. In 1991, India liberalized its economy, scrapping its system of quotas and licenses and opening whole sectors of the economy to foreign investment. Parvinder intensified his focus on expanding overseas. Within the company, his reputation was towering: "He was larger than life," one former associate remarked. "The people worshiped him. When he walked by, [employees] would bow."

However, he and his father were soon locked in conflict. Not long after gifting Parvinder his stake in Ranbaxy, Bhai Mohan began waging a battle against him. He accused Parvinder of violating the family settlement agreement by blocking his veto power over company matters. Their conflict was not just over the internal exercise of power, but rather over two different visions—one old and one new—of India. With the economy now open to outside capital, India's companies needed competent professionals with real skills, not just "liaison managers between the government and industrialists." Bhai Mohan's connections had become less important. Feeling sidelined, he openly fought with his son at board meetings.

Their biggest clash centered on a manager, Davinder Singh Brar, whom Parvinder brought into the company in 1978. Brar was a master tactician with an MBA who had become Parvinder's indispensable deputy. As father and son battled over whether to keep Brar on, Ranbaxy's board split their allegiances. Dozens of warring executives, some for Bhai Mohan, some for Brar, marched to the boardroom shouting competing slogans. An executive who led the pro-Brar delegation was fired by Bhai Mohan one morning but reinstated by Parvinder the same afternoon. On February 6, 1993, Manjit warned his father of plans to oust him at a board meeting that day. Bhai Mohan and his supporters resigned. Parvinder took over as Ranbaxy's chairman and managing director, and Bhai assumed the title of chairman emeritus. Parvinder and his father never made amends. Instead, Bhai Mohan filed a lawsuit against his son for violating the terms of the inheritance settlement, a dispute that would drag on into the next generation.

By 1995, Ranbaxy was the first company in Indian pharma to get a plant approved by the FDA to make products for the U.S. market. By then, four-fifths of Ranbaxy's sales came from abroad. Two years later, Parvinder was diagnosed with cancer of the esophagus. His father was so bitter that he wrote his son to say that he wouldn't attend his funeral. (In fact, he ended up organizing it and making use of all his connections to ensure that four hundred people attended the cremation.) With Parvinder's death in July 1999, his ownership of Ranbaxy passed to his two sons, Malvinder, age twenty-six, and Shivinder, twenty-four.

Bhai Mohan tried to draft his grandsons onto the Ranbaxy board immediately, as a way to reassert control over the company. But the spirit of Parvinder still seemed to be blocking this possibility. In his final press interview a month before his death, Parvinder insisted that his sons should only join the board once they'd gained enough professional experience to merit it. Honoring their deceased father, the brothers issued a statement saying that they would follow his wishes.

Shivinder went to work in a separate family-owned hospital business. Malvinder worked at Ranbaxy in a low-level position, making sales calls to doctors and pharmacists in small towns and villages. With his Western business education (he had attended Duke University's Fuqua School of Business) and Indian spiritual values, Malvinder rose quickly at Ranbaxy. Meanwhile, the company was being run by the professional manager Brar, whose original tenure had precipitated the ugly fight between father and son.

Parvinder had set up Ranbaxy for a globalized world. But for all his foresight and preparation, he never could have planned what would ultimately bring the world to Ranbaxy's doorstep. An immense crisis was looming, one that would eclipse the skills of even the most professional manager. It would require moral imagination on a vast scale—and summon Cipla's CEO Yusuf Hamied to action.

CHAPTER 7

ONE DOLLAR A DAY

1986

Mumbai, India

Dr. Yusuf Hamied of Cipla was a prolific reader of medical journals, with an annual subscription budget that topped $150,000. One day in 1986, he was introduced to something he knew nothing about. A colleague mentioned, "According to the Tufts report, AZT is the only drug available for AIDS."

"What is AIDS?" Dr. Hamied responded.

Just five years earlier, in 1981, the U.S. Centers for Disease Control and Prevention (CDC) had reported the emergence of a rare cancer, Kaposi's sarcoma, among clusters of young gay men in San Francisco and New York. The following year, doctors and the media gave the puzzling ailment the misleading shorthand name GRID (gay-related immune deficiency). In Africa, doctors confronting a mysterious wasting syndrome called the disease "Slim." By the summer of 1982, the CDC had connected these dots, identifying acquired immune deficiency syndrome (AIDS) and its precursor, the human immunodeficiency virus (HIV).

At the time of Dr. Hamied's question, the disease had barely surfaced in most of India. But it was brewing so forcefully in Bombay's red-light district, not far from Cipla's headquarters, that within a few years Hamied's own city would earn the moniker "AIDS Capital of India."

Within a decade, AIDS was destroying Africa. More than five thousand people a day were dying. In certain African countries, one-quarter of the population was infected. In some local communities, the biggest industry was making wooden coffins. Africa was becoming a continent of orphans, the number of children without parents doubling every year. The disease was projected to kill up to 90 million Africans by 2025.

In 1991, Dr. Rama Rao, the research head of an Indian government laboratory, told Hamied that he had developed a chemical synthesis of AZT, or azidothymidine, and wanted Cipla to manufacture it. It was the only drug that postponed the onset of AIDS. But just one company, Burroughs Wellcome in the United States, made it, and it was selling the drug at roughly $8,000 per patient a year. Hamied readily agreed to manufacture it and launched the drug in 1993 at less than one-tenth of the international price, or about $2 a day. Even that price was still well beyond what most Indians could afford. "Our sales were zero," Hamied recalled.

At that point, Hamied asked the government if it could purchase and distribute the drug. But the Indian government refused: it had money only for detection and prevention, not treatment. In total disgust, Hamied ended up discarding 200,000 capsules. With the stigma surrounding AIDS, he had no one to sell them to—or even give them to.

A few years later, Hamied read in a medical journal that a cocktail of three drugs called HAART (highly active anti-retroviral therapy) was effective in controlling AIDS. The three drugs in question—stavudine, lamivudine, and nevirapine—were made by three different multinational drug companies. The combined price for a single patient reached $12,000 a year. Not only was the treatment regimen onerous, but few could afford it. Hamied immediately set out to make the drugs in the cocktail.

In 1997, under the leadership of Nelson Mandela, South Africa altered its law to make it easier to sidestep pharmaceutical patents and import low-cost medicine. No country needed the AIDS cocktail more badly than South Africa, which had emerged as an epicenter of the epidemic. But South Africa, along with over 130 nations, was bound by an international trade agreement called TRIPS (Trade-Related Aspects of Intellectual Property Rights), which required that all members of the World Trade Organization (WTO) ensure basic protection for intellectual property.

South Africa's new law sparked a furious reaction from Big Pharma. Fearing a domino effect, thirty-nine international brand-name drug companies, with the support of the U.S. government, sued South Africa, claiming that the new health law violated the TRIPS agreement. The Pharmaceutical Manufacturers Association of South Africa, representing Big Pharma, rolled out newspaper ads with an image of a crying baby, warning that the law would allow "counterfeit, fake, expired and harmful medicines" into the market. Brand-name drug companies closed factories and withdrew from the country, with the claim that South Africa was intent on destroying international treaties.

It was a deadly global stalemate. As drug companies skirmished over intellectual property, 24 million people got sicker, with no foreseeable access to the affordable medicine they so desperately needed. On August 8, 2000, Hamied got a call from an activist in the United States whom he had never met. "Me and some of my colleagues would like to come see you," said the man on the phone. It was William F. Haddad, the expletive-spouting former investigative journalist who had campaigned so vigorously for the Hatch-Waxman Act, the law that launched the U.S. generic drug industry.

He had gotten Hamied's name from Agnes Varis, who told

him, "He's a brilliant chemist. And he's not frightened by the multinationals." The colleagues Haddad referred to were a motley group of activists who had banded together in pursuit of a single objective—to find a way to get affordable AIDS medicine to those who needed it most, free from the stranglehold of patents. Jamie Packard Love, an intellectual-property activist, had helped advise the South African government on amending and defending its new patent law. In the United States, he had set out to learn the real cost of manufacturing the drugs, and nobody seemed to know. "If it was 40 million white people that were going to die, somebody would know the answer to that question," Love recalled.

Four days after contacting Hamied, Bill Haddad, Jamie Love, and three others, including a French doctor with the group Doctors Without Borders, arrived at the elegant London duplex where Hamied sat out the brutal Indian summer. He led them up the stairs to a glass dining table. Surrounded by costly art, including a work from one of India's most famous artists, M. F. Husain, and overlooking the exclusive Gloucester Square garden, they asked him: how low could he price the AIDS cocktail, and how much of it could he make?

As they spoke, Hamied scrawled calculations with a pencil and paper. He concluded that he could cut his price by more than half, to about $800 a year. The men talked into the night, and the group vowed that they would support Hamied in the inevitable battles with the multinational drug companies that lay ahead. Together, an Indian drug maker and international activists had forged an extraordinary alliance, pledging to upend the established global commercial and pharmaceutical order in order to save millions of lives.

About a month later, in part through their efforts, Hamied received an invitation to speak at the European Commission's conference in Brussels on HIV/AIDS, malaria, tuberculosis, and poverty reduction. He readily accepted and was given three minutes for his address. On September 28, 2000, he took to the podium and looked out over the gathering of staid, skeptical—and white—Europeans, who included health ministers, ex–prime ministers, and representatives of multinational drug companies. "Friends," he told the unfriendly group, "I represent the third world. I represent the needs and aspirations of the third world. I represent the capabilities of the third world, and above all, I represent an opportunity."

He then proceeded to unveil three offers: he would sell the AIDS cocktail for $800 a year ($600 to governments buying in bulk); give the technology to make the drugs free to any African government willing to produce its own drugs; and provide nevirapine, the drug that limited transmission of the disease from mother to child, for free. He was literally slashing his own prices before their eyes. He closed with a challenge: "We call upon the participants of this conference to do what their conscience dictates."

Hamied had been expecting a pharmaceutical revolution as governments embraced his offer. The world did not often hear the word "discount," let alone "free," attached to expensive medicine. But he finished to a stone-silent room. No one took him up on his proposal. Partly, this was because Hamied was offering discounted medicine in the middle of a minefield. The global pharmaceutical marketplace was intersected by patents and trade agreements, which precluded many countries from reaching out for cheap medicine. But the other problem was credibility. Much of the world viewed Indian generics as poor-quality knockoffs, a perception that Hamied had labored against for years.

In 2000, that perception led *New York Times* reporter Donald G. McNeil Jr. to India to try to reconcile these conflicting views. He'd heard from officials at Doctors Without Borders that some Indian manufacturers were producing high-quality drugs at a fraction of the cost. Which was true? Fly-by-night counterfeiter or quality discounter? Hamied offered McNeil full access to his operations and laboratories. The resulting article, a detailed profile of the Cambridge-educated chemist, ran on the front page of the *New York*

Times and introduced Western readers to a new concept: counter to the claims of brand-name companies, costly drugs could actually be made reliably for pennies.

On January 26, 2001, one of the most devastating earthquakes ever recorded rocked the western state of Gujarat. It killed 20,000 people and injured 160,000 more. The world scrambled to respond. Bill Clinton, who'd just left office, raised money and traveled to India to offer aid. Dr. Hamied opened his warehouses and donated vast amounts of medicine. But for him, the event was a different sort of wake-up call. As the world scrambled to save the victims trapped under buildings or left homeless, Hamied realized that AIDS dwarfed the Gujarat earthquake by orders of magnitude. He decided then that he could not simply wait for governments to take him up on the offer he'd made in Brussels.

Just as he was pondering his next steps, the way forward presented itself. A few days after the earthquake, William Haddad called Hamied back, this time with a specific question. Would Cipla be able to offer the AIDS cocktail for $1 a day? After some back-of-the-envelope calculations, Hamied agreed. He would offer the price exclusively to Doctors Without Borders. It was a number low enough to be world-changing.

On February 6, 2001, at around midnight, Hamied was at a dinner party in Mumbai when his cell phone rang. The caller was the *New York Times* reporter Donald McNeil. "Dr. Hamied, is it true you offered $1 a day [to Doctors Without Borders]?" McNeil asked him. Once Hamied confirmed this, McNeil laughed aloud: "Dr. Hamied, your life will not be the same from tomorrow."

McNeil's story was published the next morning on the front page of the *Times*. According to the article, Cipla was offering to sell the AIDS cocktail for $350 a year per patient, or roughly $1 a day, as compared to Western prices of between $10,000 and $15,000 a year, but was being blocked by the multinational drug makers that held the patents, who were being backed by the Bush administration. McNeil's story "completely broke the dikes," Jamie Love recalled.

Papers all over the world picked it up. News of Big Pharma's patent protection efforts in the face of the global pandemic and the Bush administration's support of them sparked international outrage and stoked street protests from Philadelphia to Pretoria, even accusations of genocide. The result was a PR debacle for Big Pharma. Even among the industry's lowest moments—the illegal marketing of drugs for off-label uses; the payoffs to doctors who acted as promotional mouthpieces; the concealment of negative safety data for high-profile drugs—its stance in South Africa seemed uniquely horrible. As the *Wall Street Journal* summed it up: "Can the pharmaceuticals industry inflict any more damage upon its ailing public image? Well, how about suing Nelson Mandela?"

It was an outrage that William Haddad would never forget. "Big Pharma, those cock-sucking bastards," he yelled to a journalist years later. "Thirty-four million people had AIDS and every single one of them would die without the medicine. Would die and were dying. And they charged $15,000 dollars a year, and only four thousand people [in Africa] could afford the medicine."

The disgust was mutual. As GlaxoSmithKline's CEO Jean-Pierre Garnier declared of Cipla and the Indian generics companies at a 2001 health care forum, "They are pirates. That's about what they are. They have never done a day of research in their lives." Some in Big Pharma accused Hamied of trying to grab market share in Africa, to which he responded: "I am accused of having an ulterior motive. *Of course* I have an ulterior motive: before I die, I want to do some good."

On March 5, 2001, as Big Pharma's legal case against the South African government began in Pretoria, people rallied against the drug companies all over the world. In South Africa, demonstrators marched outside the high court in Pretoria. In Britain, they picketed GlaxoSmithKline plants. In the United States, AIDS activists held rallies in major cities.

Hamied and the activists prevailed. The following month, the multinational drug companies announced that they would drop their lawsuit and waive their patents so that generic fixed-dose combinations of the AIDS cocktail could be sold cheaply in Africa. In August, Cipla announced that its scientists had succeeded in creating Triomune, a single pill containing the unwieldy AIDS cocktail. To do so, Cipla had skirted Western patent laws, since the three drugs were made by different manufacturers, under separate patents, and could not legally be copied for years.

The Clinton Foundation stepped in and hammered out a deal in which Indian drug makers agreed to slash their prices even further, to 38 cents a day, in exchange for guaranteed large-volume purchases from African governments. The foundation even brought in process chemists to help the companies find ways to reduce the number of steps required to manufacture the drugs, which also brought down costs.

However, it was the $1 a day figure that changed the calculus of the West—from "we can't afford to help" to "we can't afford not to." The AIDS activists had not counted on President George W. Bush as an ally. But on January 28, 2003, he stunned them by announcing in his State of the Union address a new program that would spend $15 billion on AIDS drugs over five years. He explained that the dramatic drop in costs "places a tremendous possibility within our grasp. . . . Seldom has history offered a greater opportunity to do so much for so many." The program, which still operates today, was named PEPFAR (President's Emergency Plan for AIDS Relief). Finally, the world, in its halting way, had caught up to the revolution that Hamied had launched.

For Big Pharma, PEPFAR was a nightmare scenario: a U.S.-taxpayer-funded effort to spend billions on generics bound for Africa. Just days after Bush unveiled the plan, some CEOs of multinational drug companies petitioned the White House to undo the commitment of the $1 a day cost. The answer was no. As a concession, however, Bush allowed the group to choose the head of PEPFAR. To the dismay of AIDS activists, they installed Randall Tobias, the former CEO of the drug company Eli Lilly.

Beyond the problem of cost, another question still loomed: quality. How could the West guarantee the quality of all the AIDS drugs it was buying for Africa? The generics advocates turned to the World Health Organization, which agreed to serve as an international clearinghouse for quality generics. It would inspect companies that wanted to sell AIDS drugs internationally and, if approved, would add those names to a prequalified list. But this solution didn't satisfy everyone. Suddenly, under Tobias's leadership, PEPFAR introduced a new requirement: any AIDS drugs being purchased for sale to Africa with U.S. taxpayer dollars had to be approved by the U.S. Food and Drug Administration.

The requirement triggered an avalanche of criticism. To AIDS activists, this was the ultimate bait-and-switch. Most of the Indian companies had never gotten FDA approval for any of their drugs. Activists suspected that this was a needless safeguard whose real purpose was to funnel money to Big Pharma and keep out generic manufacturers. They gained powerful adherents. In March 2004, six senators, including John McCain (R-AZ) and Ted Kennedy (D-MA), wrote a biting letter to President Bush, arguing that FDA approval would needlessly delay access to drugs and that the WHO's standards "meet or exceed those used by respected regulatory agencies around the world." In an obvious dig at Big Pharma, the senators wrote, "We question the purpose behind the Administration's duplicative process being developed to review the safety and efficacy of generic drugs." According to a former White House aide, even President Clinton called President Bush to explain that WHO approval had satisfied his foundation.

But the requirement for FDA review had not simply been a cynical advancement of Big Pharma's interests. Inside the Bush administration, not everyone felt so sanguine about the quality of Indian generics. In a series of tense meetings at the White House and at the FDA, officials wrestled with the question of how to verify the quality of the drugs they would procure. They were bitterly divided. "We thought it would be a terrible ugly thing for U.S. taxpayers to buy subpotent, contaminated drugs for AIDS [in Africa]," recalled Dr. Scott Gottlieb, who was then director of medical policy development at the FDA and today is its commissioner. "There was a lot of pressure to buy Indian counterfeit drugs," he recalled to a journalist, years prior to becoming commissioner, adding, "I call them counterfeits, everyone else calls them generics."

Finally, in the face of intense bipartisan pressure, a compromise emerged. The FDA created an accelerated review process for PEPFAR drugs, resulting in a perceived win for public health. Affordable generic drugs available in Africa would have the benefit of FDA review. On May 27, 2005, Ranbaxy became the first Indian generics company to get approval from the PEPFAR program for one of its AIDS drugs. Many other companies followed.

Some of the world's poorest people owe their lives to Dr. Hamied. However, his revolution came with an unintended consequence. As more Indian companies got approval from the FDA to sell their drugs in Africa, this led to a realization that would in short order upend the generic drug industry and transform the American drug supply: if Indians could make affordable medicine good enough to be approved by American regulators, then the drugs were good enough for Americans to take too.

CHAPTER 8

A CLEVER WAY OF DOING THINGS

DECEMBER 2005

Canonsburg, Pennsylvania

A brand-name drug, no matter how complex or difficult to make, inevitably follows a recipe, such as: mix fifteen minutes, granulate, mist until ingredients reach 4 percent moisture content, mix again for thirty minutes. Making a generic version, however, requires figuring out a different recipe, ideally one that is faster to make but produces a similar result. That effort of reverse-engineering is undertaken by process chemists.

Among them, Rajiv Malik ranked as one of the best. Trained in the laboratories of Punjab, he held more than sixty process patents for reverse-engineering. Over seventeen years at Ranbaxy, he rose to the head of formulation development and regulatory affairs. He was also a veteran of laboratory disasters, such as when Ranbaxy's anti-acne drug Sotret yielded failing results. His colleagues' "irrationality" in choosing to continue sales of the defective formulation contributed to his decision to leave, he would later say. He submitted his resignation in June 2003.

Taking an eight-step chemical synthesis, like A-B-C-D-E-F-G and altering it to G-C-B-F is no simple task. The formulation has to produce results that can withstand scrutiny from regulators and legal challenges from patent lawyers. Malik was in the business of finding solutions to problems that eluded others. Fast-talking, with a buoyant personality, he had a warm smile, graying salt-and-pepper hair, and a tendency to curse exuberantly.

Malik left Ranbaxy just as the U.S. government came to rely on Indian companies to make low-cost drugs for Africa. With that came a notable shift. Indian companies were also moving into the U.S. market, and U.S. companies were moving their operations to India, in a turbo-charged global marketplace that prized Malik's particular skills.

Two and a half years after his departure from Ranbaxy, he became chief operating officer of Matrix Laboratories, a drug company in Hyderabad, founded by an Indian industrialist. Some of his Ranbaxy colleagues followed him. Together, they helped build Matrix into the world's second-largest producer of active ingredients, with a particular focus on the AIDS drugs being purchased for the PEPFAR program. There, Malik was essentially perched on top of a new ecosystem, one created by the promise that Dr. Yusuf Hamied of Cipla had made to the world. Indian companies could make mass quantities of effective drugs at bargain-basement prices, while still following all the good manufacturing practices required by Western regulators.

As to how Indian companies could pull off such a feat, Dr. Raghunath Anant Mashelkar, a renowned evangelist for Indian science, advanced a theory: India's scientists excelled at rethinking old processes and making them more efficient because of their engineering excellence and experience of destitution. The result, said Mashelkar, was "Gandhian innovation." One of Mahatma Gandhi's essential tenets held that the inventions of science should be for the public benefit. Because Indians had minimal resources, Mashelkar argued, they had developed a "clever way of doing things" that delivered more benefits to more people at a lower cost.

Some still viewed Indian drug companies as bottom-feeders, living off the remnants of painstaking research and innovation. But Mashelkar explained that "affordable" did not necessarily mean "worse." It could often mean "better." At Matrix, Malik's exceptional results stood out. And it did not take long for the West to come calling.

The Appalachian generic drug company Mylan Laboratories was every inch American. In 1961, two army vets launched it in an abandoned ice skating rink in White Sulphur Springs, West Virginia. The company became known for its ethos, as articulated by one of its founders, Mike Puskar: "Do it right, or don't do it at all." Its flagship plant, in Morgantown, West Virginia, which sits on twenty-two acres, became one of the world's largest. Its size and importance led the FDA to establish an almost continuous presence there. As the company-sponsored book *Mylan: 50 Years of Unconventional Success* recounted, an FDA investigator would perch on a ladder and draw a white-gloved finger across the top of the manufacturing equipment. (Regulations required that all surfaces be dust-free.) Company executives let out a sigh of relief once the gloved finger came up "white as ever."

The plant required its technicians to pay keen attention to detail. Prospective employees were shown a fifteen-minute video (unrelated to drug making) and then quizzed on what they'd seen: What came first? What came second? "In a GMP [good manufacturing practices] environment, you want them to do it the way they're instructed to do it," explained Kevin Kolar, Mylan's former vice president of manufacturing technical support. "If someone makes a mistake, an investigation has to be launched."

Any responsible drug maker had to try to minimize risk. But by late 2005, Mylan's CEO, Robert Coury, was confronting the ultimate wild card: Mylan was losing market share to Indian drug companies that made their own active ingredients in-house and operated at rock-bottom costs. By contrast, Mylan was ordering ingredients from Chinese and Indian suppliers. Mylan couldn't beat their price—unless it joined them and went global.

Coury turned to one of Mylan's ingredient suppliers, Matrix Laboratories. In December 2005, he met Matrix's chairman in a New Jersey airport lounge, and the two men hammered out a deal on a cocktail napkin. Mylan would become the first U.S. company to buy a publicly traded Indian company. The deal, which closed in January 2007, gave Mylan a global platform. But perhaps Mylan's greatest asset from the acquisition was Rajiv Malik himself, who became Mylan's executive vice president in charge of global technical operations. He brought a number of his tried-and-true team members who'd been with him at Ranbaxy.

At Mylan, he became part of an American executive team. Alongside Coury sat Chief Operating Officer Heather Bresch, the daughter of West Virginia Democratic governor Joe Manchin (now senior U.S. senator). The Indians and the Americans liked to describe the integration of Matrix and Mylan as seamless. "Bresch and Coury saw in the Matrix team a mirror image of Mylan's: they were ambitious, hard-working, and committed to quality," as Mylan's company-sponsored book recounted. By the time the deal closed, "we started speaking the same language," Malik said. At their first celebratory dinner, they all ate Indian food, which most of Mylan's executives had never sampled, reared as they were on meat and potatoes.

Yet the differences between the two teams were as real as the different worlds from which they'd come. In India, generics companies were star performers whose every flicker of stock price got breathless coverage by the business press. In the United States, generic drug companies operated with relative anonymity. When Malik eventually settled in a Pittsburgh doctors' community, he noted with surprise: "Nobody fucking knows Mylan."

But a more consequential difference between Malik's past employers and his new one was the companies' orientation toward quality. In theory, all companies that made drugs for highly regulated markets operated inside a triangle of cost, speed, and quality. Of those, quality was supposedly a fixed point, its requirements set by regulation. Manufacturing processes had to be transparent, repeatable, and investigable, without exception or deviation. But generic drug makers, under enormous pressure to reduce costs and speed up development so that their applications were first, faced a central tension: How low could they slash their costs? And how quickly could they move before quality began to suffer?

Some in the industry claim that it costs about 25 percent more to follow the good manufacturing practices required for regulated markets like the United States. That leaves companies with difficult choices. What if a sterile mop costs $4 (far more than a regular one), and in a typical day you are supposed to use nine mops? What if your customers want a vaccine for 4 cents a dose but it costs 40 cents to make? But the central problem is the generic drug business model itself. How can you maintain quality when a brand-name pill that costs $14 one day is going to cost 4 cents as a generic the next day? This dynamic "doesn't motivate you to invest" in maintaining high manufacturing quality, as Malik himself acknowledged.

In the face of this conflict, company culture mattered. Malik and his team had gotten their training at Ranbaxy, where the posters on the office walls exhorted employees to push toward the company's goal of reaching $1 billion in U.S. sales by 2015. But Malik had landed in a different culture. The posters in Mylan's conference room emphasized: "Discover why at Mylan quality isn't just a claim, it's a cause we've made personal." Mylan was so committed to its image of quality and transparency that it eventually built a glass-walled headquarters in Canonsburg, Pennsylvania, and gave its executives partly transparent business cards.

At Mylan, Malik was tasked with doing what he had always done: navigating an array of obstacles in the laboratory; filing applications swiftly enough to keep the pipeline full; and ensuring that these efforts passed muster with the world's toughest regulators. But he was doing it all from the glass-walled headquarters of an American company that anticipated, and was accustomed to, continuous oversight.

Patients tend to assume that their generic drugs are identical to brand-name drugs, in part because they imagine a simple and amicable process: as a patent expires, the brand-name company turns over its recipe, and a generic company makes the same drug, but at a fraction of the cost, since it no longer has to invest in research or marketing. But in fact, generic drug companies fight a legal, scientific, and regulatory battle, often in the dark, from the moment they set out to develop a generic. Mostly, their drugs come to market not with help from brand-name drug companies, but in spite of their efforts to stop them.

Brand companies often resort to "shenanigans" and "gaming tactics" to delay generic competition, as the exasperated FDA commissioner Scott Gottlieb put it. They will erect a fortress of patents around their drugs, sometimes patenting each manufacturing step—even the time-release mechanism, if there is one. They may make small alterations to their drugs and declare them new, to add years to their patents, a move known as "evergreening." Rather than sell samples of their drugs, which generic makers need in order to study and reverse-engineer them, brand-name companies will withhold samples, which in 2018 led the FDA to begin publicly shaming the companies accused of such practices by posting their names on its website.

To successfully launch a product, generic drug companies must tread in reverse through this obstacle course. Once a generic company zeroes in on a molecule, and its scientists figure out how it operates in the body, its lawyers get to work to establish how well protected it is legally. The next step takes place in the laboratory: developing the active pharmaceutical ingredient by synthesizing it into ingredient form. That alone can take several years of trial and error. Once successful, the finished generic has to take the same form as the brand, whether that be pill, capsule, tablet, or injection. Formulating it requires additional ingredients known as excipients, which can be different, but might also be litigated.

Then comes testing. In the lab, the in-vitro tests replicate conditions in the body. During dissolution tests, for example, the drug will be put in beakers whose contents mimic stomach conditions, to see how the drugs break down. But some of the most important tests are in-vivo—when the drug is tested on people.

Brand-name companies must test new drugs on thousands of patients to prove that they are safe and effective. Generic companies have to prove only that their drug performs similarly in the body to the brand-name drug. To do this, they must test it on a few dozen healthy volunteers and map the concentration of the drug in their blood. The results yield a graph that contains the all-important bioequivalence curve. The horizontal line reflects the time to maximum concentration (Tmax) of drug in the blood. The vertical line reflects the peak concentration (Cmax) of drug in the blood. Between these two axes lies the area under the curve (AUC). The test results must fall in that area to be deemed bioequivalent.

Every batch of drugs has variation. Even brand-name drugs made in the same laboratory under the exact same conditions will have some batch-to-batch differences. So, in 1992, the FDA created a complex statistical formula that defined bioequivalence as a range—a generic drug's concentration in the blood could not fall below 80 percent or rise above 125 percent of the brand name's concentration. But the formula also required companies to impose a 90 percent confidence interval on their testing, to ensure that less than 20 percent of samples would fall outside the designated range and far more would land within a closer range to the innovator product.

After the active ingredients are manufactured, the additional ingredients chosen, and the principal laboratory and clinical tests conducted, the formula then moves to the manufacturing floor to see if it can be made on a commercial scale.

As the manufacturing runs become larger, the processes become harder to control. If something can go wrong, it will. You can build a fortress of current good manufacturing practices around the drug-making process and still "shit happens," as Malik liked to say. Conscientious manufacturers try to protect against past disasters and prevent new ones. But because manufacturing plants are operated by humans, the systems will break down, no matter how perfectly designed they are. For example, Johnson & Johnson's epilepsy drug was fine until the company stacked it on wooden pallets that likely leached solvents into the medicine. At Mylan's Morgantown plant, one lab technician left a note for another stating that he had to "rig" a hose on the equipment to get it to work properly—a word choice that easily could have shut down the plant had an FDA investigator stumbled across it and suspected fraud instead of primitive problem-solving.

The only remedy for this variability is for plants to adhere scrupulously to good manufacturing practices and create real-time records of each drug-making step. The resulting data serve as a blueprint for finding and fixing the inevitable errors, a process that FDA investigators scrutinize. How well and how closely did the company investigate itself? The goal is to address a problem "in a way that it never happens again," as Malik explained.

As Malik confronted these challenges, he proved that he was more than a formulation wizard. He was adept at remaking himself. He rose swiftly at Mylan to chief operating officer. Bresch, in turn, became CEO, and Coury was elevated to executive chairman. Malik oversaw a swelling operation in India, where the company would soon grow to have twenty-five of its forty global facilities and over half of its 30,000 employees.

In many ways, Malik took Mylan by storm. He reoriented the company toward India and helped create competition between the research and development teams in Morgantown and Hyderabad. Within three years, the number of drug applications Mylan submitted to the FDA tripled and approvals doubled. But Malik was also careful to emphasize the importance of quality and summed up this priority for employees in typical fashion: "If you miss something in the quality space, we don't live with that shit."

His rapid ascent seemed proof that he'd mastered the tensions between cost, quality, and speed, using cleverness, as the innovation expert Mashelkar had put it. But Mashelkar made one important distinction. India's "clever way" of doing things was not to be confused with an approach that Indians referred to as *Jugaad*—taking ethically dubious shortcuts to get as quickly as possible to the desired goal. The word connoted a compromise on quality and an approach that should be "banished" entirely, Mashelkar explained to a visitor in his office at the National Chemical Laboratory in Pune.

But at Mylan, as Malik and his team seemed to hit every formulation deadline, some employees began to wonder whether Gandhian innovation was the only explanation for their success.

CHAPTER 9

THE ASSIGNMENT

AUGUST 18, 2004

Gurgaon, India

At 8:30 a.m., the heat already stifling outside, Dinesh Thakur looked around the conference table at his six project managers and saw tired faces. Some had left their homes hours earlier to beat rush-hour traffic and arrive on time. They knew the meeting was important, but didn't know its agenda. Thakur was about to give his team one of the stranger tasks in the annals of corporate due diligence, but had decided in advance to keep the true reasons for it to himself.

"We have a new assignment from Dr. Kumar," he began. "He wants to know if we can substantiate all of the data that we have provided to various countries. This is a retrospective review of our portfolio, and he wants to know how confident we are about the information we have provided to various regulatory agencies in the last twenty years."

The members of his team looked surprised, yet the assignment fell within their wheelhouse. Their job was to map all of Ranbaxy's data, so it was only logical to find out if it was accurate first.

Thakur directed their attention to a large whiteboard where he'd drawn a graph. On the vertical axis were all the regions of the world where the company sold its drugs. On the horizontal axis were myriad questions. What products were on the market? When were they registered? Where were the actual dossiers used to register the products? Where was the supporting data? How many batches were sold in that market? What facility manufactured them?

Thakur assigned each of his staffers a region of the world. He directed them to compare the company's raw manufacturing data for the drugs in those markets against the claims made in submissions to regulators. Did the data match up, or were there any discrepancies? Did the submissions comply with local regulations?

A picture of the company's entire operations had never been pieced together before. Until then, Ranbaxy had been partitioned. Specific groups worked on product development for different regions, but almost never met to compare notes. No one had a complete picture of how—or even where—the company's drugs were approved. But Thakur had directed his team to make a multidimensional assessment that spread across the entire globe and stretched back years.

His own efforts began with a visit to the associate director of regulatory affairs, Arun Kumar, who had been directed by Raj Kumar (no relation) to cooperate.

Arun, who worked in the office right above Thakur's, was waiting with a bemused air. "Everyone knows," he said, by way of greeting.

"Knows what?" Thakur asked.

"What the reality is," he said. Arun went on to describe how Ranbaxy took its greatest liberties in markets where regulation was weakest and the risk of discovery was lowest.

"Are you saying that the products on the market in those regions are not all supported by data?"

"Well, not all of them," said Kumar in a casual tone, flipping through a report on his desk. "We know where the gaps are."

Thakur was stunned by his nonchalance. "Did you bring it up to management?"

"What for?" Kumar replied. "They already know. In fact, they probably know it better than I do."

Thakur thought he must have misunderstood. If the applications included known gaps, he asked, how could Arun sign off on them when he was the one certifying their accuracy?

That was part of the problem, Arun explained. Even though he had prepared the dossiers, the regional regulatory heads, like Abha Pant in the United States, could make any changes they liked to any of the applications. They took their orders directly from the top management and signed off on dossiers without Arun's knowledge or consent.

This was incredible to Thakur. At a company like Bristol-Myers Squibb, the regulatory affairs directors had absolute control over what was submitted to the FDA, and for good reason. When regulatory executives signed submissions, they were asserting the data was accurate. It was a criminal offense to make a false statement on a government record.

"Surely you are not saying what I think you're saying, right?" Thakur asked.

"There is too much to lose in advanced markets like the U.S. and Europe if you get caught, so it doesn't make sense to take blatant risks for those portfolios," Arun explained. "However, [Africa], Latin America, India, that is a whole different story."

Thakur was dumbfounded. "Who all knows?"

"Everyone does," Arun said, adding, "Everyone knows where the directives come from."

"Is there no fear of the repercussions?"

"They are managed," said Arun. "Everything is managed."

Thakur was so stunned that he had to end the meeting and leave, just to compose himself.

As he walked through the door of his office, his executive assistant asked, "What is wrong with you? You look like you have seen a ghost." Thakur collapsed in his chair.

He knew that compliance failures and ethical lapses existed within the brand-name drug industry. After he left Bristol-Myers Squibb, the company's finance chief and the head of its worldwide medicines group would both be charged with criminal conspiracy and securities fraud for allegedly inflating sales and profits by concealing unsold inventory, charges that were later dropped.

It was one thing to game the stock market or harm shareholders, allegations that would lead the men to reach a settlement with the U.S. Securities and Exchange Commission, without admitting guilt. But what Arun had just described was another matter altogether. You *had* to test the drugs to see if they were properly formulated, stable, and effective. The resulting data was the only thing that proved the medicine would cure instead of kill. Yet Ranbaxy was treating data as an entirely fungible marketing tool, apparently without consideration for the impact on patients. It was an outright fraud that could mean the difference between life and death.

Thakur could barely wrap his mind around it. But the assignment he'd been given drove him to return to Arun Kumar's office later that day.

"There is no point digging up these things," Arun told him. "You will find yourself out of the company if you go down that path. Tell Raj you have looked into it, and there is nothing more to see."

"I can't lie to my boss," said Thakur.

"What is wrong with you guys who go to the U.S. for a few years and think you have become the moral police of the world?" asked Arun. "Do you think U.S. pharma companies never do such things?"

In his ten years with Big Pharma, Thakur had never seen or imagined such conduct. He was young and somewhat naive, but also stubborn and not inclined to back down. "Let's get on with it. Where do we start?" he asked impatiently.

Grudgingly, Arun went to a whiteboard and drew a diagram, by region, of the liability that Ranbaxy faced: the United States and Canada on the bottom; Europe next; Latin America above that; India next; and ROW (rest of the world), comprising the poorest African nations, on top. "I'd start there," Arun said, pointing to the top.

Thakur still felt that he was groping in the dark. He needed numbers. Arun called in his executive assistant to help. Thakur asked the young man what percentage of the dossiers submitted to regulators contained data that did not match what the company had on file. The assistant was evasive: "It . . . varies from region to region."

"Give me an estimate in each region," said Thakur. "How about in the U.S.?"

The assistant thought for a moment, then estimated, "Perhaps between 50 and 60 percent?" Thakur could barely breathe. Ranbaxy had faked over half its dossiers to the FDA? And that was one of the better regions?

"How about Europe?"

"About the same," came the assistant's reply.

"And India?"

After some hemming and hawing, the assistant answered, "100 percent." Testing the drugs for India was just a waste of time, he explained, because no regulators ever looked at the data. So the regional representatives just invented the dossiers on their own and sent them to the Drug Controller General of India (DCGI). What was needed for the DCGI was not real data but good connections, which they had, the assistant explained.

The scale of the deception stunned Thakur. He felt physically sick thinking about the patients. Thakur told the men he wanted a breakdown: each product by year, and the problem with each dossier.

As Thakur's project managers began their analysis—obtaining data, conducting interviews, visiting laboratories and manufacturing plants—the rigid hierarchy within the company proved to be a major obstacle. Not only was Thakur's team new, but by the unwritten rules of Indian corporate culture, its members were also too junior to question department heads. "We were unwelcome people," recalled one of the team members, who had been tasked with tracking down data from Asian and Brazilian markets. As a result, they had to operate by stealth and stubbornness. They showed up at the factories unannounced. They waited hours to speak with department heads. They drove hours to distant manufacturing plants. Little by little, as the team members stitched together small pieces of information, they stumbled into Ranbaxy's secret: the company manipulated almost every aspect of its manufacturing process to quickly produce impressive-looking data that would bolster its bottom line.

Each member of Thakur's team came back with similar examples. At the behest of managers, the company's scientists substituted lower-purity ingredients for higher ones to reduce costs. They altered test parameters so that formulations with higher impurities could be approved. They faked dissolution studies. To generate optimal results, they crushed up brand-name drugs into capsules so that they could be tested in lieu of the company's own drugs. They superimposed brand-name test results onto their own in applications. For some markets, the company fraudulently mixed and matched data streams, taking its best data from manufacturing in one market and presenting it to regulators elsewhere as data unique to the drugs in their markets. For other markets, the company simply invented data. Document forgery was pervasive. The company even forged its own standard operating procedures, which FDA investigators rely on to assess whether a company is following its own policies. In one instance, employees backdated documents and then artificially aged them in a steamy room overnight in an attempt to fool regulators during inspections.

There was little effort to conceal this method of doing business. It was common knowledge, from senior managers and heads of research and development to the people responsible for formulation and the clinical people. Essentially, Ranbaxy's manufacturing standards boiled down to whatever the company could get away with.

As Thakur knew from his years of training, a well-made drug is not one that passes its final test. Its quality must be assessed at each step of production and lies in all the data that accompanies it. Each of those test results, recorded along the way, helps to create an essential roadmap of quality. But because Ranbaxy was fixated on results, regulations and requirements were viewed with indifference. Good manufacturing practices were stop signs and inconvenient detours. So Ranbaxy was driving any way it chose to arrive at favorable results, then moving around road signs, rearranging traffic lights, and adjusting mileage after the fact. As the company's head of analytical research would later tell an auditor: "It is not in Indian culture to record the data while we conduct our experiments."

One of Thakur's team members, who'd been thrilled to join Ranbaxy months earlier, found himself in a paradoxical situation. He had no data indicating the drugs *weren't* safe, but he had no reliable data proving that they were. He'd go home after work and urge his relatives and loved ones not to buy any Ranbaxy products.

Thakur worked fourteen-hour days. He tried to build spreadsheets for each market: data the drugs had been filed with, data the company had sent to regulators for every approved product, existing data to support those claims. He would stay in the office till 9:00 p.m., preparing a work plan for the next day. At home, despite Sonal's objections, Thakur would descend to his basement office and work until midnight, trying to synthesize the data flowing in from his team. As was typical, he was vague about the project, and Sonal didn't ask. Nor did he ask himself the larger question about the professional ramifications of investigating his own company. Had he taken that single step back, he might have seen how perilous the project actually was.

Instead, he kept trying to wrap his mind around the scale of the emerging crisis. How could there be so much deception? Was there even a name for this behavior, which extended so far beyond what he'd ever imagined? Finally, after days of work, a word came into his mind, one that seemed to clarify what he was learning. *Criminal.* Yes, that was it. He was uncovering nothing short of a global crime.

After weeks of exhaustive research, Thakur brought his team's preliminary findings on the Latin American, Indian, and ROW markets to his boss, Raj Kumar.

It was 7:30 a.m., the time Kumar usually got to work, and they met in his office. The hallways were still quiet. Thakur placed some preliminary spreadsheets in front of Kumar. They showed that numerous drugs had never been tested properly, if at all, and had no underlying data to support the company's claims. Kumar perused them in silence. "This can't be right," he finally said. It seemed impossible that Ranbaxy had filed dossiers on drugs the company hadn't actually tested. Kumar had never heard of anything like it. "You must have missed data."

"We've looked, and it doesn't exist," Thakur insisted.

"You have to go back and check again," Kumar asserted. "This has got to be wrong."

To Kumar, the only plausible explanation was that Thakur had either overlooked existing test results or misinterpreted the results he'd found. Otherwise, there was no precedent for what Thakur had uncovered. In the coming weeks, Kumar sent him back to check and recheck so many times that finally Thakur organized a meeting with the team so that Kumar could hear from them directly.

They, too, had been stunned by their own findings and remained at a loss for how to think about what they'd found. "Corruption to me was more Enron, more how you fudge earnings," Dinesh Kasthuril recalled. Venkat Swawinathan had expected "nepotism and inefficiency." But doing things that jeopardized people's lives was another thing altogether.

Once Kumar heard from each member of Thakur's team, it finally sank in. The company was committing fraud and potentially harming patients on a global scale. He distilled the information into a four-page report for the CEO, Brian Tempest. Though blandly titled "Inadequate Dossiers filed in various countries for various products" and written in the gray lingo of corporate quality assurance, the report was explosive. It laid bare systemic fraud in Ranbaxy's worldwide regulatory filings. "The majority of products filed in Brazil, Mexico, Middle East, Russia, Romania, Myanmar, Thailand, Vietnam, Malaysia, African Nations, have data submitted which did not exist or data from different products and from different countries."

Kumar's document explained that while the company had slashed production costs and used the cheapest ingredients in those markets, it submitted data from the drugs that had been made for more regulated markets, a dangerous bait-and-switch that concealed the low quality. The report also noted that active pharmaceutical ingredient (API) that failed purity tests had been reblended with good API until it met requirements.

The report noted the "non-availability" in India and Latin America of validation methods, stability data, and bioequivalence reports. In short, Ranbaxy had almost no method for confirming the content of drugs in those markets. For example, the data collected by Thakur's team showed that of the 163 drug products approved in Brazil since 2000, almost all had been filed with phony batch records and stability data that did not exist.

The report noted that in a majority of regulatory filings, Ranbaxy had "intentionally misrepresented" small research and development batches (some two thousand doses) as exhibit batches one hundred times the size, and then deceptively performed crucial tests for bioequivalence and stability on the smaller, easier-to-control batches. The result was that its commercial-sized batches had not actually been tested before being sold, putting millions of patients at risk.

In an email to Tempest marked "confidential" accompanying the report, Kumar noted that lack of adherence to regulation was only part of the problem. "It appears that some of these issues were apparent over a year ago and I cannot find any documents which sought to address these concerns or resolve the issues." In closing, he made clear that his ultimate loyalties lay not with the company but with the truth. "I can not allow any information to be used for any dossier unless fully supported by data," he wrote, adding: "With your permission, I would like to take advice from legal counsel in London as to my current responsibility and indemnity with respect to the above issues."

In response, Tempest assured Kumar that the company would do the right thing.

Though the picture was grim, Kumar confided to Thakur that he believed he could fix the problems, if given the authority.

Thakur's findings were not news to Ranbaxy's top executives. Just ten months earlier, in October 2003, outside auditors started investigating Ranbaxy facilities worldwide. In this case, the audits had been ordered up by Ranbaxy itself. This was a common industry practice: drug companies often hired consultants to audit their facilities as a dry run to see how visible their problems were. If the consultants could find it, they reasoned, then most likely regulators could too.

The fact-finding mission by Lachman Consultant Services left Ranbaxy officials under no illusion as to the extent of the company's failings. At Ranbaxy's Princeton, New Jersey, facility, auditors found that the company's Patient Safety Department barely functioned and training was essentially "non-existent." The staff had no written protocols for investigating patient complaints, which piled up in boxes, uncategorized and unreported. They had no clerical help for basic tasks like mailing out the patients' samples for testing. "I don't think there's the same medicine in this medicine," was a common refrain from patients. Even when there were investigations, they were so perfunctory and half-hearted that expiration dates were listed as "unknown," even when they could easily have been found from a product's lot number.

An audit of Ranbaxy's main U.S. manufacturing plant, Ohm Laboratories in New Jersey, found that the company, though required to report adverse events to the FDA, rarely did so. There was no system to capture patient complaints after hours, and no global medical officer to ensure that any potential negative consequences for patients were being monitored. The consultants from Lachman urged Ranbaxy to address these problems globally. Ranbaxy's initial reaction to the findings was to question the number of hours, and the resulting invoice, that Lachman had sent for its work.

The picture was not a lot rosier overseas. At a plant called Mohali, in India's northern state of Punjab, auditors found so little control of records that twenty people were authorized to change test results. Over 120 different batch records had been reprinted, which Ranbaxy claimed was due to faulty dot matrix printers that had not been replaced. If the goal of good manufacturing was total control, this was about as wild and careening a picture as one could get.

The head of Lachman later sent a top company official a wide-ranging plan for corrective action. Among the suggestions was to establish a training program for workers, including a module entitled "Creating a Culture of Trust, Ethical Behavior, and a 'Quality First' Mindset." But Ranbaxy refused to implement the proposed ethics training after a company executive deemed it unnecessary.

Other employees were growing suspicious. In May 2004, three months before Thakur embarked on his research, Dr. Kathy Spreen had joined Ranbaxy's U.S. office as executive director of clinical medicine and pharmacovigilance. A fifteen-year veteran of Wyeth and AstraZeneca, she was there to help launch the company's brand products division, which planned to create new dosages and formulations of existing drugs. Spreen envisioned her job as that of a regulatory coach to help guide Ranbaxy through the FDA's intricate system.

At first, the company's manufacturing prowess seemed to exceed her expectations. She had been on the job a few months and was preparing slides for a presentation about the company's launch of Riomet, a version of the diabetes drug metformin, when she noticed something remarkable. The data showing the concentration of Ranbaxy's drug in the bloodstream appeared to match that of the brand name perfectly. *Look how good this company is,* she remembered thinking. *The bioequivalence data is superimposable on the drugs we are modeling.*

About a month later, while comparing the data for Sotret, the company's version of the acne drug Accutane, a formulation the company had secretly wrestled with, Spreen found it almost identical to the brand-name data. That's when she began to worry. *If it's too good to be true, it's probably made up.*

She knew that data is tricky. Even two batches of the same drug made by the same company at the same plant under the exact same conditions will have slight variations. Test results for a similar or copycat drug made by a different company with a different formula should look different.

With her suspicions aroused, Spreen began asking her Indian counterparts to send underlying data that supported their test results. They repeatedly promised that the information was on the way. When it didn't arrive, she got excuses: it was a "mess," they'd be "embarrassed." She begged her colleagues in India, "I don't care if it's written on the back of toilet paper. Just send me something." But no data came.

Spreen kept thinking that if only she could explain American regulations more clearly, Ranbaxy's executives would understand. But no amount of explaining seemed to change how the company did business. Indian executives approached the regulatory system as an obstacle to be gamed. They bragged about who had most artfully deceived regulators. When sales of a diabetes drug were sluggish, one executive asked Spreen if she could use her medical license to prescribe the drug to everyone in the company so they could record hundreds of sales. Spreen refused.

When she asked Ranbaxy's global manufacturing director to send documentation showing that an antibiotic acne gel was made with good manufacturing practices, he offered to send her an "impressive looking" certificate. To Spreen, it sounded like an offer to send her a forged one. She tried to explain that "the look of the certificate means nothing to me unless the FDA says it's GMP."

In New Jersey in October 2004, Thakur's boss Raj Kumar quietly confirmed to Spreen what she had already come to suspect: that crucial testing data for many of the company's drugs did not actually exist and submissions to regulators had been forged. At one point, she confronted Malvinder Singh, then president of pharmaceuticals, with her suspicions. He told her to be patient and assured her that everything would work out. But for that to happen, the company would have needed to care about compliance and feel a sense of urgency about protecting patients.

That sentiment seemed absent, and shockingly so. In a conference call with a dozen company executives, Spreen expressed her fears about the quality of the AIDS medicine that Ranbaxy was supplying for Africa. One of the company's top medical executives responded, "Who cares? It's just blacks dying."

Like Kathy Spreen, Kumar felt uncertain about what to do next. On a trip back to the United States, he went to see the company's lawyer, Jay Deshmukh. "I want to talk to you lawyer-to-client," he said. Deshmukh replied that he represented the company and couldn't act as a lawyer for him. "You don't know what's going on," Kumar told him, explaining that terrible things were happening inside the company. "I'm afraid for my freedom."

"I can't advise you," Deshmukh said. "I'm not an expert." After Kumar's visit, however, Deshmukh made some quiet inquiries and learned about the review that Kumar had launched. It struck the lawyer as extremely hazardous to undertake a corporate self-assessment without proper safeguards. Once that was launched, it would be impossible to undo, and there was no telling where such information might lead. "It's a bad thing to do with unsophisticated people," he would later say, as though Kumar had allowed children to play with matches, unsupervised.

For now, Thakur's analyses moved quietly inside the company. At an executive management meeting in Bangkok, Kumar distributed a spreadsheet prepared by Thakur that listed Ranbaxy's different markets by region and alphabetically, beginning in Algeria and ending in Vietnam. It had one column for the problems with each drug, another column headed "Risk"—by which Thakur had meant risk to patients—and a third column, "Action Plan." Thakur had filled in the column for risk to patients with the words "high," "medium," or "low," depending on how much data, and what kind, was missing from the company's records.

But at the meeting, the company executives reviewing the spreadsheet had misunderstood the column as meaning risk to the company. In the "Action Plan" column, two executives—the heads of global marketing and regulatory affairs—had scrawled notes in the margins. These made clear that the company would decide whether to discontinue the drugs or temporarily withdraw them from the market for retesting by weighing the risk of being found out against the risk of losing market share. The risk to patients did not enter into their calculation.

At the end of the meeting, Kumar took the spreadsheets with him that contained the men's handwritten notes. Whether intentionally or not, he was now collecting evidence.

On October 14, 2004, several months after assigning Thakur to dig up the truth, Kumar stood in the boardroom at Ranbaxy's corporate headquarters in New Delhi, facing members of the scientific committee of the board of directors. His audience included Brian Tempest; Malvinder Singh, then president of pharmaceuticals; the board chairman, Tejendra Khanna, who had served as the lieutenant governor of New Delhi; Dr. P. S. Joshi, a prominent cardiologist; and several others. The company secretary was asked to leave the room.

Kumar showed the men a PowerPoint of twenty-four slides that Thakur had prepared. It was entitled "Risk Management for ANDA Portfolio." To some extent, it was a work-in-progress, as it still did not contain U.S. market data. But the presentation made clear that in its race for profit, Ranbaxy had lied to regulators, falsified data, and endangered patient safety in almost every country where it sold drugs. "More than 200 products in more than 40 countries [had] elements of data that were fabricated to support business needs," the PowerPoint stated. "Business needs," the report showed, was a euphemism for ways in which Ranbaxy could minimize cost, maximize profit, and dupe regulators into approving substandard drugs.

No market or type of drug was exempt, including antiretrovirals purchased by the United States and the World Health Organization to fight HIV in Africa. In Europe, the company used ingredients from unapproved sources, invented shelf-life data, tested different formulations of the drug than the ones it sold, and made undocumented changes to the manufacturing process. The PowerPoint also noted that the fallout from the Vimta audit, which had initially taken Kumar to South Africa, was already drawing the attention of regulators and could do further damage to the company's reputation.

In entire markets—including Brazil, Kenya, Ethiopia, Uganda, Egypt, Myanmar, Thailand, Vietnam, Peru, and the Dominican Republic—the company had simply invented all the data. Noting a corporate agreement to manufacture some drugs for brand-name companies, a slide stated, "We have also put our partners (Bayer & Merck in Mexico and in South Africa) at risk by using suspect data in our dossiers."

Kumar proposed a drastic course: pull all compromised drugs off the market; repeat all suspect tests; inform regulators of every case of switched data; and create a process for linking the right data to the right drugs. A slide entitled "Guiding Principles" laid out what Kumar considered to be the company's obligations: "Patient safety is our first responsibility. Our products have to be proven safe and effective. A short-term loss of revenue is better than a long-term losing proposition for the entire business."

Kumar completed the presentation to a silent boardroom. Only one director, a scientist, expressed any surprise about the findings. The others appeared more astonished by Kumar's declaration that if he was not given full authority to fix the problems, he would resign.

"Can you not bury the data?" one of the board members turned to ask Tempest. No one responded. The silence told Kumar everything he needed to know. Tempest asked for every copy of the PowerPoint to be destroyed and for the laptop on which it was created to be broken down piece by piece. No minutes of the meeting had been created.

Kumar had been certain that Ranbaxy would have to do the right thing after seeing incontrovertible proof that it had done the wrong thing for so long. Instead, within two days of the board meeting, Kumar submitted his resignation. He had been at Ranbaxy for less than four months. "Given the serious nature of the issues we discussed," he wrote to Tempest, his only choice was to withdraw "gracefully but immediately."

But the specter of Kumar's PowerPoint, one of the most damning internal documents ever created by a company executive, would divide the executives for years to come. Inside the company, it would come to be known as "the SAR" (for Self-Assessment Report). The incriminating document was like a slow-burning fuse, headed straight toward the company's top executives.

Thakur remained behind. But with Kumar's departure, he had lost his protection. Three months after the board presentation, the company's internal auditors arrived at his department for what they called a routine review. They stayed for ten weeks, combing through his department's books and interviewing staff. To Venkat Swaminathan, a member of Thakur's team, the auditors were the "secret police of the company. We eventually figured out, this was part of the whole case, and Dinesh was the target."

In late April, the company accused Thakur of browsing porn sites from his office computer. Thakur vehemently denied doing so. Furious, he got his network administrator to pore through the computer records and found that someone in the corporate IT department had logged into his division's servers and planted his IP address on several searches.

At home, Thakur told his wife Sonal that he would be leaving the company. Though he didn't go into details and she didn't pry, he told her that it had become impossible to do his job.

"What are we going to do?" she asked.

Thakur did not have an answer for her. But given what had transpired, it was clear that he had come to the end of his time at Ranbaxy.

On a Thursday morning, Thakur drafted a resignation letter and printed out the data showing that porn websites had been planted under his IP address. The next afternoon he went to Dr. Tempest's office, where he'd managed to book a thirty-minute meeting.

Thakur showed him the evidence of computer tampering. "I can't work in an environment where people are out to get me for doing my job," he explained. Thakur gave Tempest his resignation letter.

Tempest, though pleasant enough, said, "I understand why you're leaving." He told Thakur not to return to his office that afternoon but instead to collect his belongings the following week.

Thakur returned on Monday. As the head of human resources stood watch, he plucked a few photos of Ishan from his desk. He was not even allowed to open a drawer or say good-bye to his group. He was then escorted from the building, where Vijay was waiting in the car. He had been at Ranbaxy for twenty-two turbulent months. His departure took less than ten minutes. He was done—or so he thought.

PART III

A CAT-AND-MOUSE BUSINESS

CHAPTER 10

THE GLOBAL COVER-UP

NOVEMBER 18, 2004

Little Rock, Arkansas

Dr. Brian Tempest and several other Ranbaxy executives huddled by the Arkansas River in plastic ponchos under a driving rain. They were determined to make the most of the soggy event, which included President George W. Bush, three former U.S. presidents, and numerous members of Congress. Ranbaxy had donated close to $250,000 so its directors could be at the opening of the William J. Clinton Presidential Library and Museum.

By their side was Dr. Agnes Varis, the founder and CEO of AgVar Chemicals, who had been serving for some time as their political chaperone in the United States. A major Democratic Party donor with a deep knowledge of America's drug industry, she was a longtime friend of the Clintons and had taken Ranbaxy executives to parties at their Westchester home. When Clinton's presidency ended, Varis had put a political bumper sticker on her chauffeured Bentley that read I MISS BILL. She had contributed nearly $500,000 to the Clinton Foundation for the library dedication.

Using the event as leverage, Ranbaxy did everything possible to link itself to the former president. The company released a public statement touting its "close association" with the Clinton Foundation and their shared goal of "providing pharmacotherapy to patients stricken with AIDS in financially depressed countries." A subsequent Ranbaxy newsletter described the company as an "honored guest at this momentous event." Though certainly an overstatement, it was true that Clinton took extraordinary steps to thank Ranbaxy for its role in making cheap HIV drugs for Africa.

Within six months of the library dedication, Clinton returned to India, where he spent more time with Malvinder Singh and other Ranbaxy executives on an AIDS panel and at a cocktail event. So much face time with the former U.S. president was like jet fuel for the company's public image—and its bottom line.

From a distance, Ranbaxy's ascent appeared unchecked. By early 2004, the company's global sales had surpassed $1 billion. In the United States, Ranbaxy had become the fastest-growing foreign generics maker, with ninety-six products on pharmacy shelves and fifty more applications before the FDA. Its medicine had become important to the AIDS programs of two American presidents. And the company had big future plans: to reach $5 billion in global sales by 2012, move from the world's eighth-largest generic drug company into the top five, and launch its own specialty products.

The company's marketers spelled all this out in a newsletter, *Ranbaxy World,* which emphasized the company's integrity and social commitment. It pointed to new quality initiatives, Ranbaxy's elaborate code of conduct, and its dedication to making low-cost antiretroviral drugs for poor Africans. As the newsletter stated, the company's "quest for growth and excellence goes hand in hand with unflinching commitment to integrity in all relationships with employees, customers, suppliers, government, local communities, collaborators and shareholders."

Beneath this gauzy sentiment, glimpses of a different sort of company were visible. One week after the library dedication, President of Pharmaceuticals Malvinder Singh gave an interview to an Indian website in which he attributed Ranbaxy's success, in part, to being a "very aggressive marketing company, fighting for a market share at rock-bottom prices and making that model work." He didn't explain how the model worked exactly, but went on to say, "Ranbaxy is today what it is because we took the risks." He would later explain that Ranbaxy had risked being the first Indian pharmaceutical company to set up operations and manufacturing facilities outside of India.

But inside the company, executives were grappling with a different set of risks.

The initial disclosure of fraud at Vimta, the company Ranbaxy had hired to test its AIDS drugs, was like a teetering domino, threatening to topple interconnected drug applications approved by regulators around the world. As charitable organizations asked the company for underlying data to support its claims, the problem confronting Ranbaxy executives had become almost unsolvable. Drugs that had never been tested, or whose tests revealed a failing product, were now due to be reregistered in countries around the world. Much of the raw data didn't match what the company had filed with regulators. Either it didn't exist, didn't make sense, or had been fabricated at some point. A refusal to share the data would trigger further suspicion, leaving the company with two bad options: come clean—which would have disastrous business consequences—or lie more.

The company needed to start testing drugs properly. But that not only risked exposing past fraud but often required a new set of lies. This Catch-22 played out in a torrent of confidential emails in which Dr. Tempest and future CEO Malvinder Singh were often cc'ed and also weighed in. In mid-July 2004, a UNICEF official asked why Ranbaxy had submitted only limited stability data for several of its AIDS drugs. Companies must prove that their drugs will remain stable in a range of temperature conditions. The required tests, which help to establish the shelf life of a drug and measure impurities over time, are conducted in chambers that resemble oversized refrigerators, which can replicate extremes of heat and cold.

UNICEF's question prompted a round of panicky internal emails. In one titled "Stability Studies—Urgent," an executive wrote, "According to UNICEF, if we fail to furnish the data by Wednesday evening and do not provide the information requested in the mail below, then we can forget about this tender." He added, "this tender is worth $ 5 million and we cannot take nay chances on it."

But the only data that Ranbaxy had to give UNICEF was a nonsensical hodgepodge that would raise more questions. The limited testing that Ranbaxy had done on the HIV drugs showed some impurities remaining constant or even decreasing between nine and twelve months, which was technically impossible. As an executive pointed out, these problems "will certainly raise a doubt in the mind of the reviewer . . . we need to revise this number."

Executives grappled with similar problems in dossiers for markets around the world. In February 2005, a company executive wrote to colleagues regarding the company's filing in Spain for the antibiotic cefuroxime axetil: "Please advice the way forward. This dossier was scheduled to go in Dec, 04. We have been waiting for your response for the last 2 months. We need to conclude this ASAP." This email triggered a terse reply from a senior scientist: "During our discussion in Gurgaon on 27th Jan, I mentioned clearly that the data in our Archives and that of the filed one is Differing Entirely. So, I cannot send the Data."

Months earlier, in September 2004, as the FDA evaluated Ranbaxy's PEPFAR application for certain AIDS drugs, it asked to see data the company had filed with the World Health Organization. The director of regulatory affairs, Dr. Arun Kumar, wrote to colleagues, including Dr. Tempest: "In case we do not share the data at this stage we will be under question. . . . Reasons for not sharing the data on existing batches could be difficult to explain." He added,

"We are crossroads becoz adequate data has not been generated on the products as per WHO requirements."

This was particularly problematic, as the company had assiduously cultivated its relationship with the FDA. In the intense debate over what data to share with the agency, Ranbaxy's U.S. president, Dipak Chattaraj, noted in an email that the two top officials in the FDA's Office of Generic Drugs "are so well disposed towards Ranbaxy that trying to [be] difficult with them can only cost and not help us."

The company's problems with the FDA were far bigger than just the drugs intended for Africa. Ranbaxy had not properly tested the stability of almost any drugs on the U.S. market. The most basic good manufacturing practices require continuous monitoring of drug quality. Drug stability must be tested at intervals called "stations": three months, six months, nine months, and so on. So long as a drug is on the market, that data has to be filed in an annual report with the FDA. One is never *out* of data, because obtaining it is simply part of the process.

But the company had hit an impasse, leaving its executives to confront the fact that they had virtually no thirty-six-month stability data to submit for any U.S. commercial batches. This was more than an "oops." It was the equivalent of trying to read a roadmap upside down—after you've already crashed the car into a tree.

In frantic emails, executives grappled with this seemingly insurmountable problem. Abha Pant sent a terse email to her colleagues: "This is a very serious issue. I do not know how are we going to file the Annual Reports and what reasons are we going to give to the [FDA] for not submitting the stability data. . . . We need all this data and there is no way out."

Even as company executives resolved to start testing the drugs in "right earnest," as one of them put it, a similar crisis was playing out in dossiers around the world. By 2005, twenty-two high-priority products faced reregistration in at least one country. All had been made at Ranbaxy's Dewas manufacturing plant in Madhya Pradesh, and none had been tested adequately. Arun Kumar explained to his colleagues in an email, "For most of the products the data is not available, and also the archival data is not there." In short, they had never been tested.

In assessing the task before them, a quality assurance director at Dewas wrote to his colleagues in February 2005, cc'ing Brian Tempest, "We are going to start almost on zero basis in case of majority of products where we do not have required stability data. The task seems very difficult." In other words, drugs already on the market had to be tested from the ground up. By then, Tempest had sent the director of global quality an urgent email: "There will be no business left to pay your salaries unless we get on top of the stability work for re registrations."

Previously, the company had taken data from fledgling research and development batches and falsely represented in filings that it came from much larger exhibit batches, which are harder to control. One troubling drug was the co-amoxiclav suspension, the antibiotic that is often used to treat ear infections in children, the same medication that had failed to cure Thakur's son. Ranbaxy had registered the drug with a twenty-four-month shelf life in almost thirty countries, but tests showed that its shelf life was actually closer to eighteen months. In reregistering the drug, a senior consultant noted in an email, cc'ing Tempest and Malvinder Singh, it would be "useful" to have a "plausible explanation for this reduction in shelf life" when dealing with the regulatory agencies.

If most executives were seeking guidance on how best to lie to regulators, others were concerned about the fraud they were expected to commit as part of their job. Some balked at filing false data. Others flat-out refused to participate in illegal acts. However, sometimes even the most scrupulous employees ended up being drafted, unwittingly, into the company's fraudulent schemes. Most Ranbaxy executives were expected to carry suitcases full of brand-name drugs when they traveled to India. At Ranbaxy's New Jersey headquarters, suitcases purchased at the local Walmart were kept packed with drugs, waiting for the next traveler to India. The suitcase-toting seemed innocent enough. Most executives assumed that the drugs were needed for research and development.

Generic drug companies often study small amounts of a brand-name product in order to reverse-engineer it or to reference it as a point of comparison in applications. But proper channels for purchasing and transporting such drugs are well established and became ironclad with the 2001 passage of the Patriot Act. Personal transport of drugs was technically illegal and a form of smuggling. To the dozens of employees pressed into ferrying the drugs, often on an emergency basis, it seemed like a minor shortcut, possibly to cut shipping costs, avoid quarantine, or speed up delivery.

In one year alone, seventeen executives took undeclared drugs from the New Jersey office through Indian customs, four of them doing so multiple times. The most frequent couriers included the company's U.S. president and even its U.S. executive director of regulatory affairs, Abha Pant, who was responsible for ensuring that the company followed the rules.

At Ranbaxy, top executives skirted these regulations and sometimes oversaw the illegal ferrying of drugs at the very moment when the company faced deadlines to resubmit data to regulators. Some executives came to suspect that the company was using the brand-name samples as a substitute for its own, in order to generate data showing how closely Ranbaxy's drug matched the brand it was seeking to replicate. This would explain the urgency surrounding the drug runs, especially when some Ranbaxy staffers strenuously resisted being used as drug mules.

In May 2004, a regulatory project manager refused to take French brand-name samples to India. He protested in an email, "I will NOT be bring any samples with me, not only I believe is this company policy but I personally do not feel comfortable bringing samples in this manner." An executive pushed back: "It is critical that the samples are carried by you. We cannot delay it." The employee flatly refused.

Malvinder Singh, then the company's worldwide head of pharmaceuticals, got involved. Through his secretary, he asked when the samples would reach Gurgaon. "These products have been sitting in our London office and it is a pity to see that nobody takes ownership of them."

This triggered a response from the company's president of global pharmaceutical business: "Dear Malvinder, I need to explain to you how labour laws work within Europe. As taking these samples to India is in principle illegal we cannot force people to do so. . . . Normally however we find our people willing to take the risk." So important was this to the company's business that the executive then went on to make an extraordinary suggestion to Singh: that since Tempest and Singh had been passing through the United Kingdom on a regular basis, "[I] would ask you to in future also make yourself available for carrying samples back." Other senior executives were also pressed into service.

In the event that they were caught, those who carried the drugs for Ranbaxy were given a letter claiming the products were for research and development and had no commercial value. In June 2004, one executive got stopped by Indian customs with hundreds of packs of an antinausea drug, Kytril, worth thousands of dollars that he hadn't declared. The drugs were seized. One Ranbaxy executive noted internally that "in the absence of correct documents this is considered as an illegal way of bringing the medicine in to India."

The lies and efforts to conceal them absorbed the energies of the entire company. In August 2004, the company's top executives, including Tempest and Malvinder Singh, held a meeting in the executive conference room. On the agenda, according to emails, was the "strategy for filling gap between requirement & availability." In other words, how were they going to submit data they didn't have?

In September 2004, the executives met again and opted for a solution: to move the most important manufacturing for the United States and PEPFAR from the troubled Dewas plant to the newer one in Paonta Sahib, in the hope that by severing links to the past fraudulent manufacturing, regulators would not detect it. Instead, Dewas, which had almost nonexistent quality systems, would remain the manufacturing site for the least-regulated markets, including Brazil, Mexico, Vietnam, and elsewhere.

Publicly, company executives spun this change as a response to big demand from the American market and PEPFAR. In early January 2005, at a meeting of Indian generic drug makers and AIDS activists in Mumbai, Ranbaxy's HIV and essential drugs project manager, Sandeep Juneja, explained to the meeting attendees that the company's new strategy would result in its drugs being quickly approved by PEPFAR and reinstated on the WHO list. "We wanted to harmonise everything to the U.S. market and consolidate manufacturing in one place," he said. "If a product is approved by the FDA, then it is accepted anywhere."

Two days later, Arun Kumar wrote to a UNICEF official, explaining the shift in the manufacturing site for lamivudine, an AIDS drug: "We have changed the site of manufacture of the product from Dewas to Paonta Sahib facility to facilitate handling high business requirements." Four days after making that claim, however, as the company prepared to resubmit the data for its antiretroviral drugs to the WHO, Juneja reiterated the company's *real* strategy in an email, cc'ing Tempest. "We have been reasonably successful in keeping WHO from looking closely at the stability data in the past," he wrote, adding, "The last thing we want is to have another inspection at Dewas until we fix all the process and validation issues once and for all."

But the shift to Paonta Sahib was far from a perfect solution. In some cases, the new studies weren't going to be completed in time to meet new registration dates. As one executive asked in an email, how were they going to handle the "interim situation," where regulators were demanding new data that wasn't ready yet? If the data came from Dewas, "how do we use them in the Paonta dossier?" The answer from a colleague came back: "pl speak to me over the phone. I shall be able to answer yr query." The answer was almost certainly *interim fraud,* or temporarily representing old data from Dewas as new data from Paonta Sahib.

However, as Ranbaxy geared up to start testing drugs at Paonta Sahib and redirect the world's most important regulators to that plant, the company wanted to leave nothing to chance. And legitimate testing—where drug batches can fail and formulas can become unstable—was the ultimate game of chance. Investigating the *why* is difficult and costly. Good manufacturing practices are so laborious precisely because they require innumerable steps to make an uncertain process more certain. If you have to actually test drugs, how can you best control the outcome? How do you put your finger on the scale each time in order to generate perfect data?

Ranbaxy came up with a rather ingenious, well-hidden solution. Whether it remained hidden would depend, in large part, on the next FDA investigator who walked into the plant. Would it be someone content to look merely at the well-polished surface? Or someone committed to piecing together a mosaic of clues? The company had no way to control which investigator showed up.

But two months after Raj Kumar made his ill-fated presentation to Ranbaxy's board of directors, the company had a stroke of luck. The next FDA investigator to arrive at Ranbaxy's Paonta Sahib manufacturing plant in Himachal Pradesh was Dr. Muralidhara B. Gavini.

At the FDA, Muralidhara Gavini, who was widely known as Mike, had distinguished himself in one important respect: he was one of the few investigators not only willing but happy to do foreign inspections in India.

Most of the agency's investigators did not want to travel there, especially as their colleagues returned with stories of broiling heat, incessant rain, harrowing traffic, hours spent traveling to distant manufacturing plants over pitted, washed-out roads, and the ubiquitous threat of getting sick from unclean water or contaminated food, which forced them to bring suitcases of peanut butter and granola bars. The difficult travel had contributed to a growing crisis at the FDA.

In theory, the agency tried to inspect every facility making drug ingredients for the U.S. market roughly every two years, whether the plant was in Maryland or Mumbai. But the FDA's actual rate of inspections overseas was closer to once a decade, and the backlog of applications from foreign drug facilities was growing rapidly. The FDA had no obvious solution for who to send to inspect overseas plants and how to pay for it all. The agency was so desperate that it had even explored doing remote inspections, in which the plants would supply videotapes of their facilities. The proposal for inspections by videotape cited "ever-decreasing resources."

In a system starved for volunteers, Mike Gavini was in high demand. The area around Hyderabad was beginning to develop, and his inspections took him back to his old stomping grounds. He'd grown up in Guntur, south of Hyderabad. Sending Gavini there did not violate any agency rules, in part because there weren't too many at that point. But it did violate a basic principle. As one of Gavini's colleagues would later observe, "You never send people to their native area. That's just Regulatory 101." Such an arrangement could invite corruption and compromise. But Gavini was eager for a homecoming.

He had first left India in 1972 to pursue a PhD in chemistry at the University of Arkansas. For him, America and the town of Fayetteville, were alien environments. But his work was good enough to land him a job at the Woods Hole Oceanographic Institution in Massachusetts, where he continued his research, tracing plutonium isotopes in rainwater. At Woods Hole, he uncovered traces of curium in the sediment of Lake Ontario and presented his findings at a national meeting of the American Chemical Society. Unwittingly, he had revealed runoff from a secret nuclear facility. Amid recriminations, his career as a scientist ended.

With three children to support, he spent the next ten years at a laboratory testing company until a management change left him looking for work again. In 1996, he took a 70 percent pay cut and joined the FDA's New Jersey district office, where he got paid $35,000 a year to visit dairy farms and look for evidence of bovine spongiform encephalopathy (mad cow disease). With his PhD and a decade of private-sector experience, he was an "oddball," as he recalled. By 1999, he was promoted to a position as a compliance officer at the FDA's Center for Drug Evaluation and Research (CDER).

He may have been out of his depth in Fayetteville and given short shrift at the FDA, but in India he was a bigwig. Not only did he have a PhD, which automatically commanded respect, but he also represented the world's most powerful regulator. Most of the time, he went alone to the plants—an inspectional "one-man army," as he put it. He—and he alone—could determine whether to clear a plant to export its drugs to the United States, a project he undertook without clearly defined rules.

Domestic inspections had long followed a clear formula. The FDA's investigators would show up unannounced at U.S. factories and stay as long as they needed to follow data trails wherever they led. "We walk in, show the badge, give them notice of inspection," as FDA investigator Jose Hernandez put it. The relationship with the facility was also clear: there was none. "Those guys [in New Jersey] won't even take a cup of coffee," as one investigator put it.

But the rules for overseas inspections were muddy at best. Seeking to avoid confrontations that might involve a foreign government and lead to an international incident, the FDA prioritized diplomacy over confrontation. It announced its visits to the plants weeks, even months, in advance, and relied on the companies to act as hosts and travel agents for its investigators, booking hotels and ground transportation. This suited Gavini.

The approach he chose was collaborative. He viewed himself as a partner, educating the companies as to what constituted robust quality systems. "He conducted himself more as a consultant," one of his colleagues noted. Gavini regarded the FDA and the companies as sharing the same objectives. "We're not on opposite sides of the table," he said. "Quality is our goal."

He tried to improve the companies' techniques. He taught them how to clean manufacturing equipment properly and stressed "common sense." The industry, he said, "learned a hell of a lot from me." His personal measure for each inspection was, "Have I contributed positively at this company? The net result should be positive." It certainly was. As he inspected, the pharmaceutical sector in Hyderabad grew under his watch, churning out active ingredients and ultimately becoming the bulk-drug capital of India.

Gavini put in ten- to twelve-hour days. He expressed disdain for colleagues who worked shorter hours or filed their reports months after the fact. But on the rare occasion when colleagues accompanied him, they were dismayed by what they saw: the chumminess of his relations with the companies he was inspecting; his direct communications with the executives, in person and by phone; and his tendency to let companies off the hook. He would even send draft versions of his regulatory findings, called 483 reports, to the companies for review before he submitted them officially.

Gavini rejected what he saw as needless secrecy. "I don't know why FDA investigators keep everything to themselves," he said. "I discuss every aspect I am writing [with the companies]." This gave companies the opportunity to influence his findings. If the firm said that it would fix its problems, that was good enough for Gavini. In one 2003 inspection report, he even documented how a plant's managing director had pledged improvements by phone, a promise that he deemed "satisfactory."

The honor system was a bad way to improve drug quality—at least, according to a report by the U.S. Government Accountability Office (GAO) written in 1998, two years before Gavini started his inspections. The report bashed the FDA for downgrading inspectional findings based on "foreign manufacturers' promises" to make changes. "As a result, FDA conducted fewer reinspections of these facilities to verify that foreign manufacturers had corrected serious manufacturing deficiencies."

Gavini was not inclined to treat his countrymen like criminals, he would later explain. But as the pharmaceutical sector in Hyderabad exploded—and FDA investigators returned to facilities that Gavini had previously cleared to find grievous violations—his reputation as a lax investigator spread. Among his colleagues, he came to be known as the quintessential "NAI Inspector": one whose most common finding was No Action Indicated.

On December 17, 2004, to the delight of Ranbaxy's executives, Mike Gavini, an investigator they knew well, arrived at the Paonta Sahib plant. He was there for a preapproval inspection to make sure the facility could adequately make two AIDS drugs, lamivudine and zidovudine, for the U.S. AIDS relief program in Africa.

His arrival was not a surprise to the company. As was typical for the FDA's foreign inspections, a visit had been announced weeks in advance, and the company was involved in planning it. In his summary of the inspection, Gavini noted: "The firm provided transportation to and from the plant" and "accommodations were provided."

Gavini stayed for five days. Trailing him throughout the plant were many of the same executives who'd been sending panicked messages to one another about the lack of data. Gavini saw none of that. In his inspection report, he noted some unclear instructions in batch records and unclear cleaning procedures. However, he gave unqualified praise to the company's program of stability testing, which exposes drugs to different temperatures and humidity levels to see how fast they degrade in different environments and what their expiration date should be. He noted: "Stability sample traffic in and out of the chambers was monitored and the stability inventory maintained properly."

To come to this conclusion, Gavini must have walked directly past a Thermolab stability refrigerator that had been installed seven months earlier. The large walk-in refrigerator was maintained at 4 degrees Celsius. Use of the refrigerator was not noted on any of the company's applications with the FDA. Had Gavini opened it, he would have found hundreds of bottles of samples that had not been recorded, stuffed inside cardboard boxes.

Why Ranbaxy was using the refrigerator would become one of the most contentious, bitterly fought questions for years to come. But again, Gavini didn't raise any questions. To Ranbaxy's nervous executives, he was a dream investigator, either unable to find or not on the lookout for illicit solutions to drug testing.

Not surprisingly, his findings were minimal. He gave the plant a clean bill of health and determined that it was in "overall compliance" with current good manufacturing practices. His conclusion was NAI: No Action Indicated.

CHAPTER 11

MAP OF THE WORLD

AUGUST 2005
Gurgaon, India

The heat and humidity of monsoon season rolled in, pelting the house with rain. On increasingly fitful nights, as the diesel generator rumbled, Dinesh Thakur lay awake with a map of the world in his head. It was divided into Ranbaxy's five major markets: the United States and Canada, Europe, Latin America, India, and ROW (rest of the world). Night after night, he visualized the reams of data he had prepared about the drugs in each market, every data set spelling out a hazard to patients that was almost certainly continuing.

The HIV drugs bound for Africa troubled Thakur the most. He knew they were bad. They had high impurities, degraded easily, and would be useless at best in the hot, humid Zone IV conditions of sub-Saharan Africa. They would be taken by the world's poorest patients, who had almost no medical infrastructure and no recourse for complaints. The injustice made Thakur livid.

Before he'd arrived at Ranbaxy, he assumed that a pill was a pill, manufactured identically for all the regions of the world. Publicly, the company made this claim. It stated that it keyed all its standards across world markets to the most rigorous—those of the United States. But Thakur now knew that the company reserved its worst drugs for countries with little to no regulation, where the chances of getting caught were minuscule.

The FDA, which was now monitoring the quality of all PEPFAR drugs, obviously had no idea this was going on. A month after Thakur resigned, the agency approved Ranbaxy's Paonta Sahib plant to make AIDS drugs for the PEPFAR program. And in early August, the World Health Organization had restored the company's AIDS drugs—the ones tested at Vimta—to its prequalified list, to the great relief of AIDS activists who wanted to ensure a low-cost supply of lifesaving drugs.

After leaving Ranbaxy in late April 2005, Thakur tried to convince himself that the company's medicine was no longer his problem. But his immediate relief had given way to anxiety. He was jobless and piecing together haphazard consulting work as the family's savings dwindled. Sonal was pregnant with their second child. He mentioned nothing to her about his larger concerns. But night after night he found himself wondering what, if anything, he could do about Ranbaxy's deceptions. Did he have an obligation to expose them?

The incessant questions took him back to the nighttime stories—and daytime riots—of his childhood. Thakur grew up one hundred miles north of Hyderabad in the lush agricultural town of Nizamabad, where sugarcane, turmeric, and maize grew in abundance. Three generations of Thakurs lived in the family's ancestral home. His mother was a homemaker. His father, a lawyer, was a civil litigator who did more pro-bono than paid work.

They lived comfortable if modest lives. "Money was not a driver in any sense," said Thakur. But education was. Thakur and his younger brother and sister were taught by nuns at a strict Catholic school that emphasized discipline and memorization. Each day they took a rickshaw to school along one of the town's two major roads.

His grandmother, Amba Bai—a bespectacled slip of a woman who wore simple saris—made the biggest impression on Thakur. At home, the principal entertainment for the children were the stories she told every night, drawn from India's two most famous epic poems, the *Ramayana* and the *Mahabharata*. The stories were teeming with characters facing long odds: righteous kings, scheming relatives, fantastical gods, monkey armies.

Each night the characters confronted essential questions of right and wrong and how to live their lives. Would they grab power or choose justice? Would they descend on a road to the underworld or ascend into the light? And while demons and goddesses clashed in his nighttime stories, the real world outside his door was also replete with conflict. In that postcolonial powder keg, perceived slights led to frequent strife between Hindus and Muslims that sometimes led to full-scale riots just outside his door.

His father frequently waded into these skirmishes, racing out to the street to mediate. "I would always ask him, 'Why are you going out there? It's not our problem,'" Thakur recalled. "His answer was always, 'When you see something that's wrong you have to make sure that you do whatever you can.'" Often, his father would return with cuts and bruises from his peacemaking trips, to the dismay of Thakur's mother. "She used to absolutely hate him going out there," Thakur recalled, and inevitably the same discussions ensued: "This is not your problem. They don't listen to you anyway!" But there was no stopping his father.

For young Thakur, the moral dilemmas his father faced were subtler than those playing out in his grandmother's stories. If a bad situation is "not your fight," then what is your obligation? His father's answer was clear. As a senior lawyer in town, he believed that he had a duty to intervene and serve as a liaison between the people and the justice system, however flawed.

Thakur absorbed this lesson but found it difficult to implement. In eighth grade, one of his friends got suspended from a soccer game for an infraction that wasn't his fault. Thakur appealed to the physical education teacher on his friend's behalf but made no headway. So he took his appeal directly to the headmaster. In front of the entire team, Thakur told him that the physical education teacher was being punitive and his friend did not deserve the suspension. The headmaster slapped Thakur across the face and told him never to complain about any of his teachers. "The headmaster did not want to see a teacher challenged by an eighth-grader," Thakur realized afterward.

The school's value system, and that of the culture at large, prioritized deference to authority above simple fairness. Thakur rejected this doctrine. He maintained his own value system as he grew into adulthood. His father had gone into the streets to reason with hostile neighbors, with little thought of the dangers to himself. But for a problem that sprawled all over the world, where were Thakur's streets? With whom could he reason? Thakur knew that others had detected fraud at Ranbaxy, but he believed he was one of the few insiders who understood its full extent. And among that small circle, none had expressed any inclination to address the crime.

Some had even profited from it. His old boss, Barbhaiya, had been given a hefty compensation package on his departure, which Thakur and Kumar believed was to ensure his silence. "The systems in India are so corrupting that even if you are not corrupt, you get corrupted," Thakur would later say. The middle course for the honest man was to do nothing, which seemed to Thakur to be a version of complicity. There was the option to speak up, but honesty came with peril.

Whistleblowers in India faced mortal risks. Just eighteen months earlier, a project director at the National Highways Authority of India had exposed massive corruption in a highway building project. He was found shot to death by the side of a road. While his death provoked national outrage, it was not all that unusual—especially since whistleblowers in India had no legal protection. As for Ranbaxy, the powerful Singh family had a reputation for being bullies. Their internecine family feuds had even involved gangs of paid thugs, according to published news accounts.

It would have made the most sense for Thakur to do nothing. Being a good Samaritan could backfire in ways that were hard to predict, as it had when he'd rescued the drunk pedestrian from the Mehrauli-Gurgaon Road. That had led to a police officer attempting to frame him as part of a shakedown. The lesson was to keep moving and mind his own business, as his driver Vijay had urged him to do. But Thakur rarely considered the societal norms when deciding whether to act in a way he considered just. He had an absolute set of ethical coordinates that dictated his actions.

On the morning of August 15, 2005, four months after submitting his resignation, he woke up determined to do *something*. It was India's Independence Day, a national holiday celebrating the day fifty-eight years earlier when the country had gained its freedom from British rule. Thakur wanted his own liberation from the concerns that had dogged him for months.

He descended to his basement office and opened a Yahoo email account that he had created while weighing his options. Posing as a low-level company scientist and using intentionally broken English, he wrote to officials at the U.S. Agency for International Development (USAID) and the World Health Organization. He stated, "Ranbaxy Laboratories in India is fooling you to get their product on the market with fake data." He claimed that Ranbaxy was forcing him to falsify data: "I cannot sleep at night knowing that these drugs will be used to cure sick patients in Africa. At best, this drug are ineffective and at worst, they cause adverse reaction and kill people." He had sent his email using a pseudonym. In choosing one, he reached for a name that invoked power and prestige, in the hope of drawing attention to his cause. He had created an email account using the name of Ranbaxy's heir apparent, Malvinder Singh.

Each night after that, he went down to his basement computer to check his email, expecting *some* response. The wait was excruciating. And each day that passed without a response brought a fresh round of pain and self-doubt. He suspected that his email had not been authoritative or detailed enough to penetrate the bureaucracy. So he wrote again, this time more pointedly, to half a dozen FDA officials: "I fear for the poor people in Africa who buy these fake medication from WHO and PEPFAR in the hope of getting better, but no, they do not get better, they get died." Silence.

He continued to write, adding details and even attaching documents. The silence persisted. A week later, he wrote back to FDA officials, still in the voice of a lowly bench scientist, but this time with more detail. "Ranbaxy management, including the CEO, Head of business and Head of QA systematically ask people in the lab and plant to fabricate data to support stability. There is no data to support shelf life and the formulation you now approve will be degrade in Zone IV conditions in Southern Africa before it gets to the patients. The formulation is worthless. It produce no results."

Thakur was relentless—and disappointed. He'd thought that if he could only overcome his fear of speaking up, the world would respond and regulators would descend on the company. But no one seemed to care. Over the course of several days, polite but vague responses trickled back. A secretary from the World Health Organization wrote that the officials he'd contacted were out of the office, but his message had been received "and will be dealt with in due course."

After two weeks of waiting, he put aside the ruse of writing in broken English and sent a message directly to FDA commissioner Lester Crawford. In a forceful and urgent email, he alleged that

Ranbaxy was selling "untested, spurious, ineffective medication." He noted that he'd written repeatedly to Crawford's subordinates and had included "documents, e-mail messages exchanged between the senior management of this company, including the CEO." He finally used the word that had come to him months earlier, as he unearthed the company's misdeeds. "I plead with you," he wrote, "to put a stop to this crime."

This time his message broke through. Two days later, Thakur got a detailed email back from Edwin Rivera-Martinez, then chief of investigations and preapproval compliance in the FDA's Center for Drug Evaluation and Research. Rivera-Martinez stated that he had Thakur's "e-mail communications of August 15th, 17th, 27th and 31st" and asked Thakur if he would consent to a conference call. Thakur had planned to stay hidden. He had initially expected to set regulators on the trail but limit his own involvement. It had not actually occurred to him that he would have to do more than that.

The two went back and forth. Thakur declined the call, citing a risk to his family, but attached more documents. Rivera-Martinez responded, reassuring him that he could remain anonymous. "From your emails, we got the impression that you believe that people are being killed by fake medications and you want to stop this from happening," Rivera-Martinez wrote. ". . . without opening a telephone dialogue with you, our investigation may be significantly hampered."

Thakur was wary. He wrote back: "Will this conversation be recorded? Who will be with you when we talk? Is there any personal liability for me if I talk to you? What protection do I have from prosecution? What we are discussing here is a crime that was committed by the company." Rivera-Martinez wrote back that it was far more effective to gather relevant officials for a conference call than to go back and forth on email. He also assured Thakur that his identity would be held in strictest confidence, unless the matter went to court and the FDA was required to divulge his identity.

Hesitantly, Thakur consented to the call. But still wanting to maintain control of the interaction, he tried to school Rivera-Martinez on the best and most secure way to set up the conference call. "Do you have access to publicly available VOIP applications?" he asked in one email. "I mean apps like GoogleTalk or SKYPE? In order to use either of these apps, you only need a microphone and speakers on your computer." He then sent links for the FDA official to download. But the FDA had its own technology—and its own way of operating.

Thakur ended up phoning in to the agency, as instructed. The conference call lasted about ninety minutes. Rivera-Martinez, who was formal and persuasive, gently asked the questions while staffers from different divisions of the agency listened in. There was a special agent, Douglas Loveland, from the FDA's Office of Criminal Investigations. Dr. Mike Gavini sat in, as did a compliance officer, Karen Takahashi. They wanted to know where Thakur had gotten his information, how confident he was about it, and what some of the documents he'd shared meant.

In an email afterward, Thakur wrote to Rivera-Martinez, sounding almost disappointed in advance: "If I am successful in proving to you that the medicines that this company sells worldwide are not of the quality the US FDA mandates, I will be satisfied. Whether you choose to take this up further is completely your prerogative. I personally hope you would."

But in the coming months, Thakur was not satisfied at all. To him, the wrongdoing was black-and-white. He had given proof and expected action. And yet ten days after the conference call, the FDA announced that it had approved Ranbaxy's application for the first generic pediatric AIDS drug for the U.S. market, zidovudine. "Given all the data you have in your possession today about the criminal activities of this company in registering ARVs with fabricated data, I am confused how the USFDA could give such an approval," Thakur wrote to Rivera-Martinez. "Does this mean that you have concluded your investigation and decided that Ranbaxy is not guilty of these crimes?"

The bureaucrat wrote back that because the drug had been approved before Thakur made contact, only actual proof of fraud could reverse the decision. Thakur was staggered. He'd emailed the agency reams of internal data and communications that clearly showed top-level executives conspiring to alter test results. If that didn't count for proof of fraud, what did? It was a question with no easy answer, as he would come to learn.

Over the weeks that followed, the communications between Thakur and Rivera-Martinez unfolded as a cordial, slow-motion, arm-twisting duel, fought almost entirely over email. Rivera-Martinez coaxed a reluctant Thakur to give up more information and be more patient; Thakur goaded and needled Rivera-Martinez to make the FDA take action more quickly and aggressively.

On October 6, Rivera-Martinez sent an email asking Thakur "urgently to call regarding several Agency press releases in the works." Thakur did not get the message in time. The agency announced two more approvals for Ranbaxy: a diabetes drug, glimepiride, and an antiseizure drug, gabapentin. Despondent, Thakur emailed Rivera-Martinez back, noting that he'd provided as much information as he could, "at great risk to me and my family." He went on: "The actions of the agency really worry me, since it appears to me that you have chosen to largely ignore the evidence I have given you. . . . If you have already concluded that this company has done nothing wrong, please tell me and at least I would have the satisfaction that I have done my duty."

He wanted to give up and several times declared as much to Rivera-Martinez. In the wake of the October approvals, he sent the resignation letter written by his old boss Raj Kumar, and his last communications with the CEO, Brian Tempest. Thakur called it "a final gesture. . . . The ball is in your court now Mr. Rivera-Martinez, I cannot do anything more for you or the Agency. I will wait and watch your actions . . . before initiating any further dialogue with you."

But the dialogue continued. Thakur had remained anonymous. Those in the agency referred to him simply as "M," or "Mr. M," short for the name he'd used to first make contact with them, Malvinder Singh. And though Thakur had shared many documents, he had withheld the one that mattered most: the Self-Assessment Report (SAR) that Kumar had presented to board directors. That document was radioactive inside the company and would lead directly back to Thakur. On November 2, Rivera-Martinez wrote back, "During our teleconference . . . you mentioned a risk assessment document that you were asked to prepare while you were at Ranbaxy. . . . A copy of this document may be what we need to focus our investigation."

The request sparked another concern. Though Thakur had done nothing wrong, he had no lawyer and no immunity. "As you have seen from our two-month long association, my only interest in this is to protect the people who pay for the medicines this company makes," he wrote to Rivera-Martinez. Before he could share more documents, "I need to have immunity from prosecution," Thakur warned.

Rivera-Martinez tried to explain that the FDA had no power to confer immunity. But he arranged a teleconference with an FDA criminal investigator who was able to reassure Thakur. With that remaining obstacle removed, Thakur sent Rivera-Martinez the document that Ranbaxy's CEO had been so intent on destroying: the PowerPoint Kumar showed to a subcommittee of the board of directors. The agency now had its fullest picture of a company that was fraudulent to its core.

Though Thakur didn't know it at the time, the FDA had found his information credible and had been moving to confirm it. In October 2005, less than two months after he first contacted the agency, Rivera-Martinez's division had sent a request to the Division of Field Investigations to perform high-priority inspections at two of Ranbaxy's main manufacturing plants, Dewas and Paonta Sahib.

The five-page assignment memo set out a host of alleged frauds for which the inspectors needed to look. Rivera-Martinez requested a face-to-face meeting with the chosen investigators before they left on their trip. The memorandum also recommended that the investigators collect any documents on the day they requested them, noting that the "informant said that the firm has fabricated documents overnight during inspections."

The agency needed an unvarnished view of the company. In January 2006, however, Thakur urgently relayed to Rivera-Martinez what he had learned from former colleagues: the senior leadership of the company was "camped out in the plant locations, both at Paonta Sahib and at Dewas," he wrote, warning of "a massive cover-up effort underway to 'produce'" any documentation that agency investigators might request. Given what Ranbaxy seemed to know about the upcoming inspection, Thakur demanded to know whether the investigation had been compromised in some way. The answer stunned him: Ranbaxy had been notified months in advance that the regulators were coming, because overseas companies always were. It's the way things were done.

CHAPTER 12

THE PHARAOH OF PHARMA

JANUARY 19, 2006

Gurgaon, India

Five months after Thakur first contacted the FDA, Malvinder Mohan Singh succeeded Brian Tempest as managing director and CEO of Ranbaxy, returning the company to the founding family's leadership. Malvinder was just thirty-three and knew little about the science of medicine. But his own personality and upbringing, as well as his family's background, made him seem well suited to a competitive industry with a social mission at its core.

His father, Parvinder, had been austere, restricting everything from candy to conversation. He had indoctrinated his sons into their maternal grandfather's spiritual organization, Radha Soami Satsang Beas (RSSB), from an early age. Parvinder often took his sons and his wife, Nimmi, to the Beas community, where they volunteered as laborers. Though Parvinder sent his children to elite schools, they had few of the indulgences of their peers. At college, the children of the wealthy drove fancy cars, got extravagant allowances, and ate dinner at five-star hotels. By contrast, Malvinder took public transportation, commuted to college by way of Delhi Transport Corporation buses, got an allowance of less than $10 a month, and ate street food.

Though the Singhs emphasized the ascetic values that had long been central to the family, Malvinder had always been a corporate-titan-in-training, who developed expensive taste and an air of mastery as he grew older. He had been introduced to Ranbaxy's inner workings while he was a boy. His father would let him browse company reports, hoping he would glean industry trends. During school breaks, Malvinder would tag along with Ranbaxy drug representatives on sales calls to doctors and chemists, riding pillion on the back of their scooters.

He assumed his new role as CEO with ease. His management style was brash, competitive, and ambitious. Malvinder immediately looked around the globe for opportunity. The fawning Indian business press dubbed him "the Pharaoh of Pharma" and hailed him as an "out-of-the-box decision-maker." Inside Ranbaxy, some viewed him as petulant and immature. He was preoccupied with his own ranking on the *Forbes* list of India's 40 richest people. He and his brother Shivinder, with $1.6 billion in assets combined, had fallen from tenth in 2004 to nineteenth in 2005. This year was shaping up to be even worse, which Malvinder seemed to blame on a lack of employee loyalty. When told that a division wasn't making its numbers, he would yell at employees, "I want profit!" He later explained to a journalist that his passion for Ranbaxy's collective mission, to become an international research-based pharmaceutical company, drove his conduct, adding, "We too had a sharp focus on topline and bottomline, which is true for any business organization."

Malvinder and Shivinder both drove $100,000 champagne-colored Mercedes S-class sedans and collected art and photography. They liked fine clothes, and were prized clients of one of Delhi's most upscale tailors, Vaish at Rivoli, which boasted of serving the "maharajas of business." Each morning, Malvinder and his brother, who worked in a different part of the family's businesses, coordinated their outfits, making sure not to wear the same thing to the same meetings. The listeners of radio station Fever 104 FM voted Malvinder among Delhi's most stylish, and he even sat as the judge of a Miss India competition.

Interacting with employees, Malvinder liked to quote from his favorite book, *The Art of War,* the 2,500-year-old Chinese military treatise that he viewed as required reading for anyone in business. Perhaps this was not surprising, since the disputes within his own family over real estate and corporate assets resembled warfare.

For years, the extended Singh clan had lived in the heart of New Delhi at one of the world's most exclusive addresses: Aurangzeb Road, where mansions—or "bungalows," as the residents call them—sit on acres of planted gardens behind impregnable walls. In 2006, Malvinder's mother, Nimmi, and his uncle, Analjit, both of whom lived on the multi-acre family property in separate bungalows, filed complaints with the police against each other. Nimmi alleged that she'd caught Analjit erecting an unlawful wall; in response, he'd hired "huge and bulky" men armed with axes and hammers to intimidate her. "I was attacked, threatened, physically abused by goons, who said that not one member of the family of Parvinder Singh, including grandchildren will be alive," she alleged to the police. Analjit filed his own complaint against Nimmi and Malvinder, alleging criminal intimidation and assault.

The "Ranbaxy Family Feud" hit the Indian press as the next sensational chapter in the almost two-decade saga of claims and counterclaims between family members over Bhai Mohan Singh's division of his empire and property. In desperation, Nimmi turned to her sons for help. In less than a month, Malvinder and his uncle announced that the family had reached an amicable settlement. Both sides withdrew their lawsuits, of which there were over thirty going back to the 1990s. Malvinder came off as the family diplomat and peacemaker.

He appeared to be a "spiritual young man," said an Indian journalist who knew the family well. Eight times a year, Malvinder returned to the small town of Beas in Punjab to visit his guru from the RSSB organization, which focused on reuniting man's soul with God through meditation. His spiritual upbringing, and his family's core values of austerity, became part of his brand. Reflecting on his childhood in an interview for Duke University's business school alumni magazine, Malvinder spoke for himself and his brother: "We agree we were born into an illustrious family, however, we were fortunate to have an environment that was simple, pious, and spiritual. . . . Our family values centered on hard work, high ethics, equity of relationships and humility." Just six weeks after assuming the leadership of Ranbaxy, Malvinder would need to bring all the enlightened values he liked to promote into an increasingly fraught conflict.

On February 20, 2006, two of the FDA's most experienced investigators, Regina Brown and Robert Horan, arrived at the Paonta Sahib manufacturing plant in the northern state of Himachal Pradesh. Though they had only six days, they were armed with a confidential assignment memo from Edwin Rivera-Martinez's division: five pages outlining the frauds alleged by Dinesh Thakur.

Despite Ranbaxy's lead time and preparation, the investigators found troubling lapses. Raw data was routinely discarded. Patient complaints went uninvestigated. But their most important finding was the one that Dr. Mike Gavini had seen and walked right past fourteen months earlier: the unregistered walk-in refrigerator, set to 4 degrees Celsius, and another one like it that had been added since then. The contents of the refrigerators made no sense. Inside were cardboard boxes stuffed with bottles of unlabeled drug samples.

"There were no counts for the items, no test status and no reason for storage given for the over 1000 items on the refrigerator inventory and over 150 items on the second refrigerator inventory," the investigators noted in their report.

What were the refrigerators for? One drug vial inside was labeled 30 degrees Celsius. Was it in the wrong refrigerator by accident, as Ranbaxy would later claim? Or was every sample in the wrong place on purpose? The investigators asked for a log of the fridges' contents but were told that none was kept. Later, the company claimed that it did maintain a list of the drugs but had not provided it to the FDA "because we did not understand FDA to have requested it."

The fridges seemed an unlikely hub of fraud. But in trying to explain away unlabeled samples in unregistered refrigerators, the company got tangled up in shifting explanations. Ranbaxy first claimed that the samples were for "regulatory filings globally," but its refrigeration had no impact on stability tests. In a seeming contradiction, company officials separately claimed that the bottles were "'test-on-demand control samples' used only for 'reference purposes' but not used for creating any [official] data." The FDA noted that "it remains unclear to us what these 'stand-by' samples are actually used for." The investigators also took samples of Sotret, Ranbaxy's generic version of the anti-acne drug Accutane. The FDA found that the drug degraded far in advance of its expiration date and had lower-than-expected potency.

The following week, Brown and Horan set out to inspect Dewas, one of Ranbaxy's most troubled manufacturing plants. Mike Gavini had last inspected it in December 2004 and found nothing wrong. This time dozens of Ranbaxy officials trailed behind Brown and Horan, scrambling to respond to their questions. The investigators discovered that Ranbaxy had been discarding original electronic data and had changed the policy to retain it only weeks before their arrival. The investigators even found themselves explaining the basics of good manufacturing: that raw data should not be changed "after the fact or outside of the laboratory operation by someone who was not performing the test."

Horan and Brown had done their jobs. They'd found critical deficiencies that pointed to much larger issues. The findings were serious, and the company's explanations were too inconsistent to fend off the FDA. In June 2006, the agency issued a warning letter to the Paonta Sahib plant that looked to the world like a severe rebuke. It chronicled a list of failings: not retaining "analytical raw data, undocumented stability sample test intervals, the unclear purpose of 'standby samples' [the drugs in the fridge], the inadequate staffing and resources in the stability laboratory," and the FDA's lab results for the anti-acne drug Sotret, which revealed that the drug degraded and lost potency. The agency said that it would not consider any new applications for Paonta Sahib's drugs until the company demonstrated corrections.

This should have been a serious checkmate, since the company had moved manufacturing of its most lucrative and important products for the United States and PEPFAR to Paonta Sahib. But the FDA's action did nothing to stop all the drugs that were already on the market, drugs from the plant that had already been approved, or drug applications submitted from other sites. Just weeks before the FDA issued the warning letter, Rivera-Martinez sounded almost plaintive when he wrote to Thakur: "We are under a lot of pressure to approve Ranbaxy's generic version of Pravastatin [a cholesterol-lowering drug] when the patent exclusivity runs out this Thursday." In short, the bureaucrat's hands were tied by the agency's inexorable machinery: to keep approving drug applications almost no matter what.

Thakur had done as much as he could. Despondent that the FDA was still greenlighting Ranbaxy's drugs, he tried to focus on his family. Sonal had just given birth to their second child, a girl they named Mohavi. She would be their good luck charm, Thakur told his wife. Yet even as Sonal lay in a private hospital in Gurgaon with her arms around her new daughter, both of them receiving excellent care, she worried. Four days in the hospital would cost a major sum. The Thakurs no longer had health insurance and were living from savings. They had gone from being well situated in the corporate world to a fragile, uncertain existence.

Thakur had not been able to find sufficient employment in India. The consulting work, and the income, came in dribs and drabs. Within a few months of Mohavi's birth, he was finally offered a job with Infosys Technologies that would require him to travel heavily and relocate to the United States, almost full-time. He felt he had no choice but to take it and tried to convince Sonal to come with him, even petitioning her mother, but Sonal resolved to remain in India with the children.

She had always prided herself on her independence. She was neither helpless nor without resources. Yes, she was in an arranged marriage, had two small children, and wasn't working. But she had been raised by a mother with a master's degree in Sanskrit, who had insisted that her daughter get a superb education and be able to make her own way in the world. Sonal had done just that. Once she'd gotten her master's degree in computer engineering and worked as a software engineer at the Carrier Corporation, she enjoyed the most companionable span of her marriage. She and Thakur had worked and provided for Ishan, side by side.

But the day her husband prepared to leave India for the United States felt like the darkest of her life. Though she had chosen to stay in India, she felt alone nonetheless. Her mother and father lived over eight hundred miles to the south, in Raipur. That morning, unable to find Ishan, she went down the basement stairs to Thakur's office. She found her husband weeping, with his arms wrapped around Ishan, trying to conceal his heartbreak from his wife. She had never seen Thakur cry. Nor had Ishan, who was asking in a puzzled voice, "Why are you crying, Daddy?"

Taking in the scene, Sonal called out, "Don't! Dinesh, don't do that in front of Ishan." Among all her inchoate fears, the one she fixed on was that the boy might somehow be scarred by her husband's sorrow. Sonal still had no idea what her husband had embarked on with the FDA, yet the stress of it had settled over their marriage like a fog.

There had always been a divide between them. Even under the best of circumstances, Thakur kept a wall around himself. He was not accustomed to having confidants and had revealed his anonymous contact with the FDA to only one person, his friend and former employee Dinesh Kasthuril. Kasthuril, who assumed that Thakur had just tipped off the agency, thought that was "the right thing to do." He had no idea that his friend was in ongoing dialogue with the FDA. Had he known, he might have expressed his doubt that anyone in India could personally take on a corporation and succeed.

Thakur was doubting the wisdom of what he'd done. He'd put his family at risk, and he feared for their safety. The Singhs had a reputation for threatening people, and he knew too many of their company's secrets. Thakur expanded the hours of the security guard outside their house to twenty-four hours a day. He told Sonal that he'd done it because he'd be away. Meanwhile, his disappointment only intensified as he waited for definitive action from the FDA, which seemed either helpless to act or uninterested in doing so.

In all of this, there did seem to be one person who understood what Thakur was going through, shared his goals, and believed in the rightness of what he'd done. Her name was Debbie Robertson, and she was an agent in the FDA's Office of Criminal Investigations. She wrote to Thakur in January 2006 to explain that she would be his new point of contact at the FDA.

Though Edwin Rivera-Martinez's division continued to examine whether Ranbaxy was complying with regulations, Robertson's involvement signaled a new dimension to the case. Her job was to investigate whether Ranbaxy had broken any laws or should face any criminal liability.

Robertson was new to the agency, having arrived in October 2005. But she was a seasoned law enforcement professional who had spent ten years as a criminal investigator at the IRS. As she was junior in the agency, she was handed the matter of Ranbaxy to take another look at it, with the implication that the case was a dog. Robertson was not so sure. As soon as she got in dialogue with "M," as Thakur was known to the FDA, she was struck by his seriousness and evident intelligence, as well as the significant risks he'd taken in contacting the FDA. In turn, her kind and reassuring emails offered Thakur something that no one else at the agency had given him yet: hope.

Nonetheless, Thakur felt disheartened when he returned to the United States. He'd heard initially from former colleagues that the FDA's inspections in India had been uneventful, which he feared would spell the end of the road for the agency's inquiry. But Robertson was reassuring on this point too. "It is actually a good thing that [Ranbaxy] thinks the inspection was uneventful," she wrote to Thakur. "It means they are not suspicious about anything."

It had been months since Thakur first contacted the agency. He had watched as the FDA announced one new drug approval after another for Ranbaxy. Debbie Robertson tried to ease his frustration. "Imagine, if you will, that we were able to prove even half of what you have told us," she wrote to Thakur. "This would bring down the entire corporation. One of the largest in the world." She added, "To lose on a technicality would be a crime in itself." As she advised him, "Think of the U.S. corporation Enron and how long that took."

Thakur wrote back, "Do you believe there will be any concrete action against this company ever? . . . It makes me wonder if all my efforts and troubles were worth anything at all."

She urged him not to lose hope. "The wheels of justice turn slowly," she wrote, "but they do turn."

CHAPTER 13

OUT OF THE SHADOWS

OCTOBER 11, 2006

Princeton, New Jersey

On a bright, crisp morning, Debbie Robertson brought two of her colleagues to the AmeriSuites Hotel off U.S. Highway 1 to meet the man who was referred to in the agency only as "M," short for Malvinder Singh.

It had been fourteen months since Thakur first reached out to the FDA. He'd been reliable and accurate. He did not appear to have shaded the truth or overblown his information. Nor did he seem to have any discernible ulterior motive, other than to stop his former employer from making dangerously inferior drugs. But the investigators were preparing to take a fateful step, and it was Robertson's responsibility to confirm that her agency and U.S. prosecutors were not being led into some sort of internecine feud or ambush, and that "M" was as credible as he seemed.

As FDA agents and prosecutors prepared to serve a search warrant at Ranbaxy's New Jersey headquarters, Thakur had become the agency's eyes and ears. Robertson had turned to him for answers to question after question: Who sat where? How many entrances and exits? Where was the WAN access? Where were the domain controllers? Did the MS exchange email server use encryption? Could the Princeton servers be shut down remotely from India? Was there a way to prevent this once the India office got wind of the intrusion?

In his deliberate and detailed manner, Thakur helped Robertson navigate a virtual maze. He created a digital schematic of the New Jersey offices, down to the location of the bathrooms. His emails were full of minute details.

But in September 2006, after nine months of intense email contact, Robertson wrote to him with a different question: Could they meet? As she explained:

> I am not trying to deceive you in any way, I just think it would be the most judicious use of time and would help clarify some issues. To be completely honest with you, the U.S. Attorney's Office is totally supportive of this case, but because of the complexity and political ramifications, the Assistant U.S. Attorney that is prosecuting this case is being very conservative and wants to ensure we understand exactly what all the information means.

This wasn't exactly true. So far, federal prosecutors at the Maryland U.S. Attorney's Office barely seemed to know the case existed and certainly didn't understand it. There was only one prosecutor assigned to it, the supervisor of the Greenbelt office, who had a full caseload and limited attention. Nonetheless, it was incumbent on Robertson to know her source and to make sure she understood his motives.

Thakur, who was now working in New Jersey, was willing to meet and said so immediately. Still, the idea concerned him. He had no legal representation and had been working with her as an "honest individual," he wrote back. "Since this is a more formal meeting, do I need a lawyer representing me? I have to tell you that I am not all that wealthy and getting U.S. legal representation would be very prohibitively expensive and unaffordable for me." But Robertson was reassuring on this point too. The meeting would simply be a face-to-face version of all their email contact to date.

Thakur was in the lobby of the AmeriSuites Hotel at 9:30 a.m., impeccably dressed in a blazer and slacks. There was Robertson, with dark wavy hair down to her shoulders, warm brown eyes, and a kind and practical demeanor. Her Sig Sauer .357 was holstered under a flowy blouse. With her were two other women: another agent from the Office of Criminal Investigations and the compliance officer, Karen Takahashi, who had been on the original conference call. They sat in the lobby, the fall sunshine spilling through the windows.

The group remained there for two hours as Thakur retraced the history of events inside the company, how he'd gotten the information, which U.S. products he believed were compromised, and how Ranbaxy had gamed the FDA's previous inspections. Robertson asked most of the questions. When she asked what risks he foresaw for his family, Thakur relayed how the Singh family had hired thugs in previous disputes and how whistleblowers were typically dealt with in India.

A hush settled over the group. Though Thakur was doing everything possible to control his nerves, Robertson could see he was shaking.

By the time the trio of FDA agents left, the man they'd referred to only as "M" had come fully to life as Dinesh Thakur. But his choice of aliases had been apt. He had intuited that, in bringing Ranbaxy to justice, Malvinder Singh would become one of the biggest obstacles.

On November 29, 2006, Singh led a delegation of five top company executives, Abha Pant among them, as well as the company's intellectual property attorney, Jay Deshmukh, Ranbaxy's longtime outside lawyer, Kate Beardsley, and a representative from the consulting company Parexel, to FDA headquarters. They had come on a mission vital to Ranbaxy's survival: to convince agency officials to lift their freeze on the review of drug applications from Paonta Sahib. Across a conference table, they faced ten skeptical FDA regulators, including Edwin Rivera-Martinez, who had agreed to hear out the company. Ranbaxy had asked for the meeting, and the executives had prepared for it.

Abha Pant, then Ranbaxy's associate vice president of regulatory affairs, wearing a serious expression, began: "Thank you for the opportunity to meet."

Malvinder Singh, gazing calmly at the regulators, explained that Ranbaxy was committed to quality manufacturing. "We take the FDA's observations very seriously, and have taken immediate action to resolve them," he explained. "I've authorized all resources to ensure that we are fully compliant." No one spoke. He added, "Compliance is important to me personally. I am the third generation of the family to manage Ranbaxy." He then offered, "If for any reason you feel that you're not getting full cooperation, I would invite you to contact me personally."

The regulators seemed skeptical as the executives took turns describing the company's new quality improvement plans. This included hiring an outside auditing firm, Parexel, which was staffed by former FDA veterans, and a new management review committee.

The executives explained that they had added eighteen new analysts in the stability laboratory and eliminated its backlog of samples. The company had also stopped refrigerating its pending stability samples, they claimed. The consultant, Dr. Ron Tetzlaff, chimed in, explaining that his company, Parexel, had made extensive recommendations and Ranbaxy had addressed each one.

But the conversation circled back to the mysterious samples in the refrigerators. The regulators wanted to know whether any had been used to generate stability data for the U.S. market. A Ranbaxy executive denied that. The executives also pushed back against the FDA's findings that the Sotret had lower-than-expected potency. The company had pulled samples from the market, tested them, and was unable to replicate the FDA's findings, they claimed. This was, the Ranbaxy executives suggested, because the FDA's testing method was less precise than their own.

The regulators looked displeased. They dug deeper. One of the more senior bureaucrats there said that the FDA wanted to see the audits done by Ranbaxy's consultant, Parexel. The company argued that those were confidential. The two sides volleyed back and forth on the audits.

Ranbaxy's outside lawyer, Kate Beardsley, piped up. Given the company's concerted effort to address the FDA's concerns and the devastating impact of the agency's finding of Official Action Indicated, would the agency be willing to lift its hold on Ranbaxy's applications from Paonta Sahib? The regulators' answer was no.

Though Malvinder, at thirty-four, knew how to resolve problems in India, solving them in the United States was another matter entirely. By meeting's end, the standoff had only intensified. Regulators would not lift the stay on applications from Paonta Sahib until the plant cleared another inspection. And they wanted to see the Parexel audits.

That was bad enough. But there was something potentially even worse that Ranbaxy executives took away from the meeting. One of them thought he'd spied Raj Kumar's PowerPoint—the infamous Self-Assessment Report (SAR)—among a stack of documents in front of one of the regulators.

Three months earlier, Abha Pant had shared important intelligence with Jay Deshmukh and Ranbaxy's U.S. chairman during the men's cigarette break outside the company's Princeton headquarters. An FDA investigator, who was Indian, had warned her confidentially that regulators had a "bombshell document" that could potentially destroy the company. At the time, Pant had no idea what document the investigator meant, but apparently it had slowed the agency's approval of Ranbaxy's drug applications. Deshmukh, following clues inside the company, was able to unearth the calamitous document, only to have it surface again on an FDA conference table.

Two months after the tense meeting at FDA headquarters, the agency returned to Paonta Sahib. As was customary, the inspection was announced in advance. The FDA's ostensible purpose was to monitor the manufacture of the active ingredient pravastatin sodium, used in the generic version of Pravachol, to lower cholesterol. But inside the agency, the assignment sheet for the inspection left little doubt as to what regulators were looking for: "We continue to have concerns about the integrity of the data in the laboratory. . . . We are concerned about the deletion of records and contradictions in the account of statements between the inspection team and the firm." The assignment also noted, "Be mindful that there may be two sets of books for [active pharmaceutical ingredients] not manufactured at Paonta Sahib."

More telling than what the agency was looking for was the investigator they sent to look for it: Jose Hernandez, now the top compliance officer for the Baltimore District Office, and one of the investigators most likely to detect a fraud. Hernandez was the investigator who had smelled dog meat in the crab-processing plant and found the smoking pile of drug bottles in the woods behind the pharmaceutical plant in Louisiana. What would he be able to find at a plant that was on notice that the FDA was coming and whose survival depended on impressing the agency?

Hernandez arrived on January 26, 2007. A group of officials trailed him as he surveyed the plant carefully. The facilities were immaculate. The units seemed well staffed. Officials retrieved records promptly while he watched. On full alert, he studied the raw data from the pravastatin sodium batch records. Something wasn't right, but he couldn't put his finger on it.

He turned up an unauthorized notebook in the desk drawer of a warehouse supervisor, with notations about possible use of an active ingredient, made by a company that Ranbaxy had not registered with the FDA. It was a tantalizing clue. But company executives explained that Ranbaxy had never used that material, which was why they hadn't reported its use to the FDA. Hernandez asked an employee to run the name of the unapproved company through the computer system, as he watched, but he could see no trace of it.

Hernandez stayed for just three and a half days, all the time the FDA gave him. Although his inspection was tough and yielded some findings, he hadn't nailed it, despite his legendary nose for malfeasance. But he knew that something was wrong, and he vowed to find it on his next visit.

CHAPTER 14

"DO NOT GIVE TO FDA"

FEBRUARY 14, 2007

Princeton, New Jersey

As a hail-and-snow storm pelted the Princeton area, elaborate Valentine's Day floral bouquets arrived for employees that morning at Ranbaxy's U.S. corporate headquarters. Then, around 9:30 a.m., chaos erupted as federal agents, led by Debbie Robertson, swarmed the reception area. Vincent Fabiano, the vice president of global licensing, was in his office when a man he had never seen before walked in and told him, "Step away from your desk."

"Who the hell are you?" Fabiano asked.

"I'm an FDA criminal investigator," the man responded. Fabiano noticed the gun on the man's hip and stepped away from his desk as directed.

On the second floor, an employee heard a voice boom behind her: "Don't touch your computer. Don't touch your phone. Step away from your desk." Her first thought was that there was a bomb threat. She turned around to see FDA criminal investigators in bulletproof vests, guns strapped to their waists, surging across the floor, along with local New Jersey police.

Panic spread as the building was surrounded by police cars. "People were freaking out, crying," a former employee recalled. "They took every computer. There were people with guns." Employees dove under desks, not sure whether the action was an immigration raid or something else. The agents were carrying out boxes of documents and herded employees to a conference room, where they began interviewing each one: they asked about their citizenship, how long they'd worked for the company, even their height and weight. No one could use the bathroom without being escorted by a federal agent.

Jay Deshmukh, the company attorney, who was out of the office that morning, got a panicked call from his assistant, begging him to come. Deshmukh arrived to a scene of bedlam, with federal agents lugging out computers and interrogating staff. He dove in, trying to get the agents to stop interviewing employees.

Even as Deshmukh tried to restore calm, he didn't feel at all composed, especially after he studied the search warrant. Federal prosecutors seemed to be looking for a world of incriminating material. As the news ricocheted from New Jersey to New Delhi, Ranbaxy issued a statement: "This action has come as a surprise. The company is not aware of any wrongdoing. It is cooperating fully with the officials."

By late afternoon, traumatized employees were finally escorted from the building, past piles of drooping flowers in the reception area, which smelled like a funeral parlor. They would later call it the "Great Valentine's Day Raid." It lasted so long that the FDA agents who participated called it the "Valentine's Day Massacre." It killed their evening plans.

Federal agents emerged from the raid with around five terabytes of data—about half the contents of the print collection of the Library of Congress, they estimated. But even amid this avalanche of records, one document in particular stood out. It was the company's own secret report on its Sotret formulation problems that Abha Pant had filed away. Investigators found it in her office with the cover page that read in bold letters, "**Do Not Give to FDA**." The document made clear that the company's defense of its Sotret drug at the FDA meeting two and a half months earlier—and its claim that regulators had tested it improperly, thus getting a bad result—had been a bald-faced lie. The company knew its drug was failing. It had known for years.

Thakur was in India during the raid, celebrating Mohavi's first birthday. It should have been a peaceful family time, but the raid devoured his attention. Former colleagues were contacting him to explain what was happening inside the company. He'd told no one about his role in the search warrant. Ten days later, in late February, he learned what he feared most: Malvinder Singh and board chairman Tejendra Khanna had apparently held a meeting with top company executives in which they reviewed a list of people who might have given the FDA the information that had led to the raid. Thakur's name was on it. So was that of his former boss, Raj Kumar.

What if the company tried to harm him or his family? What if something happened when he was away in the United States? He shared his fears with Robertson, who got him the name and direct contact information for the regional security officer stationed at the U.S. embassy in New Delhi. He mentioned to Sonal, as though in passing, that the FDA was reaching out to former employees and had also reached out to him. If she encountered any trouble, there was someone at the embassy who could help.

Sonal felt increasingly ill at ease. A pall hung over the family: their money worries, her husband's preoccupation, his long absences in the United States, living in a freestanding home with a solitary guard post outside, alone with an infant and a toddler. The thought never occurred to her that her husband had played some pivotal role in the investigation of his former company. Nonetheless, she was concerned enough about the family's safety to paste the information he'd given her on the inside of her closet door.

In the weeks after the search warrant, as the FDA's investigators fanned out, soliciting both those who might help them and those who might implicate themselves, employees began choosing sides. Some remained loyal and hired company-paid lawyers. Ranbaxy promoted, and even relocated to India, several senior executives—including Abha Pant—thereby moving them out of easy reach of the FDA and American prosecutors. Others broke off contact with the company and cooperated with investigators. Dr. Raj Kumar received two messages at home from the company lawyer, Jay Deshmukh, urging him to call.

To Deshmukh, Kumar was a bomb that could detonate at any moment. He'd left on principle and had never agreed to remain silent. He might still have the devastating SAR that the company had worked so hard to destroy. The company's top leadership had been shown the document, and it implicated them directly.

Kumar was "basically in hiding," Deshmukh recalled. "We were trying to get him out of hiding. If he had good facts, we wanted to know. If he had bad facts, I wanted to know." He added, "They were mostly bad facts."

Kumar, who had returned to London, didn't call Deshmukh back. He hired his own lawyer. Deshmukh, increasingly desperate to keep Kumar under wraps, called Kumar's lawyer to offer legal assistance. He also left a chilling message: "Make sure that Raj is careful in what he tells the FDA, because he also has exposure in this case."

Kumar read it as a clear threat.

On the morning of March 16, about one month after the search at Ranbaxy's New Jersey headquarters, Dinesh Thakur arrived at the FDA's Office of Criminal Investigations and was led to a conference room, where numerous investigators and prosecutors from the U.S. Attorney's Office in the District of Maryland, whom he'd never met, were seated around the table. Robertson was there and explained that they now had enough evidence to proceed against the company.

One of the prosecutors told Thakur bluntly, "You need to get a lawyer."

He was flabbergasted. "Why? I told you everything I know. You're a lawyer. Isn't it your job to protect people like me?"

"No, no, no . . . ," the prosecutor began.

"You are telling me you verified all my allegations? I have barely worked for two years. How am I going to be able to afford a lawyer?" Thakur shot back.

"That's not my problem," said the prosecutor. "Now this is a formal investigation. None of the government attorneys can actually represent you."

Robertson, noting Thakur's distress, asked if he needed to use a restroom—which gave her the opportunity to speak with him privately in the hall. Robertson gave him contact information for an organization called the Taxpayers Against Fraud Education Fund (TAFEF), an organization that helped potential whistleblowers, and urged him to contact the group. They could help him find an attorney, she explained.

Thakur felt at his lowest ebb on the drive back to his temporary apartment in Belle Mead, New Jersey. He was seven thousand miles from his children, living on cereal and salad in a cheerless rental, while barely staying afloat financially. Now he had to find an attorney to protect him from the process he himself had set in motion. All evening he thought about the piece of paper Robertson had given him. He felt that calling the number would lead him even further away from the life he knew and expected. Yet he was already in a place he couldn't recognize and getting in deeper by the day.

Thakur went to sleep thinking about the phone number Robertson gave him, and it was his first thought when he woke. That morning he called and left a message. By the afternoon, he had received a call back from TAFEF, which supplied him with a lawyer's name and phone number.

At age thirty-seven, Andrew Beato was finding his way at the prominent firm of Stein, Mitchell, Muse & Cipollone LLP in Washington, D.C. The young lawyer had thinning brown hair, wore steel-rimmed glasses, and spoke in hushed tones so low that visitors had to lean forward to hear him. He typically wore a neutral expression, which did little to mask his air of intensity. When he smiled, which was infrequent, slight dimples emerged. He had been representing whistleblowers for about five years and was told by TAFEF to expect a call from a potential client.

His firm had a history of representing whistleblowers. In 2002, it had represented Cynthia Cooper, the internal auditor at WorldCom who'd uncovered an almost $4 billion accounting fraud. One of the firm's founding partners, Jacob Stein, had represented Monica Lewinsky during the Clinton impeachment hearings. But Beato was still proving himself, and his judgments were subject to review.

It was late in the day on a Friday, and Beato was packing up to leave the office, running late (as usual) to meet his wife, who had also worked as an attorney and was now a stay-at-home mom. He had just gotten his coat on and was almost out the door when the phone rang. He was certain it was his wife, calling to ask where he was. The voicemail picked up. Instead, he heard the soft polite voice of a man with an Indian accent and a formal manner: the whistleblower.

Beato called him back and explained that he was just running out and asked for a "very short" overview. Thakur did not know where to begin, so he just started talking. As Beato listened, trying to hold on to the fragmentary pieces, he became certain that the man was crazy.

Thakur's story sounded so implausible as to be impossible. The top executives at India's biggest pharmaceutical company had committed intentional global fraud. The company's drugs were being taken by U.S. consumers. And to hear the man tell it, the scope was not just limited to one manufacturing plant, or one drug. It was numerous plants and dozens of drugs being distributed around the world. The scope of the fraud the man was alleging made him sound deranged. He must not know much about how businesses operated, or even about how to make a drug, Beato figured. Fraud was typically limited and select—a rogue employee, a single incident, or a poorly managed plant. How could *everything* at a company be fraudulent?

It made no sense that hundreds of employees were allegedly participating in this fraud, as though it were just business as usual. That couldn't happen. If it had, how could it have gone undetected for so long? As Thakur talked and talked, Beato grew more concerned about being late for his wife and increasingly dubious about the whistleblower.

"I have to go now," Beato told him. He did not want to be unkind. "Why don't you send me an email instead, with details about what happened?"

Less than twenty-four hours later, after exchanging about half a dozen emails with Dinesh Thakur, Beato had reversed his earlier view. Though he was far from clear on what had happened at Ranbaxy, he was beginning to think that some version of Thakur's insane story just might be true.

In many ways, Beato was the perfect lawyer to make sense of the Ranbaxy case. Health care had long been his family's business, and illness its fate. Beato had grown up in St. Louis, Missouri, the youngest of seven children. His mother died of breast cancer when he was two. His father, a dedicated internist, was the rare physician who would make house calls late into the night. Unflaggingly committed to his patients' welfare, he was a poor businessman and refused to limit time with his patients in an era of managed care.

As some of his siblings entered health care, Beato turned to the law; he had arrived at Stein Mitchell right out of law school. At first, he represented companies in their dealings with the Federal Trade Commission. But the work left him cold. He didn't want to get up every morning figuring out a way to get a company off the hook. How could he help people, not just corporations, from inside a corporate law firm, where he had to remain to pay off student loans? While Beato was still struggling to answer this question, his father died from a rare brain tumor. Though his father left the family with no savings, he had "wealth of character," Beato recalled. Hundreds of his grateful patients turned out for the wake, proof that his father had lived a meaningful life dedicated to helping others.

Beato's desire to do the same led him to a fledgling area of the law: representing whistleblowers. The practice dated back to the Civil War and fell under a part of the federal False Claims Act known as *qui tam,* an abbreviated Latin version of the phrase translated as "he who brings this action for the king as well as for himself." It allowed whistleblowers to sue those defrauding the government and collect a portion of any recovered funds. The law had initially been intended to stop profiteers from selling defective supplies to the Union army. In the 1940s, after an amendment to the False Claims Act reduced the amount that a whistleblower could recoup, the law fell into disuse. In 1987, however, after reporting of widespread fraud by defense contractors (like the infamous $640 toilet seat sold to the Pentagon), the law was again amended to increase a whistleblower's reward, creating a new incentive to report fraud, and for lawyers to take such cases.

Though Beato lost his first whistleblowing case, which involved a cardiologist doing unnecessary procedures, he saw the righting of wrongs against the people and their government as an essentially moral pursuit.

Weeks after their first conversation, Thakur sat across from Beato and several of his colleagues, including a senior partner, at the law firm's conference table. Though smartly dressed and articulate, Thakur seemed exhausted and anxious, with dark circles beneath his eyes and sagging shoulders. Painstakingly, in a quiet voice, he began to describe the complex web of deceit that Ranbaxy had engaged in and his role in penetrating it.

Barely ten minutes into the meeting, Thakur broke down and began to sob. "What did I do?" he asked repeatedly. "What did I do? I just wanted to do the right thing." He'd stupidly put his family in jeopardy, and now there was no going back. The FDA had started issuing subpoenas to Ranbaxy officials, and Thakur was fearful about how the company would respond. These problems were sorted out in India very differently, he explained. In considering the physical risks to Thakur and his family, the lawyers realized that they were dealing with something they had never confronted in the United States.

Thakur's case was also inordinately complex. The lawyers had to consider the scope of the fraud. Was it even a false claims case? What violations had Ranbaxy committed, and had those materially altered the drugs? How could they prove that? And how could they best protect the man sobbing in their conference room? The case also posed a major financial risk for the firm: they could incur huge expenses, lose, and wind up with nothing. But after meeting with Thakur over the course of two days, the lawyers were almost certain, despite the daunting complications, that the firm would represent him. "This was a public health issue," Beato recalled. "You were not going to find anyone at this firm backing down."

In a quiet moment, Thakur asked Beato, "How do I pay you and what do I pay you? Obviously, you are not doing this for free."

Beato's answer came as a revelation. Thakur would pay them nothing. Instead, the firm would roll the dice on representing Thakur for free and work to help build the government's case against Ranbaxy, with Thakur's evidence as its primary guide. The firm would then file its findings in a confidential lawsuit against the company, which would remain under seal. While the government investigated, Thakur's identity would remain secret. If there was a settlement, Thakur could net up to a third of the government's recovery and Beato's firm would get a portion of that. Under the arrangement, Thakur would become a legally protected whistleblower. Until that moment, he had not even known such a protection existed.

CHAPTER 15

"HOW BIG IS THE PROBLEM?"

2007

Rockville, Maryland

As the regulatory case against Ranbaxy moved slowly through different FDA divisions, it landed on the desk of Douglas A. Campbell, a thirty-seven-year-old officer with the agency's international compliance team. Some of his superiors found little of interest. "We don't really think there is anything there," one of them advised. The case looked like a protracted fight over a single manufacturing plant. Ranbaxy's lawyers were insisting that the FDA lift its temporary ban on new applications from the Paonta Sahib plant, claiming that the company had made all needed corrections. Regulators were demanding that Ranbaxy hand over full copies of the audits done by its consultant, Parexel. Stalemates like these littered the FDA's offices.

But as Campbell read the paperwork for the case, he became intrigued. In its responses to the FDA's inspection findings, the company blamed most of the problems on transcription errors, lost data, or mismatched internal systems. But in excerpts from the audits conducted by Parexel, which Ranbaxy's lawyers had refused to release in full, the consultants had noted "inconsistent entries" in some of the data from stability tests. The company was now sending *corrected* data. But some of the discrepancies seemed awfully big to be mere mistakes, such as a forty-five-day discrepancy as to when a test was conducted. How could the company have been so confused? Or sloppy?

Companies had to continue testing their drugs at predetermined intervals long after they launched them on the market. Every year a company had to file the test results in annual reports to the FDA. Few at the agency actually read these reports, and they piled up in back offices. But the information in the reports still needed to be true.

On July 3, 2007, Campbell drove over to the Office of Generic Drugs, which was headquartered at the Metro Park North campus, and dug out the annual reports that Ranbaxy had filed for three anti-infective drugs: fluconazole, ciprofloxacin, and efavirenz. From the reports, the company appeared to be testing the drugs at appropriate intervals. But as Campbell compared that data to the corrected data on his desk, he was startled by the differences. The annual report for fluconazole stated that Ranbaxy had conducted the three-month stability test on September 26, 2004. But in its response to the warning letter, the company stated that it had performed that same test on August 17, 2005. There was almost a year's difference between the two dates.

To Campbell, it looked like Ranbaxy had either lost control of its manufacturing process or lost track of its lies. "Once you matched up [the dates], it was blatant," he recalled.

His discovery mattered—at least to him. The whole reason to test drugs so frequently was to ensure that unsafe products were detected quickly and didn't stay on the market any longer than necessary. But Campbell's superiors didn't seem particularly excited by the mismatch of dates. Some seemed satisfied by the company's claims that its problems stemmed largely from "transcription errors" or simple failures to update data. Campbell didn't buy it. This was not just ".54 vs .45 type of stuff," as he recalled. Ranbaxy's data was so inaccurate that he couldn't even "find a place where the data actually meant anything." And if the data was meaningless, then there was no proof that Ranbaxy's drugs were safe and effective.

Doug Campbell's boss, Edwin Rivera-Martinez, the first FDA bureaucrat to have corresponded with Thakur, sided with Campbell. He was not mollified by Ranbaxy's claims of correction. As he wrote to a colleague in March 2007, he was convinced that the agency should continue to withhold approval of new applications from Paonta Sahib until Ranbaxy had "completely addressed all issues" from the FDA's warning letter. As Campbell and his colleagues continued to review the documentation from Ranbaxy, each thread they pulled, each strand of testing data they followed, seemed to lead to something bigger—some new revelation that the regulators had not anticipated.

By October 2007, the company reported that its gabapentin, a sensitive drug used to treat epileptic seizures, had shown a spike in an impurity called compound A. Under good manufacturing practices, abnormal spikes or troughs in data were known as out-of-specification (OOS). The company had been required to report the gabapentin finding to the FDA within three days of discovering it, but instead it had waited four months. The regulators soon discovered that the company had not just failed to report the gabapentin impurity spike. It had failed to report irregular test results to the FDA's New Jersey district office for six years. For any vigilant drug company with a large volume of products, such reports should have been a regular occurrence.

The company blamed the troubling lapse on a cascade of small internal problems. But when the FDA investigators went to Ranbaxy's New Jersey headquarters, which should have been issuing the reports, they stumbled on an even bigger finding: for the 600-milligram gabapentin tablets, the company had conducted its three-, six-, and nine-month stability tests in a span of four days, and all on the same day for the 800-milligram tablets. But the test dates had been falsified and documented as if they were performed at the proper intervals.

Inside the FDA, this news was regarded as slam-dunk proof of fraud. Campbell emailed Rivera-Martinez, "Bullseye!!!" Rivera-Martinez reported the finding up the chain of command: "We hit a goldmine!" He laid out the details of the phony testing. Deb Autor, director of CDER's Office of Compliance, responded with one word: "Wow."

Suddenly, the lapses, irregularities, and omissions looked like something very different. Campbell scrawled in his government-issued notebook, "How Big Is the Problem?" And then, "What do they do to come into compliance?" Below this he wrote, "Present the chance to come clean or lie again. Will they continue to jerk us around?"

And below that: "Can we trust them?"

Campbell, a brawny former football player, spent eight years in the army, three on active duty. He had started his FDA career in 1998 at the Roanoke, Virginia, resident post, inspecting everything from infant formula to a tilapia aquaculture farm. By 2006, Campbell had moved to the Division of Manufacturing and Product Quality's international compliance team, nested deep within CDER. His group performed about one hundred overseas inspections a year. Campbell himself had inspected cheese in Nicaragua and stuffed grape leaves in Greece.

Almost overnight, it seemed, the agency was hit by a wave of globalization. The number of pending inspections for Campbell's division spiked. The foreign applications were "all stacked up in our offices," Campbell recalled. From 2002 to 2009, the number of facilities overseas that required inspection by the FDA skyrocketed from around five hundred to over three thousand. As Campbell noted in a review, "The responsibilities related to our mission have exploded. The burden is immense, and the resources have not been readily allocated."

As the need for inspections grew, FDA policy became an all-out scramble to keep up. At one point, Campbell wrote to a colleague to ask about the agency's travel policy: "Do we send teams to India during Monsoon season? Do we send teams to India when it's going to be 110+ degrees in the shade?" The answer came back: "In the past we would routinely defer travel to India during the monsoon, however, due to increase workload we no longer adhere to this."

Before a company was cleared to manufacture a drug at a plant overseas, Campbell's division was supposed to conduct a preapproval inspection to determine if the facility could safely and competently make the drug. But just keeping track of the facilities, and the respective codes that identified them, was a fearsome task. Were the manufacturing plants in the system those really making the drugs? Were the plants that were actually making the drugs the ones being inspected? Campbell understood that he and his colleagues held an enormous responsibility, which weighed on him as he sifted through the Ranbaxy paperwork: "[Should] we let these drugs come into the U.S. or not?"

In October, nine months after Jose Hernandez had inspected Paonta Sahib and come away with suspicions but little else, he received a remarkable email. It was from an employee, writing under the pseudonym "Sunny," who said that he'd seen Hernandez at the plant in January and had finally summoned the courage to write. Sunny explained that Hernandez had been deceived, like so many FDA investigators before him. "At last my conscious did not allow me to keep quiet as its a matter of health of people. Ranbaxy has been hiding many facts till now."

Sunny went on, "What you see at these locations is not the real stuff. It requires a minimum of one month to find out the real matter." He described how the company's top managers exerted relentless pressure on lower-level employees and forced them to clear crucial medicines—isotretinoin, gabapentin, flucanozole, metformin—for release. All had "issues but they are dispatched from the site by QA [quality assurance]." He named senior executives who had orchestrated the scheme: "Over the years they have all bluffed and fooled the FDA."

Sunny explained that before Hernandez arrived at Paonta Sahib, a team of twenty people from research and development had descended on the plant to review and alter data. "This cleanup was done just before the FDA inspection," he wrote. "Such thing is done routinely at all the plants at Ranbaxy." Executives stage-managed the deceptions at the direction of top company leaders, he explained, with lower-level employees carrying out their orders under duress. To support the claim that the company had increased staffing at its stability lab, it had shifted employees from around the plant to that lab during Hernandez's inspection.

As the whistleblowers' revelations spread among FDA officials, regulators were surprised by Ranbaxy's duplicity, in part because it was difficult to fathom the extent of it. The FDA was confronting a system in which data had been so artfully altered that everything seemed perfect. Until then, despite the evidence that had emerged from the search warrant, they had thought of fraud as discrete acts committed by individuals. But what if the way the entire company operated was fraudulent? How could they pierce a scheme of lies that all the employees were in on?

As the FDA descended on Ranbaxy facilities, investigators suddenly noticed everywhere the clues they had previously walked by: the use of unapproved materials; the secret changes in formulation; the use of unregistered active ingredients; the plagiarism of already published data, down to copying chromatograms, or the graphs measuring impurities from brand-name drugs, and passing them off as the company's own. The company seemed to have sprung a leak, with at least half a dozen whistleblowers now writing in to the agency with their own examples of fraud and misconduct.

Sunny continued to write, and Hernandez soon put him in touch with Debbie Robertson, who was leading the probe at the FDA's Office of Criminal Investigations. Sunny revealed that Ranbaxy, unable to find a legitimate fix for its troubled Sotret drug, had secretly altered the formulation while it was on the market, adding oil to the wax base to try to improve the drug's dissolution. None of this had been reported to the FDA, making it an egregious violation. Companies were strictly prohibited from making any changes to approved formulations without the FDA's permission.

Another whistleblower also flagged the Sotret and urged the agency to check the difference in the formulations of the drug before and after December 2006, when the switch was made. "Some scientists wanted to address the issue earlier during 2005 and early 2006 but the commercial staff was ruthless with them," the whistleblower wrote to CDER's ombudsman. "They really have no appreciation for the honest people. I don't know the harm that must have been caused to U.S. users but being part of the [manufacturing] team, I thought as a good world citizen, I must bring this to your notice."

It seemed there was almost nothing the company wouldn't do or say, no excuse it wouldn't make, no claim too far-fetched, to trick the agency into approving its drugs. Instead of investigating its own out-of-specification results, as required, Ranbaxy claimed that its own lab had mishandled samples, which had erroneously led to poor test results. Sometimes it seemed that Ranbaxy was better at making excuses than it was at making drugs. As one FDA regulator urged her colleagues, the company's "discrepancies, inconsistencies, mistakes, oversights and poorly executed investigations" were not to be taken at face value. The lies were so audacious that Debbie Robertson would later say that she had never encountered anything like it: "In all my years [investigating] drug dealers, I had never had that blatant disregard for the law. They lied right to your face. I was told, it's a cultural thing. They understood, but they knew they could get away with it."

The only answer to this global malfeasance seemed to be global punishment. And the agency had a rarely used remedy, one of the most onerous punishments it could inflict: an Application Integrity Policy (AIP), which it had imposed on only four drug companies since the policy was first established in 1991. The AIP allowed the FDA to halt all review of a company's applications, to be resumed only after outside auditors—hired at the company's expense—certified that the data was legitimate. The sanction would flip the regulatory dynamic. No longer would the FDA have the burden of proving fraud if it wanted to block a Ranbaxy product. Instead, in order to get its products approved, the company would have to prove that its products weren't fraudulent.

The agency imposed an AIP only when it had found criminality or "untrue statements of material fact," which it clearly had in Ranbaxy's case, or so Doug Campbell believed. He had drafted a memorandum lining up the various falsehoods that regulators had been able to document. He proposed that CDER impose the AIP for "all approved and pending applications related to Ranbaxy Laboratories Ltd." In short, bring the hammer down on the entire company.

But as drafts circulated and meetings gave way to more meetings, Campbell began to doubt his own ability to communicate. Nothing seemed to rise to the requisite level of proof. No one seemed exactly sure if the AIP was merited or even defensible. The FDA was crowded with lawyers who seemed more intent on heading off court challenges from companies than on protecting public health. The agency's lawyers debated the FDA's requirements. Did it say anywhere that companies could not keep drugs in unregistered refrigerators? If the company claimed that it had lost raw data, was there any written requirement that they keep it forever?

Even the FDA's own role seemed unclear. Was it the agency's job to help Ranbaxy comply? Or to cut off the company for failing to do so? Shocked by the internal dithering, Campbell jotted in his notebook, "Our goal cannot be to make it easy on Ranbaxy!" But in fact, that seemed to be the case. There was intense congressional and public pressure to find more cheap drugs. Ranbaxy also played a crucial role in providing AIDS drugs to Africa under the government's PEPFAR program. Was Ranbaxy simply too big, or too important, to fail?

The FDA's confusion over its own enforcement mission was not the only problem. "There was some force holding this case down," Campbell would later conclude. Did it have to do with money? With connections? With political sway? He grew suspicious, particularly when some of the FDA's Indian employees who'd never come to his office before seemed to find any excuse to visit him.

Then there was Deb Autor. As an attorney who headed CDER's Office of Compliance, she sat a number of levels above him in the FDA bureaucracy and oversaw the work of some four thousand employees. Before she'd joined the federal government in 1995, she'd worked for three years at the law firm that became Buc & Beardsley, which was now representing Ranbaxy. As such, she was on a first-name basis with the partner who served as Ranbaxy's outside counsel, Kate Beardsley.

Beardsley turned to Autor as her direct conduit to the workings of the case, calling and emailing to get insight into the government's progress and trying, whenever possible, to move forward decisions that would be helpful to her client. Autor prodded the bureaucracy at Beardsley's behest, an effort that she viewed as part of her FDA role: to respond to lawyers whose clients had matters pending before the agency. Autor had left the firm thirteen years before the FDA's compliance office began considering the Ranbaxy case. But to Campbell, it seemed that sometimes Autor was more focused on assisting her former boss than on helping to advance the agency's case against Ranbaxy.

Beardsley wrote to Autor in March 2007: "Deb, I left you a phone message, but thought it would be sensible to send you an email too asking if you could give me a call about Ranbaxy. We're still trying to sort things out on the civil side, as they are also trying to cope with the criminal side."

Autor responded, "Hi Kate. I'm happy to call you. But it seems like I know enough about your concerns to follow-up around here first and get back to you. Okay?"

In December, in another email, Beardsley contacted her again, to explain the delays on releasing the Parexel audits in their entirety and asking that Autor call her. Two hours later, Autor told her colleagues via email, "Ranbaxy needs to think through the implications for the criminal case of providing the audits. Please consider whether you can provide Ranbaxy with Doug's 12/6/07 list of questions in order to give them more information about what GMP issues are still troubling us." Given the overlapping connections and bureaucratic foot-dragging, Campbell found it hard to distinguish who was working for the FDA from who was working against it. As the case became increasingly fractious, Campbell hesitated to disclose things to Autor.

Even if agency regulators had not dragged their feet, they faced a problem without an obvious solution: how could they verify that a company was making the changes it claimed to be making in factories that were over seven thousand miles away?

In an email to Robertson, the whistleblower Sunny described how Ranbaxy used hidden areas of the plant to store and cover up testing machines that were not connected to the company's main computer network. He was referring to the crucial high-performance liquid chromatography (HPLC) machines, the workhorses of any good testing laboratory. The bulky machines looked like a stack of computer printers. Once a drug sample is mixed with a solvent, injected into the machine, and pressed through a column filled with granular material, the machine separates out and measures the drug's components, including impurities. It displays them as a series of peaks on a graph called a chromatogram.

In a compliant laboratory, HPLC machines would be networked with the main computer system, making all their data visible and preserved. During a recent inspection, Sunny wrote, the unauthorized HPLC machines were kept in two ancillary labs: "Ranbaxy creates small such hidden areas where these manipulations can be done."

Sunny estimated that some thirty products on the U.S. market did not pass specifications and advised Robertson that the agency needed to raid Paonta Sahib and Dewas, just as it had done in New Jersey, to find the evidence. He warned, "The move has already started in Ranbaxy to share such details of problematic products personally and not on emails or letters."

But because the U.S. Attorney had no jurisdiction in India, the FDA couldn't execute a search warrant there. Robertson felt thwarted: "People said, 'You need to go to India.'" But her response was, "What am I going to do [over there], knock on people's doors and hope they talk to me? I don't have authority over in India. It's all a voluntary, good-faith system." The case had crashed like a wrecking ball into the overtaxed agency, exposing the fact that the FDA had no effective way to police a foreign drug company.

In November 2007, as the FDA prepared to inspect a factory at Ranbaxy's Dewas site that made sterile injectable products to consider whether to approve its drugs for the U.S. market, Sunny emailed Robertson his most important tip yet. The FDA's regulations required the facility to have one of the highest levels of sterility possible. But, Sunny warned Robertson, "the microbiological data is not actual and is manipulated to show less microbial counts," adding that the plant had had several unreported sterility failures. He advised, "Be careful before approving this facility."

A month before the FDA's inspection, Sunny wrote again to Robertson to alert her to a cover-up in progress: "all the actual failure data of Environment monitoring and sterility failures" were being moved out of the Dewas plant to a warehouse fifteen miles away, in Raokheri. Sunny advised that those inside the plant "will try their best to confuse the auditors and they are trained for the same. People in QA have been told to keep their mouth shut about the deviations."

The international compliance team members knew that if they wanted to get the evidence that microbial test results had been manipulated, they needed to go, unannounced, to the Raokheri warehouse. Rivera-Martinez submitted the team's request to do that. But the answer he got back was not what he had expected. In a terse email marked "confidential," Patricia Alcock, then deputy director in the Division of Field Investigations, wrote: "Please check your voicemail message. The warehouse is off the table."

Alcock explained in her voicemail that an unannounced inspection could jeopardize ongoing diplomatic efforts between the FDA's parent agency, the U.S. Department of Health and Human Services (HHS), and Indian health authorities. The two sides were negotiating a written statement of cooperation that might one day enhance the quality of FDA-regulated products made in India. To accomplish even a draft statement, HHS officials wanted to avoid angering the Indians.

Rivera-Martinez was outraged. He prepared an email to Alcock that he shared with colleagues, challenging the refusal. He stressed the seriousness of the allegations. If "true [they] raise significant questions regarding the corporation's compliance attitude, adequacy of the firm's quality management system, and manufacturing state of control." He reminded Alcock that just last month, in a briefing to the staff of Senator Charles Grassley (R-IA) on what it was doing to strengthen foreign inspections, the FDA had even pledged to conduct them unannounced. In their congressional briefing, FDA officials had acknowledged that there was no legal requirement to notify foreign facilities in advance.

Rivera-Martinez's office deemed an unannounced inspection of the warehouse to be "both warranted and necessary." Armed with another email from Sunny about up-to-the-minute manipulations of data, Rivera-Martinez wrote to his superiors: "FYI, here's another e-mail from the informant regarding Ranbaxy's keen ability to manage our inspections and investigators. It's quite clear to me that we have to consider different and bold inspection/investigative strategies and techniques in dealing with Ranbaxy to increase our chances of uncovering evidence of data falsification/manipulation. . . . That's why I'm insisting on an unannounced inspection of the warehouse."

Rivera-Martinez was insisting on common sense. But this time he heard back from the deputy associate commissioner for compliance policy in the Regulatory Affairs Office, who also nixed the warehouse inspection. The deputy explained the thinking of senior FDA officials: "We agreed that any unannounced inspections in a foreign country [need] to be well thought through and planned with all parties to ensure investigator safety and that we minimize the potential that any adverse international incident occurs. Since no such planning had taken place by the time the inspection team left, we agreed that no unannounced inspection would take place this trip." So because of the risk of international conflict that might scuttle a diplomatic effort, the FDA would not fully and unreservedly investigate foreign facilities that presented potential threats to U.S. public health. The needs of patients came last.

Instead, while at Dewas, the investigators requested a visit to the Raokheri warehouse. They were taken there the following day and spent eight hours rummaging through drawers and boxes. As Alcock informed Rivera-Martinez, they "didn't find anything pertinent or as described by the informant." The whistleblower Sunny later told Robertson that "just before the inspection somebody inside alerted to move unaccounted material from the Raokheri warehouse."

But the company could not conceal a more startling problem at the supposedly sterile facility. As Alcock conveyed to her colleagues, "The building . . . is surrounded by pig farms. [The investigator also observed] a lack of instructions/procedures advising personnel to wash their hands/feet before entering the sterile core. (Many of the workers wear sandals. . . . There are also a tremendous amount of pigs scattered on/near the facility???)."

The FDA did not approve the sterile facility. But as agency regulators continued to approve other Ranbaxy applications, their permissive approach seemed increasingly unsustainable. Toward the end of 2007, Deb Autor learned that a different federal agency, the U.S. Agency for International Development (USAID), was considering terminating Ranbaxy as a provider of low-cost drugs to Africa. Instead of applauding the aggressive action, Autor expressed concern that such a clear rebuke of the company would make the FDA's regulators look bad. It would raise "questions about why FDA has not shut down Ranbaxy," she warned her bosses. She offered to take up this question in person or by phone. One of those copied on the email advised colleagues, "Keep this information to yourselves."

On December 12, 2007, USAID sent a stern letter to Ranbaxy. It accused the company of delaying reporting negative test results to the FDA. "I am very troubled by this apparent lack of business integrity and honesty by a company performing on a subcontract funded by the U.S. Government," the letter stated. It was signed by the acting director of the Office of Acquisition and Assistance and stated that USAID was considering suspending or debarring

Ranbaxy from the program. By contrast, the FDA seemed more comfortable to continue its accustomed approach: "Regulators were perfectly happy to stall and stall," Campbell recalled.

By early 2008, Ranbaxy had submitted an application for approval of a plant called Batamandi, in Himachal Pradesh. It proposed to make sensitive drugs there, including tacrolimus, an immunosuppressant used by transplant patients to prevent organ rejection. The application caused immediate suspicion among Campbell and his colleagues. Batamandi was close enough to Paonta Sahib to possibly be a part of it. Was Ranbaxy trying to pass off a part of Paonta Sahib as a new plant as a way to evade the agency's restrictions there?

As was customary, the agency ordered a preapproval inspection for the manufacture of tacrolimus at the Batamandi facility. But there was nothing ordinary about the assignment. Jose Hernandez was tapped to go. In early March, he arrived at Batamandi, armed with his "broad scope of thinking," as he liked to put it, and a determination not to be fooled again. As was typical, Hernandez began his inspection outside, taking in the grounds. By standing at the edge of the property, he could see the Paonta Sahib plant, about two and a half miles away. He noted an eight-foot-high fence that surrounded almost all of the Batamandi plant. There was only one entryway, secured by a well-staffed guard post. The guards there, former military policemen, seemed proud to confirm that they logged all employees and visitors in and out through the gate. No one got past them.

This gave Hernandez the opportunity he needed. By examining checkpoint records at the gate, he learned that the supervisors who had signed off as present for the manufacturing of key tacrolimus batches had not actually been at the plant on those days. They had not signed in to the security gate logs. The dates, times, and signatures of the batch records were fake and had been filled in after the fact. One evening at his hotel, where other Ranbaxy executives who had flown in for his inspection were also staying, Hernandez cornered one of them and said, "When Catholics do something bad, they go to the priest and confess. Now, think that I am the priest and take this opportunity to tell me who was involved in creating the fraudulent data."

They did not confess then and there. But by the end of his inspection, while at the closeout meeting at the plant, the managers essentially admitted that the company was trying to rush Batamandi into operation to circumvent the FDA's restrictions on Paonta Sahib.

Hernandez explained to Ranbaxy's senior executives that he was planning to recommend against approving the plant. The company's senior vice president for global manufacturing, who'd arrived to oversee the inspection, took him aside. He became "agitated and desperate," Hernandez documented. The vice president admitted that the plant's construction had been rushed and mistakes had been made. Over and over, the man pledged to correct anything Hernandez wanted, but begged the investigator not to use the word "falsified" in his report.

The FDA refused to certify Batamandi as a separate plant, and Ranbaxy withdrew its tacrolimus application. It was yet another round in a global game of whack-a-mole. Each time the FDA found fraud inside Ranbaxy, it responded with a small regulatory restriction, only to find fraud pop up somewhere else. The agency had done nothing yet to clamp down on Ranbaxy's entire method of operating. But the game was about to change.

CHAPTER 16

DIAMOND AND RUBY

OCTOBER 2007

New Delhi, India

Malvinder Singh's year had gotten off to a bad start. Unlike his father—a visionary who built institutions and legacies—Singh saw his primary role as creating value for shareholders. As he told India's business press, "I am an entrepreneur at heart and value-creation is the ultimate objective of any true-blue entrepreneur."

But creating lasting value was not as easy as it seemed. When he took over the company in 2006, the young CEO quickly ran through his MBA playbook—aggressively seeking acquisitions and alliances. But even though he succeeded in generating buzz, Ranbaxy's bottom line was "sagging," according to *AsiaMoney*. Ranbaxy had to withdraw its bid to buy Germany's biggest generic drug company. Then there was the FDA. In his view, the American regulators were a sour lot—unmoved by his pledge of reform and the company's face-saving gestures.

In Malvinder's India, you could almost always make a problem go away, whether through strategic payments or the threat of force.

Just five months earlier, Malvinder's brother, Shivinder, blocked a high-profile cardiac surgeon who opposed him in a business deal from reentering a New Delhi hospital the Singh family owned. The surgeon arrived at work to encounter almost one hundred policemen and a full battalion of Rapid Action Force servicemen, armed with tear gas and a water tanker used for riot control.

But in the United States, there was no vigilante army to summon and the FDA was refusing to budge. As Ranbaxy's largest shareholders, the Singh brothers found that the problems had begun to take a toll on their bottom line. Amid these concerns, Malvinder paid close attention to an intriguing message from one of the company's advisers in New York. A fellow named Tsutomu Une, from the Japanese drug company Daiichi Sankyo, wanted to speak to him about a strategic partnership. Malvinder sensed opportunity.

In Tokyo, four thousand miles away from New Delhi, Dr. Une looked around the world for new revenue streams. As senior executive officer of global corporate strategy for Daiichi Sankyo, Japan's second-largest drug company, Une wanted to penetrate far-flung markets, like India and eastern Europe, in which his company had never set foot. He needed a low-cost, big-volume partner. His eye kept landing on Ranbaxy.

Une, a sixty-year-old microbiologist, had come up through the ranks of pharmaceutical innovation. He had been at Daiichi Pharmaceutical Company for more than thirty years and continued his rise after the company's 2005 merger with the Sankyo Company. Gentlemanly and formal, Une maintained a careful journal of his work life.

More than other countries, Japan was a brand-loving society that viewed generics as untrustworthy. The nation was devoted to quality and cleanliness. Though once scorned for poor quality, its modern-day pharmaceutical industry enjoyed a worldwide reputation as a leader in quality controls. The pills even had to be white, or patients regarded them with suspicion. Though Une was a product of that trademark caution, he recognized that the insular nature of Japan's drug industry and its focus on high-cost research had led to a slowdown in growth.

From his perspective, Ranbaxy—with manufacturing plants in eleven countries and sales in 125 countries—was enticing. It had numerous first-to-file drug applications pending in the United States, including for generic Lipitor, which would be the most lucrative generic drug launch in history. Daiichi Sankyo wanted a stable of low-cost drug products to help create a flow of revenue in between the blockbusters. To that end, Ranbaxy might be the perfect partner for an alliance. If an acquisition could be done quickly, it could help prop up Daiichi Sankyo's battered share price before the next quarterly investors' meeting.

In Japan's boardrooms, decision-making tended to be consensus-driven, so Une would have to convince his colleagues. But as Daiichi Sankyo's president, Takashi Shoda, had told his colleagues, "India will be the trump card allowing a Japanese pharmaceutical firm to go global." The idea was in the air.

Une, sensitive to cultural implications, knew the matter would be a delicate one. Ranbaxy was not just any Indian company. It was a cultural institution, built by the eminent Singh family and handed down for three generations. It was now in the hands of a Singh scion, an American-educated MBA little more than half his age. But Une's standing was high within Daiichi Sankyo, his vision respected, and his business acumen proven. And so, in early October 2007, he made his first approach, contacting one of Ranbaxy's external advisers in New York.

Within three weeks of their first conversation, Une and Malvinder met. Une, his silver hair combed neatly back, spoke in thickly accented English. Malvinder—with his polished diction, bespoke suit, and elegant turban with matching handkerchief—

was the picture of composure. Their negotiations moved swiftly, and the men scheduled a follow-up meeting in New Delhi. Une and Malvinder proceeded with utmost secrecy, using the code names "Diamond" for Daiichi Sankyo and "Ruby" for Ranbaxy in internal reports and correspondence. They also agreed on a cover story for their upcoming meetings. If the press inquired, they would claim to be in discussion about a contract manufacturing agreement. Those were a dime a dozen and wouldn't attract the scrutiny of the business press.

Little more than four months after Une's first contact, conversations had escalated well beyond a strategic partnership to negotiations over an outright sale. The Japanese microbiologist and the young Indian billionaire were haggling over share price and terms. But Une was concerned by the regulatory cloud that hovered over Ranbaxy. In emails back and forth, Une, guided by his lawyers, kept pressing Ranbaxy to make typical warranties, representations, and indemnities explicit, which would allow Daiichi Sankyo to sue for breach of contract if Ranbaxy's financial health was not as Malvinder warranted. Malvinder kept deflecting his requests. Finally, on a call, Une told him, "My colleagues are frustrated by your 'Don't worry' type answers to our questions." In a second call the same day, Malvinder, in his melodic and steady voice, moved to reassure the scientist: "There is no fear or fault in Ruby."

But Une had unresolved questions. It was public knowledge that federal agents had raided Ranbaxy's U.S. headquarters in February and issued warning letters against two of its key plants, Paonta Sahib and Dewas. But no one knew how serious the investigation was, or how deep the company's potential liability went. Malvinder had asked Une on the phone, "What are your concerns?"

Une could only guess what prosecutors working for the formidable U.S. Justice Department and the world's toughest regulators at the FDA were probing. It would be the job of his advisers to find out. In less than a month, Malvinder and Une met secretly in

Delhi, this time with their lawyers and a small coterie of top executives. Une made clear that whatever lay at the bottom of the various U.S. investigations into the company would be critical to whether Daiichi Sankyo moved ahead. Malvinder appeared unruffled and pledged help and candor with the process. He agreed to set up a due diligence meeting in which all documents relevant to the investigations would be disclosed to Daiichi Sankyo.

The young CEO then appeared to take Une into his confidence and explained what was really driving the investigations: Pfizer was retaliating against Ranbaxy for having prevailed in the Lipitor patent litigation and, through sleight-of-hand, had unleashed the investigators. As Une considered this explanation, Ranbaxy's intellectual property lawyer, Jay Deshmukh, among a handful of top advisers present, sat stone-faced and silent.

Despite Malvinder's blithe assurance that there was no "fear or fault" in Ranbaxy, the company was essentially a powder keg. As Malvinder well knew, the company's dark secrets had been memorialized in the potent document known inside the company as the Self-Assessment Report (SAR), the blistering PowerPoint that Rajinder Kumar had shown to the board.

Had the document been destroyed, Malvinder wouldn't have been in this position. But he believed that the U.S. government had it, and it had served as a magnet for prosecutors and regulators. Ranbaxy's external lawyers had made it vividly clear to Malvinder that the company would not move past its troubles with the U.S. government until it righted the wrongs described in the SAR. And though the SAR did not address the authenticity of the data used for U.S. drugs, government prosecutors had quite logically inferred that, if so many of the company's drugs were fraudulent, there was no reason to trust *any* of them.

Just weeks after Malvinder's disastrous meeting with the FDA, lawyer Kate Beardsley, who had helped Ranbaxy navigate the FDA for years, got on a plane to New Delhi. She spoke with Malvinder

PART IV

MAKING A CASE

specifically about the SAR. Until the company dealt with it, she told him, its FDA problems were likely to persist. There was only one way forward: to withdraw each drug and each dossier flagged as having false data and then retest and resubmit them, just as Kumar had proposed two years earlier at the ill-fated board meeting. The situation required global remediation, Beardsley told him.

If there was any lingering doubt as to whether the government had the SAR, the February 2007 search warrant in New Jersey erased it. Ranbaxy's lawyers learned that it had been in the trove of documents that FDA agents hauled from Abha Pant's office. The next month, in a meeting near Heathrow Airport in London, Beardsley again raised the topic of the documents, this time with Deshmukh and former CEO Brian Tempest, who remained a company adviser. A new outside attorney, Christopher Mead, a partner with London & Mead, which had been hired by Ranbaxy to deal with the growing Justice Department investigation, also attended the meeting. A veteran former prosecutor, Mead had immediately understood the significance of the SAR. Not only had this confession of fraud by executives in the company probably ignited the Justice Department investigation, but the document also opened the door to prosecution of individual company executives. The alarm was such that, in a phone call, Deshmukh warned Tempest against traveling to the United States, where he could be subject to arrest.

In June 2007, Mead sat down with Malvinder and told him that he'd gotten two letters from the Justice Department, asking for documents related to the SAR. Mead was nothing if not emphatic. The SAR was so serious and reflected such a culture of corruption that the company had to address the concerns it raised head on. Until it did so, the government was unlikely to relent.

But as Ranbaxy and Daiichi Sankyo got closer to a deal, Malvinder had a different concern about the SAR. It had attracted regulators and prosecutors but would almost certainly repel the Japanese and jeopardize the deal with Daiichi Sankyo. He would need a way to make the documents vanish in plain sight. Malvinder's trusted deputies fabricated the minutes of the 2004 board meeting, so they bore no reference to Kumar's presentation.

Malvinder also managed to wear Une down in his insistence on contractual safeguards. With the argument that contracts were usually different in India, he'd gotten Une to accept a single representation and warranty that Ranbaxy was as sound an investment as he'd portrayed. That guarantee would come from Ranbaxy, not from Malvinder personally. However, Une had been warned by advisers not to buy the company without proper due diligence regarding the U.S. government's investigations. It was advice he intended to heed.

In general, Une's approach, as a Daiichi Sankyo lawyer summarized it to a consultant at the time, was, "if we show our best efforts, sincerity and reasonableness on the coming negotiation, they also will show their best efforts, sincerity and reasonableness to us." The consultant wrote back, "We have been too trusting of Ruby, and they have taken full advantage of that." Jay Deshmukh would later observe that the two cultures were like "oil and water." Indians succeeded in business by being "ultra-aggressive," he said. "Ethics are not important." By contrast, the Japanese were "very trusting. Babes in the woods."

Inside Ranbaxy, Malvinder put Deshmukh in charge of the due diligence process. As they prepared for a major meeting with Daiichi Sankyo, Singh gave him explicit instructions: do not mention the SAR or any of its repercussions. Deshmukh balked, but ultimately made clear that he would follow orders, and he advised the external lawyers not to mention the SAR to Daiichi Sankyo under any circumstance. Other Ranbaxy executives were instructed not to speak with Daiichi Sankyo executives at all, but to communicate with them through Malvinder's executive assistant. Malvinder would later deny any misrepresentation or concealment of information from Daiichi Sankyo, saying that all relevant information about Ranbaxy was in the public domain.

Though Abha Pant was tapped to speak about the FDA warning letters at the upcoming meeting, Malvinder's personal attorney advised her to say nothing about the SAR. This lockdown was possible, in part, because Malvinder had surrounded himself with a handpicked group of loyalists who had ties to his family and to the spiritual organization so interwoven in the family's affairs. They sat in on almost every early phone call Malvinder had with Une.

On May 26, 2008, Malvinder, Deshmukh, and Une met in New Delhi, accompanied by top executives and lawyers from both sides. To ensure that Deshmukh wouldn't waver in his pledge, he was given a script that Malvinder had signed off on to follow throughout the meeting. It made no mention of the SAR or its connection to the government investigations. The script put Deshmukh in a box. He could only tell Daiichi Sankyo that the FDA and DOJ investigations were routine and dealt with different, unrelated issues and that the allegations Ranbaxy faced were unlikely to result in major liability.

By agreement, Ranbaxy had set up a data room where Daiichi Sankyo's lawyers could review documents that included communications from U.S. prosecutors. But documents that mentioned the SAR had been removed from the stacks of materials. The two letters from the Justice Department asking for related documents had also been excluded.

Weeks later, in an executive meeting with Malvinder in London, Deshmukh was thunderstruck to learn that the Japanese were to become Ranbaxy's majority shareholder. Until then, he believed the two companies were negotiating a manufacturing agreement, rather than an outright sale. He'd been kept in the dark. Once Deshmukh understood that an acquisition was being negotiated, suppressing information not only felt unjust. It seemed fraudulent.

Afterward, Deshmukh sought out Kate Beardsley for legal advice about his responsibility for deliberately withholding information from Daiichi Sankyo. Distraught, he told her that he was suffering a "crisis of conscience." Beardsley said that she couldn't advise Deshmukh personally, as her client was Ranbaxy. But Malvinder's directive to withhold information about the SAR was so serious that she consulted with her law partner about whether it was even possible to continue representing Ranbaxy. They concluded that if Daiichi Sankyo contacted them with questions about the government's investigation, they would have to resign rather than lie.

But Daiichi Sankyo never contacted them. Within weeks, the Japanese company had signed the agreement to become Ranbaxy's majority shareholder and retain Malvinder as CEO for five years. At a press conference in June 2008, Malvinder stunned the Indian business world by announcing that he and his brother, Shivinder, had agreed to sell their 34 percent stake in Ranbaxy to the Japanese for an astonishing $2 billion. The Japanese would then purchase additional shares to bring their controlling ownership to just over 50 percent of the company.

Malvinder called the sale an "emotional decision," but one that would enhance the value of both companies and leave Ranbaxy debt-free. Just because he was no longer a shareholder, he said, "the visions, the dreams, the aspirations, none of that is going to change." He faced fierce criticism that he'd betrayed national entrepreneurial pride by selling Ranbaxy to the Japanese. "Ranbaxy was the all-conquering Indian hero and should have been the last man standing instead of being the first to capitulate," a former Ranbaxy executive told the *Economic Times*.

Malvinder had capitulated just in time.

One month later, on July 3, 2008, the Maryland U.S. Attorney filed an explosive motion in district court, shocking both FDA bureaucrats and Ranbaxy executives. The motion, which was public, demanded that the court force Ranbaxy to turn over the Parexel audits, which had been subpoenaed seven months earlier. In one sense, the motion was procedural. But the language was a knockout punch. The motion described a "pattern of systemic fraudulent conduct by the company" and noted that ongoing violations by Ranbaxy "continue to result in the introduction of adulterated and misbranded products into interstate commerce with the intent to defraud or mislead."

Inside the FDA, the mood among frustrated bureaucrats was jubilant. Even if their own agency hadn't yet taken action, prosecutors had. Edwin Rivera-Martinez emailed his team members: "This is the best news I've received in a long time! Finally, after waiting for so long and so much time and effort dedicated to this case . . . we are seeing the fruits of our efforts." The usually understated bureaucrat signed off: "Enjoy the rest of your weekend if you can calm down."

The motion set off a firestorm of questions from public health experts, congressional investigators, and foreign regulators, who were suddenly reevaluating some of their own Ranbaxy inspections. Even Ranbaxy's own consultants were confounded. If the FDA had so clearly identified a company-wide culture of corruption, why did it continue to approve the company's applications, product by product, and the company's manufacturing plants, site by site?

Deb Autor, the director of CDER's compliance office, sent around a blog post written by a former FDA employee with thirty-eight years of experience at the agency, headlined, "It's Time for FDA to Block All Ranbaxy Products." The post stated, "Until [Ranbaxy] can prove to FDA that these record problems are not part of a corporate culture, I feel that FDA should just inform Ranbaxy that all of their products offered entry into the U.S., regardless of plant involved in the actual production, will be refused entry." It's what the FDA compliance officer Doug Campbell had been saying all along. The post concluded that it would be a "start for the FDA toward regarding the American consumer as a 'customer' deserving protection, rather than continuing to reserve this policy for industry." Autor appended a terse note, written in bold, to her email: "**Further discussion if any should be in person, not by e-mail.**"

Ultimately, it was neither the risk to public health nor the company's ongoing obstruction that motivated the agency to act. Two weeks after prosecutors filed their motion, an email with the subject line "Heads Up xx" landed in the inboxes of senior FDA officials. It came from an attorney in the chief counsel's office and noted: "I have been advised that Congress may soon be vigorously inquiring why FDA did not seek to bar the importation of Ranbaxy drugs made in India that are alleged to have been made under fraudulent conditions."

The prospect of congressional questions set off alarm bells within the agency. Doug Campbell was in Poland, inspecting a drug plant, when his cell phone rang. It was Deb Autor. She had never contacted him directly before. Now she was focused on guarding the agency's flanks and wanted to know: "What are we doing with Ranbaxy?" Campbell told Autor the truth: "We're not doing anything." Three days later, Campbell got new marching orders from an official by email: "Be prepared. Ranbaxy is your only priority right now."

As though in preparation for questions from Congress, an internal summary of steps the agency had taken related to Ranbaxy began circulating among top officials. It played down the findings at Paonta Sahib, explaining that they did not justify restricting the importation of products made there. The summary noted that the agency had refused to approve some of Ranbaxy's applications—but made no mention of the twenty-seven others it had approved since Thakur had first reported the fraud in 2005. It also noted that five inspections "have not found evidence to substantiate the informant allegations relating to bioequivalence testing," a shading of the truth at best.

The memo concluded that appropriate next steps had been a difficult judgment call and had triggered scientific disagreement within CDER. This made it sound as though the agency had been actively wrestling with the case as a prelude to taking action, rather than simply sleeping on it. But in its own ponderous manner—facing the specter of congressional oversight—the agency was finally gearing up for action.

CHAPTER 17

"YOU JUST DON'T GET IT"

JULY 2008

New Delhi, India

As the Maryland U.S. Attorney's Office filed its motion, alleging a "pattern of systemic fraudulent conduct," Malvinder Singh faced his sharpest crisis yet. Questions loomed about Ranbaxy's integrity, and the company's stock price fell. The deal with Daiichi Sankyo had not been completed yet, and Malvinder needed a way to salvage it.

In a cagey call with reporters, Malvinder went public with his allegations of a conspiracy that he had previously shared with Tsutomu Une. "People are trying to create confusion and obviously somebody is trying to bring our price down so that they can [buy] at a lower price," he told investors. "A multinational [company] and a leading Indian company are working in concert to bring our share price down," he claimed, without offering any proof. Malvinder said Daiichi Sankyo was "aware of these issues while conducting due diligence. There is no change in the deal and there is no exit clause in it."

Inside Ranbaxy, the principals huddled. They attended a meeting in Malvinder's office marked "VIMP" (very important) in their calendars. The first agenda item: "Self assessment." Almost everything aside from the SAR could be explained away, but that document alone would expose Malvinder's claims as a lie.

The sense of crisis, and the phalanx of outside law firms to help manage it, was growing by the day. In late July 2008, company lawyer Jay Deshmukh traveled with the outside lawyer Chris Mead to a meeting at Ranbaxy headquarters that included Malvinder and other senior management, as well as two other external lawyers: Beardsley, the company's longtime FDA lawyer, and Raymond Shepherd from Venable, brought in to deal with the new problem of scrutiny from the U.S. Congress. They needed a strategy to respond to the increasing tide of problems. In the car on the way to the meeting, Deshmukh confided in Mead that Daiichi Sankyo had not been told about the SAR because Malvinder would not allow it. Mead was furious.

At the meeting, the SAR again dominated the conversation. It became infuriatingly clear to Mead that the company had done little to respond to the allegations laid out in the SAR. He had learned, from a report Pant created, that the company continued to ship over sixty products that had been approved based on fake data. He was so frustrated that he pounded his fist on the conference table and shouted at Malvinder, "You just don't get it!" He insisted that Ranbaxy stop shipping such products immediately, advice that Beardsley seconded. Mead explained that until the company fully dealt with the SAR, the government would not yield. Malvinder appeared to acquiesce and agreed to withdraw all affected products from global markets. Both Mead and Deshmukh urged him to disclose the SAR to Daiichi Sankyo. Malvinder responded, "I will deal with the Japanese."

Among Ranbaxy executives, this came to be known as the "You just don't get it" meeting. Malvinder did take one clear step after that meeting: he let go Mead's firm and brought in another to deal with the Justice Department.

As Ranbaxy executives battled their own lawyers, FDA officials were preparing to take action. On September 16, 2008, just two months after prosecutors filed their motion, the agency announced that it was stopping the import of more than thirty drugs from two Ranbaxy plants, Dewas and Paonta Sahib. It also unveiled new warning letters for each plant. "With this action we are sending a clear signal that drug products intended for use by American consumers must meet our standards of safety and quality," said Dr. Janet Woodcock, CDER's director, in a press release.

As the compliance staff paused for a moment to celebrate, FDA officials held a media briefing to explain their new get-tough stance toward Ranbaxy. They were blocking the importation of the drugs from two of Ranbaxy's plants. However, they were not demanding that the drugs be recalled from U.S. pharmacy shelves. Characterizing the action against Ranbaxy as "proactive," Deb Autor told reporters, "FDA has no reason to believe the drugs from these two plants already in the U.S. drug supply pose a safety problem." Stating that FDA tests showed that Ranbaxy drugs met specifications, she added, "FDA has no evidence that these Ranbaxy products are actually defective. But the manufacturing process and control problems the FDA has found could impact product and for this reason FDA has taken these proactive steps."

Autor had served as an agency representative. Her statements, as she later said, "were simply a reiteration of FDA's official position, which was, as always, the result of a lengthy internal deliberation process involving many people throughout the organization." Nonetheless, her statements sparked fury inside the agency. The criminal investigators and compliance staff convened to discuss their veracity and impact. The FDA knew that two of Ranbaxy's drugs, Sotret and gabapentin, had failed quality tests and posed a potential danger to patients. If the drugs were not defective, then what was the problem, and what leverage did the FDA have? Campbell didn't know what to think. "Apparently there's an idea, we don't want to create a panic. It's a lawyer thing," he said. "Lawyers don't have the same kind of conscience."

The statements had floated the company a life raft and made the work of agency investigators that much harder. For years to come, Ranbaxy would invoke them in its efforts to fend off the FDA. In the short term, however, the FDA's announcement of import restrictions and warning letters triggered a crisis inside the company that played out on three continents.

On a conference call the day after the FDA's get-tough announcement, Malvinder Singh continued to insist to Une that Ranbaxy had done nothing wrong. He told Une that he was "shocked" by the developments. He claimed that Ranbaxy had been fully cooperative and there was "no sign or omen" that the FDA had been dissatisfied with its efforts.

Remarkably, despite every warning sign, the deal between the two companies remained on track. Ranbaxy agreed to turn over to the Justice Department the audits prepared by its consultant, Parexel, in their entirety, and the U.S. government withdrew its motion to enforce the subpoena. Une continued to express confidence in Malvinder, noting in his journal, "[Malav-San] responded to our requests in good faith on each matter and prepared accordingly. Things have started to work well." He even counseled himself to be culturally sensitive and not to harbor negative thoughts. In one entry, he wrote, "Malav-San's response to requests in connection with the FDA are slow. I understand that it is in the middle of the most celebrated festival but . . . Patience!"

On November 7, 2008, Daiichi Sankyo and Ranbaxy closed

their transaction. A month later, the Japanese took majority control of Ranbaxy's board. In the blink of an eye, Malvinder and his brother had added over $2 billion to their own bottom lines, and Malvinder went on to become an employee of the Japanese. It was an odd arrangement fraught with risk. But Ranbaxy lawyer Jay Deshmukh had new hope. Now that Ranbaxy had agreed to cooperate with the Justice Department and turn over the Parexel audits, perhaps the company could swiftly reach a deal with prosecutors that would make concealment of the SAR irrelevant. His guilt over lying to Une might lift. It would all be behind them. In the meanwhile, Deshmukh was in a strange new world. There was no more shouting—or decisions by fiat—in interactions with the Japanese. They operated by consensus, he found. It seemed to him that ten people got on the phone and little got done.

Despite Ranbaxy's lurching difficulties, Une was pleased enough. It was his deal, and he documented it closely. "Mr. Singh fits in well," he noted in his journal. But he also noted, as he delved into the work for his monthly Ranbaxy meetings, that the discussions on how to resolve the company's FDA issues were "all just the beginning of the beginning." On February 19, 2009, Une traveled to New Delhi and met with Malvinder. He was sorting through a host of issues, including how to balance Malvinder's autonomy with Daiichi Sankyo's governance and trying to understand the chronology of Ranbaxy's troubles with the FDA. Something didn't quite add up, but he wasn't sure what.

He was back in Tokyo on February 25 when he got news that came as a body blow. The FDA announced that it would level the harshest punishment available on Ranbaxy—an Application Integrity Policy. An AIP was the drug regulator's version of a scarlet "A." The agency imposed it only when it deemed a company's applications to be largely fraudulent or unreliable. Now the company would have to prove its products weren't fraudulent in order to get them approved.

The AIP covered all the products for the U.S. market that Ranbaxy manufactured at Paonta Sahib. The action left no doubt as to the depth and extent of the problem. The stock market responded accordingly. Ranbaxy shares fell 18 percent and took Daiichi Sankyo's down 9 percent with them. "Crisis!" Une noted in his journal. A gentleman to the core, however, he did not at that moment second-guess his most prominent employee, as another executive might have done. Instead, Une wrote, "even though it is the problem of the management of Malav-San, now is the time to cheer him up. We can criticize later." Both Ranbaxy and Daiichi Sankyo formed crisis response teams. Within three days, Une was on a plane to New Delhi. There, he made continued efforts to understand why the FDA had clobbered the company.

As Malvinder continued to play dumb, Une was left with a Delphic riddle of sorts. Why would the FDA deem the entire company dishonest if, as Malvinder claimed, Ranbaxy had done everything right? All Une could do was parse the clues in front of him. He was a careful observer and wrote in his journal about his meetings with Malvinder and his deputies: "What I found out is that they also do not understand the reason AIP was invoked. They are not aware that the entire system was deemed suspicious even though it was partially fixed. Malav-San joined [the meeting] at the end. Members' tone changes. They are very much afraid of him."

Inside Ranbaxy, the AIP had caused tensions to boil over. Jay Deshmukh was at his wits' end. He'd participated in a fraud, under direct orders from his boss. Now that Daiichi Sankyo needed to take charge of the FDA and Justice Department problems, it had every right to know the truth about their origins. The ongoing suppression of the SAR and the deceit perpetrated against Une seemed more indefensible with every passing day. And though Deshmukh had been put nominally in charge of cleaning up the trail of fraudulent dossiers, Malvinder's loyalists inside the company were blocking that effort at every turn. In meetings, Deshmukh openly began to voice his dismay, on occasion even threatening to personally disclose the SAR to Daiichi Sankyo, airing a view that others shared but refused to say aloud.

Finally, in early March 2009, the growing conflict between Deshmukh and Malvinder broke out into the open. "We don't think you're doing things the right way," Deshmukh told Malvinder at an operational meeting in India, with over a dozen people present.

"I'm going to stop you right there," Malvinder interrupted him. "Who's *we* and who's *you*?"

"*We* are the lawyers," Deshmukh said.

"Jay, I am very upset with you," Malvinder responded. "You're now looking at *we* and *you*, *us* and *them*? You're not one of us?"

"Take it the way you want," Deshmukh retorted. "I want this thing cleaned up. And you're not doing it. Your guys are refusing to do things that the compliance lawyers and I are insisting be done."

Not long after that clash, Malvinder called Deshmukh, who'd returned to the United States, and told him, "I want you in India." He directed him to take the next flight over. Within seventy-two hours, Malvinder was glaring at Deshmukh across his desk and gave him three choices: he could leave voluntarily with a negotiated settlement; he could relocate to India and work on intellectual property but nothing else; or he could be fired and leave on bad terms. Deshmukh asked for a day to think it over.

That night Deshmukh went out for a drink with trusted colleagues and blew off steam. He complained that the company's fall was all due to Malvinder, and that he was screwing over Ranbaxy's twelve thousand employees and all their dependents. "There's not too many people who are the worst people on the face of the earth, but this guy is one," he said. He threatened to disclose the SAR to Daiichi Sankyo.

In bitterness, he had said way too much. The next day Malvinder hauled Deshmukh back into his office. The head of human resources was seated beside him. The meeting got instantly ugly. "I heard what you said," Malvinder said, meaning that the lawyer's remarks over drinks the night before had been relayed to him. Malvinder then directed the human resources director to leave. He told Deshmukh that if he chose to release the SAR to Daiichi Sankyo, Malvinder could take matters into his own hands. "I know where you live," Malvinder said.

Deshmukh refused to cower: "Of course you know where I live, you idiot. You've come to my house." He added, "Why don't you get your HR person back in here, so he can hear your threats?" He then pointed out that if Malvinder tried to harm him while he was in the United States, "it's going to come back and bite you very hard." He added, for good measure, "I think it's safe to say my Shiv Sena connections won't appreciate that." Deshmukh's second cousin had helped form the Shiv Sena, a fearsome right-wing Hindu nationalist party linked to political violence.

"You're a womanizer," Malvinder shot back. "I have all the records on you."

The men continued to trade insults. "If your father was alive today, he'd be ashamed," said Deshmukh.

It was the lawyer's last day at Ranbaxy. Malvinder forced him to resign. As he later reflected on Ranbaxy's conduct, "Honestly, once you get to the point where you actually wholesale make up data points, hundreds and thousands . . . of data points, what's to keep you from doing anything?" The effort to deceive Une continued, over the protest of Ranbaxy's external lawyers. Had the microbiologist known about the SAR, he could have instantly decrypted the company's woes. But without that crucial information, all he could do was keep on sifting the clues.

Around the time that Deshmukh left the company, Une flew to New York for a meeting at Giuliani Partners, yet another external adviser that Ranbaxy had hired, in the hope that the former mayor of New York, Rudolph Giuliani, could use his political heft to get the FDA to back off. The SAR was still not mentioned at the meeting, and Une was left to grapple with shadows. "I was told that Jay D-san of IP [Deshmukh] was the barrier of information," he wrote in his journal. "I still do not know why Malav-san placed so much importance on Jay. (Jay has something on Malav?)"

Though Une was inching closer to the truth, a group of Malvinder's top deputies and helpers still worked to keep the information from him. On March 16, 2009, Une was in New Delhi for a critical Ranbaxy board meeting. Before the meeting, Warren Hamel, an external Ranbaxy lawyer from the firm Venable LLP, emphasized to a company lawyer that the SAR had likely triggered the U.S. government's investigation. But once Une arrived at the meeting, no one mentioned the SAR. After the meeting, Une noted in his journal that his "understanding of the background of the dealings with the FDA is still insufficient."

As Ranbaxy's problems with the FDA worsened, despite Malvinder's claims that he was doing everything possible to resolve them, Daiichi Sankyo grew unsatisfied with his management. One Ranbaxy adviser urged Une to think of Malvinder "from the perspective of a parent raising a child." At a compensation committee meeting on March 26, Une proposed to Malvinder that he forgo his bonus, owing to the deteriorating relationship with the FDA and the poor earnings result. Not only did Malvinder object. He broke down in tears. As Une noted in his journal, Malvinder "did not clarify problems of his own management, or his responsibility. In addition, he cried, exposing his vulnerability."

By mid-April 2009, Une and his top executives had concluded that Malvinder wasn't able to perform as a professional CEO. Une wrote in his journal: "Malav-san is getting more confrontational than defensive . . . we have no choice but to fire him." At that point, Une blamed the problems on poor management and noted a "difference in attitude towards quality." He pondered whether the real problem wasn't "Malav-san's secrecy."

On May 8, Une finally told Malvinder that he could not continue as CEO. Within two weeks, the Indian CEO resigned. Months later, Une was still in the dark. But from his new vantage point, he had begun to see a different side of Malvinder. Une noted in his journal, in a tone as mild-mannered as ever, "Although Malav-san seemed like thinking about the company, he actually prioritized his family profits as a private shop owner after all. I was disappointed." Even after Singh's departure, executives remained loyal to him and continued to conceal the SAR. Lavesh Samtani, who had replaced Deshmukh as the company's general counsel, instructed Hamel that an upcoming presentation to Une should not delve into the facts of the case and, specifically, should not mention the SAR.

On November 17, 2009, federal prosecutors invited Ranbaxy's attorneys and outside lawyers, including Hamel, to Justice Department headquarters in D.C. It was time for the prosecutors to show their hand. They gave a hard-hitting presentation, emphasizing that the misconduct at Ranbaxy had spanned years and involved all of the company's facilities and all of its drugs. In the sixty-seven-slide PowerPoint they displayed, the prosecutors tracked dozens of false statements. They showed that the company's top executives had known about, and were complicit in, the fraud. The prosecutors made clear that, in their view, any unexplained misstatements in Ranbaxy's FDA filings were likely instances of criminal conduct rather than innocent mistakes or oversights. Their evidence included excerpts from the SAR. At that, the Venable lawyers rose from their chairs. One walked closer to the slide presentation, as though to see if the image of the SAR was real.

It was a devastating meeting. But to Hamel, it had one silver lining: he could finally right the profound wrong done to Tsutomu

Une. He asked Justice Department officials for permission to share their presentation with Ranbaxy's board, which Dr. Une chaired. They agreed. Two days later, at a conference room in Venable's New York office, Hamel showed Une the Justice Department presentation. The microbiologist was stunned. Suddenly, the bewildering problems he'd been confronting for more than a year made sense. The next day, in an email to a company consultant, Une tried to piece all the information together. He wrote that there had been a "whisperer" from inside the company, who'd taken documents that had been presented to board members and given them to the U.S. government. Brian Tempest and Malvinder Singh had tried to conceal "all the documents relating to the issues. This is the reason why the cases could be criminal," Une wrote, and why the "FDA insists the issues could be caused by corporate culture."

Finally, Une understood the true nature of the company that Daiichi Sankyo had bought. He felt driven by the need to make his company whole for the disastrous purchase he'd championed. Within three years, Une's carefully written journal would become Exhibit A when Une and a team of lawyers brought their claim of Malvinder's deception before an international court of arbitration in Singapore.

PART V

DETECTIVES IN THE DARK

CHAPTER 18

CONGRESS WAKES UP

JULY 2008

Washington, D.C.

David Nelson, a congressional investigator for the U.S. House of Representatives Energy and Commerce Committee, which oversaw the FDA, was at his desk in the Ford House Office Building on Capitol Hill when he read the startling twenty-eight-page motion filed by Maryland prosecutors. It laid out how Ranbaxy's "systemic fraudulent conduct" was bringing "adulterated and misbranded products" into the United States.

Studying the motion, Nelson's first question was, why had he been lied to? In February 2007, during the FDA's raid at Ranbaxy's New Jersey headquarters, he'd called the agency and demanded to know whether the action had anything to do with drug quality; if it did, Congress should have been informed. *No,* an FDA official had told him. *Nothing to do with drug quality.* So Nelson figured that the raid must have been related to financial irregularities and put it out of his mind. But now, many months later, it was clear that the raid had everything to do with drug quality. Something "reeked," he would later say.

Why would the FDA continue to allow the sale of drugs in the United States made by a company that it knew was committing fraud? The agency had cause and authority to challenge every single application that Ranbaxy filed. But there had been no discernible action to remove Ranbaxy's drugs from pharmacy shelves. Confronted with fabricated data clear enough for prosecutors to put in a court record, the agency appeared to have done what it historically did: almost nothing.

As Nelson pondered the massive and seemingly lawless Indian company that had tried to stonewall prosecutors, he had another thought: *It's happening again.* Nelson had spent his year engulfed in another overseas drug catastrophe. But the Ranbaxy crisis seemed to echo one from decades earlier: the generic drug scandal of the 1980s.

David Nelson's journey into a world of dangerous and badly regulated drugs began in 1988, four years after the Hatch-Waxman Act led to the creation of the modern-day generic drug industry. On a hot July 4 weekend, a prominent Beltway lawyer, a political operative, and a private detective arrived at his modest home carrying a large bag of trash. With his permission, they emptied the bag's filthy contents onto his dining room table.

Nelson, a big, bluff Texan, had seen a lot of things in his career as a congressional investigator: venal corruption, epic incompetence, inexcusable lapses. But he'd never had anyone dump trash onto his table. As he studied the dirty, sodden papers, the visitors watched his face for a reaction. The trash had come from the Maryland home of an FDA chemist, Charles Chang, who helped oversee the review of generic drug applications. In the debris, Nelson saw tickets for a round-the-world airline trip and receipts for expensive furniture. The evidence suggested that the FDA chemist was on the take from generic drug executives seeking favorable review of their applications.

The people who had come to see Nelson had been hired by

Mylan, the respected generic drug company from West Virginia. For months, Mylan had found itself inexplicably stymied in its dealings with the FDA. Its executives watched as less-experienced competitors secured lucrative first-to-file approvals while its own applications remained in bureaucratic limbo. They had heard rumors that Chang was getting his reviewers to slow down reviews or to fabricate excuses to block certain applications. Finally, the Mylan executives hired private detectives, who found a motive in his trash. Corrupt generic drug companies appeared to be bribing Chang with trips and furniture in exchange for approving their own applications and blocking those of competitors.

The evidence was enough for Nelson's boss, Representative John Dingell (D-MI), to immediately launch a full-scale investigation and send the trash to the inspector general at the Department of Health and Human Services. In the months that followed, Dingell's committee uncovered corruption that seemed to have no bottom. Generic drug executives had roamed the FDA's halls, dropping envelopes stuffed with thousands in cash onto the desks of reviewers. Chang had accepted numerous bribes. A generic-drug trade association had subsidized the appearances of FDA reviewers at conferences with hotel bills they never saw. Representative Ron Wyden (D-OR) called the generic drug industry "a swamp that must be drained."

Congressional hearings in 1989 revealed an FDA totally out of control, with no ability to adequately review the tsunami of applications that flooded its offices. Even the FDA commissioner, Frank Young, admitted that the agency was "drowning in a sea of paper." Witnesses described wild disorder in the FDA's document room, with applications lost amid teetering piles. The Office of Generic Drugs was a "horrible world of overwhelming work," said its director, Marvin Seife, who led the office from 1972, when generics were at a trickle, through the explosion in applications after the Hatch-Waxman Act. Seife, a longtime public servant who got to work at 6:30 a.m. every day, became essential to Nelson's investigation.

The hearings revealed that the generic drug companies had resorted to bribery and fraud in pursuit of a single goal: securing the coveted first-to-file status for their applications, which would give them six months of exclusive sales at a price a little below the brand-name price. Clearly, the original drafters of the Hatch-Waxman Act had not contemplated the frenzy this incentive would unleash.

One of the companies involved in the scandal was Quad Pharmaceuticals, an Indianapolis-based company whose CEO had given Chang $23,000. Gretchen Bowker worked as a bench scientist at Quad and knew that the company was required to make three sequential lots of a drug for approval, and that these had to meet certain testing criteria. With the very first drug Bowker worked on, her boss asked her to make one batch look like three by splitting it into three and giving each a different lot number. Bowker was appalled. She documented the fraud in her laboratory notebook so that she would have proof for regulators if they showed up.

Like Quad, the other implicated companies had done whatever it took to move their applications to the front of the FDA's line. This rush left honest companies at an obvious disadvantage. The scandal shattered public trust in generic drugs. The FDA was forced to form an inspection team that went from company to company comparing claims in the drug applications with the actual manufacturing. Congress used those findings to publish a "clean list" of the companies that had actually made the drugs they claimed to. Nelson was often asked when he gave speeches, "Do you trust the drugs?" His answer was, "No—unless they're on the clean list." Even Mylan executives, who had launched the inquiry, were stunned by the extent of the corruption. They found their own company—and their industry—locked in a "life or death struggle" for legitimacy, Nelson recalled. A total of forty-two people, including multiple executives and ten companies, pleaded guilty to, or were convicted of, fraud or corruption charges.

In the scandal's wake, the Dingell committee worked to ensure that corruption like the kind exposed never happened again. It backed legislation, the 1992 Generic Drug Enforcement Act, that gave the FDA the power to withdraw any application containing false data or to debar corrupt companies entirely, if need be. New regulations required companies not only to manufacture three commercial lots of a drug but to undergo a preapproval inspection, which would ensure that the companies actually had the capability to manufacture the drugs they'd applied to make.

Despite the corruption that had saturated at least half of the nation's nascent industry, most medical experts and consumer advocates defended generic drugs. Joe Graedon and his wife, Terry, who wrote the syndicated newspaper column The People's Pharmacy, assured readers that "a few bad eggs" shouldn't spoil their confidence in the generic drug industry.

But even as the House Energy and Commerce Committee succeeded in strengthening the FDA's regulatory arsenal and put up more roadblocks to thwart companies that flouted regulations, the problem of dangerous and fraudulent drug manufacturing was slipping beyond the nation's borders. Within roughly fifteen years, the amount of pharmaceutical ingredients, by weight, that the United States imported from China would grow by over 1,700 percent, from roughly 5 million kilograms in 1992 to over 90 million kilograms by 2008. This meant that the FDA, which had struggled to regulate companies within driving distance of its headquarters, now had to regulate companies halfway across the globe. With the agency's oversight precarious at best, foreign drug supplies became "a string of ticking time bombs," as former FDA associate commissioner William Hubbard later told Congress.

A decade after Mylan's private detectives had fished through Chang's garbage, Americans started dying after taking a contaminated antibiotic, gentamicin sulfate, that contained cheap active ingredients imported from China. As it investigated these deaths, the House Energy and Commerce Committee learned that the FDA was barely monitoring the drug ingredients, known as bulk drugs, pouring in from overseas. The committee unearthed a 1996 memorandum from the agency's own Forensic Chemistry Center that stated, "We literally have no control over bulk drugs that enter the U.S. . . . These drugs can reach anyone, including the President."

By then, the FDA was inspecting about one hundred foreign facilities each year—an inspection rate of about once every eleven years for each overseas plant, according to a devastating U.S. Government Accountability Office report in 1998. Even when investigators found problems, the FDA would often skip a follow-up inspection in exchange for a promise from the manufacturer to correct them. The FDA barely knew what plants overseas had been inspected, or needed to be, as it was relying on a patchwork of fifteen different databases, most of which didn't interface with one another. It was a system built on wishful thinking and infrequent scrutiny, which yielded disastrous results. The United States "turned a blind eye," said Heather Bresch at Mylan, asking rhetorically of manufacturers overseas, "What do you do when your odds of getting caught are near zero?"

By 2007, as poisonous pet food and children's toys made with lead paint flowed from Chinese factories into the American market, the U.S. government hammered out a "cooperation agreement" with Chinese regulators to improve the safety of food and drugs. In August 2007, the Committee on Energy and Commerce sent staffers to accompany FDA investigators as they traveled to China and India. They saw a threadbare inspection program that relied on the very companies being scrutinized to help arrange the inspections.

At a hearing that November, Representative Joe Barton (R-TX) begged the new FDA commissioner, Dr. Andrew von Eschenbach, to bolster the agency's foreign inspection efforts: "I stand ready to give you the support you need to heroically improve the FDA's interception of tainted drugs from abroad." But an even graver crisis was already brewing.

That very month, at the St. Louis Children's Hospital, two young patients experienced strange and alarming symptoms. As they underwent dialysis, a lifesaving procedure to filter blood for those whose kidneys don't work properly, the patients' eyes started swelling, their heart rates escalated, and their blood pressure dropped. These were signs of a life-threatening allergic reaction. Dr. Anne Beck, the director of the nephrology unit, directed her staff to wash out the tubing with extra fluid before hooking the children back up to the dialysis machines. For the next two months, everything seemed fine. But in January 2008, the symptoms struck again.

Beck contacted an epidemiologist specializing in children's infectious diseases who immediately assembled a command center where a team worked around the clock to uncover the cause of the strange reactions. But as more children succumbed and the staff grew frightened, the epidemiologist notified the Centers for Disease Control and Prevention. The CDC immediately contacted dialysis centers in other states and learned of similar reactions elsewhere.

As the CDC and the FDA began a joint investigation, their efforts pointed to a common denominator: all the sickened patients had been given heparin made by the brand-name company Baxter, the nation's biggest heparin supplier. It was a drug that patients took intravenously during dialysis to ensure that they didn't suffer blood clots. Within weeks, Baxter—at the FDA's urging—began a sweeping series of recalls, until finally the allergic reactions stopped.

Yet the mystery was far from solved. Nobody understood why heparin—which is made from the mucosal lining of pig intestines, most of which come from China—was suddenly making patients sick. In February 2008, the FDA discovered the likely source of the contamination: a Chinese plant supplying crude heparin to Baxter. In a clerical blunder, the FDA had completely overlooked and failed to inspect the facility, Changzhou SPL, located about 150 miles west of Shanghai. Instead, it inspected and approved a plant with a similar-sounding name.

Predictably, once FDA officials finally traveled to Changzhou in February 2008 to make an on-the-ground inspection, they found serious problems. The facility had dirty manufacturing tanks and no reliable method of removing impurities from heparin, and it acquired the crude heparin from workshops that had not been inspected.

Chinese regulators were no help at all. A loophole in Chinese regulations allowed certain pharmaceutical plants to register as chemical plants, which made them subject to far less oversight. For U.S. congressional investigator David Nelson, whose committee was now immersed in the heparin crisis as well, the situation laid bare the "classically good reason to be suspect of production coming from any country that doesn't have competent regulatory authority." The FDA issued an import alert in March 2008, meaning that Changzhou SPL's shipments would be stopped at the U.S. border.

Though investigators had identified Baxter's heparin as the source of the contamination, and the Changzhou plant as deficient, neither the FDA nor Baxter could find any contaminant in the heparin. Urgently needing help to figure out what was wrong with its own product, Baxter reached out to Dr. Robert Linhardt, a chemist at Rensselaer Polytechnic Institute in Troy, New York, who had been studying heparin for years. He promptly sidelined his other work to dig into the mystery, and his laboratory joined several others working on the crisis.

Stumped, the research teams finally turned to sophisticated nuclear magnetic resonance spectroscopy machines, which revealed evidence of a contaminant: a synthetic substance called oversulfated chondroitin sulfate (OSCS). The ingredient mimicked heparin, was almost impossible to detect, and produced life-threatening reactions. The FDA formally named OSCS as a likely contaminant in March 2008 and concluded that it had been added, somewhere along the supply chain, to increase the yield, and profitability, of the drug. The contamination exposed perilous gaps in the FDA's oversight and intensified the long-simmering conflict between Congress and the agency.

In April 2008, tensions erupted during a highly publicized hearing, with regulators, manufacturers, and victims' families all in one room. By then, David Nelson had pieced together the missteps that had contributed to at least eighty-one deaths from the adulterated heparin, a number that would continue to climb. The hearing laid bare the FDA's blunders in crippling detail: the lack of facility inspections, the poor risk assessment, the woeful technology. Nor was Baxter spared. The company's own audit of Changzhou SPL, conducted a few months before Americans started dying from heparin contamination, was "incomplete, bordering on failure," Nelson testified. He concluded that corporate America could not be trusted to do the FDA's job—nor could the FDA be trusted to do its own job.

Under questioning, Dr. Janet Woodcock, director of the FDA's Center for Drug Evaluation and Research, acknowledged that the FDA had little idea how many overseas firms shipped pharmaceutical ingredients to the United States. "It's most likely between three thousand and seven thousand," she said. The FDA also attempted to pin the blame on Baxter, adopting a common refrain that it was a company's responsibility to ensure the quality of its own products. The hearing intensified David Nelson's longtime fury at Woodcock, whom he felt promoted the idea that manufacturers could be motivated to do the right thing without the threat of an inspection. "Woodcock—may her soul burn in hell for all eternity—did not believe in inspecting plants to ensure the safety of the drug supply," he said years later.

The issue is far more complex than Nelson allows, Woodcock later told a journalist. "Inspections are not a panacea," she explained. "You always have to do inspections. I'm no fool. But it really is important to make this industry responsible for quality itself," she said.

"People need to own that as a critical part of what they do, not just trying to fool inspectors."

At the hearing, the most powerful testimony came from the family members of dead patients. Leroy Hubley of Toledo, Ohio, testified about losing both Bonnie, his wife of forty-eight years, and his son, Randy, to adulterated heparin within one month of each other. "Now I am left to deal not only with the pain of losing my wife and son, but anger that an unsafe drug was permitted to be sold in this country," the seventy-one-year-old widower testified. "The FDA and Baxter have not done their job. Somebody sure as hell didn't." Members of Congress expressed anger on Hubley's behalf. At one point, a congressman interrupted a witness midsentence to exclaim, "This is thuggery. This is thievery. This is high crime and a direct assault on the American public. . . . Someone did this deliberately."

Assuming the culprits were in China, how could the United States hold them accountable? The FDA was a disorganized, weak-kneed agency with limited powers. It faced a foreign country with little effective regulation and a government with every incentive to suppress bad news. How were America's investigators supposed to track down, and prosecute, those responsible? In the decade-long investigation that ensued, pitting the United States against China and the FDA against Congress, no one has yet been held accountable. Nevertheless, the FDA's webpage on heparin still states that the agency is "continuing to aggressively investigate the situation."

Three months after the heparin hearing, with the vulnerability of the American public still fresh in his mind, Nelson read the prosecutors' motion about Ranbaxy and realized that something else linked the 1980s generic drug scandal to the Ranbaxy case. Although the earlier scandal had implicated American companies, those companies had largely been run by South Asians, such as Quad Pharmaceuticals CEO Dilip Shah. Whether fairly or not, those investigating and prosecuting the initial cases had referred to the corrupt executives as the "Bengali mafia." At the time, some of the defense lawyers had tried to justify their clients' crimes by explaining that they were viewed as acceptable business practices in their homelands. "I was insulted by the notion that people would be considered innocent because it was culturally okay with them to give bribes," Nelson recalled.

At Quad Pharmaceuticals, Bowker learned that, in some Indian companies, "it is seen as an asset to be able to do something in a creative different way that goes around the arduous tasks and gets you there quicker and cheaper. What we would view as cheating would be viewed in their culture as creativity." This was the model of aggressive shortcuts, the ability to dodge onerous rules and get to the desired results by the shortest means possible, known as *Jugaad*, which the innovation expert Dr. Raghunath Anant Mashelkar condemned. As Dinesh Thakur put it, "There's a saying in India. 'We don't have a system. We have a way to work around the system.'"

Jugaad developed as a survival mechanism in response to failing systems. In *Maximum City: Bombay Lost and Found,* the Indian journalist Suketu Mehta studies just the sort of workarounds, or alternative systems, that govern daily life in Mumbai (which was called Bombay until 1995). As Mehta concludes:

> You have to break the law to survive. . . . I dislike giving bribes, I dislike buying movie tickets in black [illegally]. But since the legal option is so ridiculously arduous—in getting a driving license, in buying a movie ticket—I take the easy way out. If the whole country collectively takes the easy way out, an alternative system is established whose rules are more or less known to all, whose rates are fixed. The "parallel economy," a travelling partner of the official economy, is always there, turn your head a little to the left or right and you'll see it.

The parallel economy operates wherever the official state lies in tatters. In drug making, too, India's manufacturers developed an alternate set of rules in part because the actual rules had been ignored by the country's regulators. The Ranbaxy case defied the imagination of U.S. regulators and investigators for so long because the fraud was so all-encompassing. The company's intricate system for faking data involved hundreds of people. And the U.S. government all but volunteered to be fooled by announcing its inspections in advance.

The congressional investigator David Nelson knew nothing about *Jugaad*. But he knew a lot about the industry he had investigated. Paging through the prosecutors' motion on Ranbaxy, Nelson saw a dangerous junction between the "get rich quick" schemes of the generic drug industry and the FDA's "see no evil" approach to foreign drug regulation. The result, he feared, was a public health disaster in the making.

CHAPTER 19

SOLVING FOR X

MAY 25, 2007
Cleveland, Ohio

Dr. Harry Lever, a sixty-two-year-old cardiologist, always listened to National Public Radio as he drove his half-hour commute into work at Cleveland Clinic. His drive took him down suburban back roads into Cleveland's early-morning traffic. But this morning he was so engrossed in a radio segment that he barely noticed the streets going by.

The story described how the United States was importing huge amounts of adulterated food and ingredients from China that the FDA was struggling to inspect. The list was harrowing: toothpaste with ingredients found in antifreeze; fish bred in polluted waters and fed banned veterinary drugs; herbal tea leaves dried with exhaust fumes from trucks using leaded gasoline. The FDA inspected only around 1 percent of all the food and ingredients that entered the United States, the story explained. Imports from China were by far the most likely to be seized as unfit for human consumption. A former FDA deputy commissioner, William Hubbard, explained that investigators often blocked products that looked rotten or smelled of decay. "Filthy" was the official term. The FDA was so understaffed, however, that "only a fraction" of "filthy" food was caught and stopped at the border, said Hubbard. The rest was "slipping through the net."

Lever parked in the hospital garage with the radio still on. His pager beeped, but still he sat there. Until that morning, he'd had no idea that so many products—apple juice, garlic powder, honey, hot dog casings, vitamin C—came into the country with such perilously weak regulation. Pets had died from tainted food. To Lever, it sounded evil.

He got out of the car but couldn't get the radio broadcast out of his head. That night he started doing research in his own pantry, where he found a bottle of garlic powder made in China. Then he noticed that it had been certified as kosher by the Union of Orthodox Rabbis. His first thought was, *a counterfeit certification*. He went to the Union's website. Sure enough, rabbis were certifying Chinese imports. But how could they know what was safe and what wasn't? He called the Union and got a rabbi on the phone who confirmed the certification. Lever even tracked down William Hubbard, the former FDA commissioner who had been on the radio program. Hubbard verified the details, down to the tea leaves coated in lead exhaust.

Lever's nature was "almost too passionate," as his cousin, who was also a patient of his, commented. "He just gets furious when he knows something is not being done right." But Lever didn't dwell for too long on the problems of "filthy" food. He'd soon turn his attention to adulterated medicine.

Not long after hearing the NPR radio segment, Lever noticed that some of his patients were suffering from a low platelet count after taking heparin. He began raising concerns with colleagues. When it emerged that the heparin had been contaminated in China, one doctor called him a "prophet." In his mind, he was just connecting the dots. As he did, his concerns grew. He realized that he could no longer assume that the drugs his patients were taking would work as intended.

Lever specialized in treating hypertrophic cardiomyopathy (HCM), a condition in which the heart's muscle tissue thickens and potentially restricts blood flow. The disease can strike without warning and is the leading cause of sudden cardiac death among young athletes. Over the years, Lever developed one of the nation's biggest practices in treating the disease and helped uncover new methods to identify it early. At Cleveland Clinic, he worked from a small ground-floor office, where mementos from grateful patients lined the walls. Medical records lay in heaps on his desk, and Post-It notes were stuck to the cabinets. Most days he took lunch at his desk, eating the salads his wife had prepared, as he reviewed his patients' echocardiograms.

Lever liked to visualize each patient's case as an algebraic equation. A new symptom put an unknown variable, an "X," into the equation. Solving for X could be relatively straightforward if the other variables were known. The brand-name drugs that Lever had prescribed for years were known variables: they had almost always worked as expected. Lever's best defense for his patients was medication: beta blockers and calcium channel blockers that could offset his patients' irregular heartbeats or help lower their blood pressure. Many of his patients also needed diuretics, pills that reduced swelling and fluid buildup. After the radio segment, Lever started to realize that some patients he had stabilized on those drugs developed symptoms again when they were switched to certain generics. Those generics seemed to be a new X that threw off the whole equation.

He started running Google searches. If a drug roused his suspicion, he researched the drug maker and the locations of its plants—basic information absent from the drug's packaging or product label. With increasing frequency, he also contacted Cleveland Clinic's high-level pharmacists, who routinely gathered data from manufacturers and the FDA as part of their research into which drugs the hospital should use. He developed a sense for which drugs, or which companies, to avoid. Ranbaxy was among them.

One night his cousin, who suffered from heart disease, called him and complained that he was feeling terrible. One of the medications he took was generic furosemide, a diuretic that helped him shed excess fluid. Lever immediately asked, "What are you taking?" His cousin read the label on the pill bottle: Ranbaxy. Lever switched his cousin to what he considered a better generic made by Teva, an Israeli drug company. His cousin lost about fifteen pounds of fluid in a week.

Another of his patients, Martin Friedman, a theater professor, was having trouble keeping off excess fluid, despite taking a diuretic. His ankles were swollen, and the only way he could sleep was in an upright position, propped on pillows. Lever soon identified that his diuretic, torsemide, was manufactured by the Croatian company Pliva. Lever switched Friedman to Demadex, the brand version originally manufactured by Roche (and later purchased by the European firm Meda). Again, Friedman immediately lost excess fluid. "It was pretty weird," Friedman noted. "It started working right away."

The list of sick patients and suspect drugs continued to grow. One patient, Karen Wilmering, had obstructive HCM, which blocked the blood flow out of her heart's left ventricle. For years, to control her cholesterol, she'd taken brand-name Pravachol, made by Bristol-Myers Squibb, until she was switched to a generic, pravastatin sodium. When Lever ran another test, Wilmering's cholesterol results were alarmingly high.

Lever asked Wilmering which company made her cholesterol drug. When she told him it was Glenmark, Lever "almost jumped out of the chair, because Glenmark is from India," Wilmering recalled. Lever dialed Wilmering's pharmacist and insisted that she be switched to Teva's version, which he believed would work better. In less than a month, her cholesterol dropped back to the normal range. Even Lever was stunned by the new numbers. "His jaw dropped," Wilmering said. The following year, Glenmark recalled thousands of bottles of its pravastatin sodium, due to patient complaints that the medicine gave off a strong fishy odor. A Glenmark spokesperson said the recall was "voluntary, proactively initiated by Glenmark, and not related to product efficacy."

Some of Lever's patients became as outraged as he was. Christine Jones, fifty-four, who had previously run a consumer relations division at PepsiCo, retired early owing to her HCM. She was put on the beta blocker Coreg, made by GlaxoSmithKline. But after she was switched to the generic version, carvedilol, which cost $10.87 for a four-month supply, her health deteriorated. "I was having way more shortness of breath and irregular heartbeats that were waking me up at night," she recalled. "This went on for months." She attributed the problems to her diet and lack of sleep.

Lever immediately fingered the generic. It was made by the Indian company Zydus. "If he hadn't put two and two together," said Jones, "it would never have entered my mind." Lever switched her back to the GSK brand, which cost $428 for a six-month supply. She felt better almost immediately after taking it. She began to research Zydus and found a trail of aggrieved patients posting to online forums. Alarmed by these complaints, she called the company "multiple times" to ask if its generic was just as effective as the brand, but could never get what she felt was a clear answer. The fact that the Zydus drug worked so poorly, and yet the FDA had approved it, seemed "unconscionable" to Jones.

As Lever made life-altering adjustments—switching patients from bad generics to better ones or back to the brands—he was no longer just diagnosing his patients. He was following the trail of the global economy and trying to diagnose the drug supply. The preponderance of problematic drugs seemed to be manufactured in India, but some were American-made. Many of the ingredients used in the finished drugs came from China. Lever was discovering that a complex supply chain made it extremely difficult to know where drugs had been made, who made them, and which ones worked best.

With some drugs, the problems were stark. When his patients were switched to certain generic versions of metoprolol succinate, a beta blocker, Lever noticed that they frequently developed chest pain and their heart rate and blood pressure became harder to control. The brand version, Toprol XL, made by AstraZeneca, had a time-release feature that helped the active ingredient stay in the blood, but that was under a separate patent from the drug itself. In 2006, as generic companies started selling versions of Toprol XL, they had to engineer their own mechanism for releasing the drug into the bloodstream.

That year, Sandoz, a subsidiary of Novartis, in Switzerland, started marketing the first generic version. The following year, Ethex, a subsidiary of KV Pharmaceutical in St. Louis, Missouri, also began selling a version. At the time, only Par Pharmaceutical, based in New York, had been authorized by AstraZeneca to sell a generic version, so it used the same time-release mechanism as the brand. Lever became certain that there was something wrong with the Ethex and Sandoz versions. "You get this feeling like you smell a rat," said Lever. Sure enough, in March 2008, the FDA inspected one of Sandoz's plants and found "significant deviations." By November 2008, after a stark warning letter from the FDA, Sandoz quietly recalled its drug.

Around the same time, a vigilant Cleveland Clinic pharmacist sent the package insert for Ethex's metoprolol succinate to Lever, with one phrase out of the nine-page document highlighted: "Does not comply with the dissolution test of the USP monograph." Lever was amazed. If the drug did not dissolve according to the agreed-upon standards set by the USP, why had the FDA approved it? Together, he and the pharmacist called the FDA's Office of Generic Drugs from Lever's office. They wanted to know: Were patients getting too much of the drug? Not enough? They couldn't get a straight answer. Then, on January 28, 2009, Ethex announced a sweeping recall of more than sixty products, and metoprolol succinate was among them. Ethex later pleaded guilty to two felonies and agreed to pay a fine of over $27 million.

Lever had no official data to support his observations. He couldn't see inside the drug plants. But using his skill as a physician and years of experience, he had figured out that the drug supply was "sick," as he described it. Whenever Lever switched his patients' drugs, he tried to keep the dose the same. This helped to eliminate another X in the equation.

Sometimes, even the same doses of different versions became dangerous, as Lever found when treating Kevin Parnell. Parnell was diagnosed with HCM at age thirty-one. Told it was essentially a "death sentence," he sought a second opinion at Cleveland Clinic. In 1998, at age thirty-nine, the very sick Parnell found himself under Lever's care. In 2003, open-heart surgery earned Parnell six years of good health. But as time went on, he took more medication and higher doses of the diuretic furosemide.

By 2012, his health was failing. As Parnell sat with his wife in Lever's office and described the problems with his swollen legs, Lever immediately suspected a poor-quality generic diuretic. Though Parnell had not bothered to bring the medication, his wife called their daughter and instructed her, "Go check the label on Dad's bottle." The pills were made by Ranbaxy. Lever switched the prescription to a version made by the American company Roxane and kept the dose the same. Parnell started taking his new medication right away and immediately started losing fluid. But three days after the switch, he got up in the middle of the night, then passed out and hit his head on the bedside table. He awoke to the sight of three paramedics hovering over him. He was rushed to a local emergency room and was later admitted to Cleveland Clinic. He had suffered ventricular tachycardia—an overwhelming, rapid heartbeat.

Lever suspected that once Parnell changed to a more effective drug, the dose was suddenly too high. The diuretic had rapidly depleted Parnell's potassium levels, causing his heart to beat irregularly. Parnell finally received the heart transplant he desperately needed. After his surgery, Lever "was in the ICU constantly," Parnell recalled. "He checked on me at all different hours." He credited Lever with saving his life. Lever was particularly infuriated by Parnell's case. He felt certain that drugs of varying effectiveness had harmed Parnell and had reduced the science of medicine to guesswork.

As Lever regularly paged Cleveland Clinic's pharmacists and alerted colleagues to problematic drugs, he began to reshape the list of drugs the health system used. Cleveland Clinic's vigilant pharmacists essentially ran a mini-FDA: they sought out bioequivalence data from generic drug makers, investigated the origin of active ingredients, submitted Freedom of Information Act (FOI) requests to the FDA for additional data, and even visited manufacturing plants, all to figure out which drugs the hospital should use and which they shouldn't. As they heeded Lever's warnings and reached for a larger set of data points—scouring FDA inspection reports and warning letters and fielding anecdotal reports from doctors—they developed a confidential black list of drugs the hospital would no longer buy, dominated by generic drugs manufactured in India.

Dr. Randall Starling, a member of Cleveland Clinic's Heart Failure and Cardiac Transplant Medicine Section, was stunned to learn in late 2013 that a generic version of tacrolimus, made by the Indian company Dr. Reddy's Laboratories, was on the black list. Tacrolimus is a crucial drug for transplant patients, because it suppresses the immune system to prevent organ rejection. The Dr. Reddy's version was by far the cheapest, but the number of recalls of

Dr. Reddy's drugs had made the health system's pharmacists uncomfortable.

Over the next six months, Starling worked with his staff to make sure that none of their patients was taking the Dr. Reddy's version of tacrolimus and that the hospital's inpatient and outpatient pharmacies no longer carried it. Though he told patients to use only Prograf, the brand-name version of tacrolimus, Starling knew that once his patients left the clinic's grounds, he could not control what an outside pharmacy might dispense. It did not take long for his fears to be realized.

In October 2014, about eighteen months after his heart transplant, Cedric Brown, a forty-eight-year-old patient, was admitted to the cardiac medicine service with symptoms of acute organ rejection. Brown swore that he had taken every pill. Up until that point, Brown's post-transplant recovery had been remarkably successful. He had been up and walking after a few days, and left the hospital within two weeks. The Prograf, which suppressed his immune system to prevent rejection, cost about $3,000 a month, and Brown would have to take it for the rest of his life. By the time he was readmitted to the hospital, he'd gained fifty pounds and felt terrible. He was released after a month, but wound up back in the hospital's intensive care unit within a week. He didn't know if he was going to die: "I just prayed to God."

On a Monday morning, Dr. Starling came in to consult on his case. Standing by Brown's bedside, he asked him, "Did you get a new prescription?"

Brown had brought in his medicine. "Yes, I got something new at Marc's pharmacy," he said. It was a different size and color than the usual Prograf drug.

"Gee, I'd like to see that prescription," said Starling.

Brown said, "Sure. Open that closet, there's a bag in there with it."

Starling fished out the bag, and there was the pill bottle of tacrolimus, made by Dr. Reddy's. "You are never going to take this one again," said Starling. And Brown never did. Starling set about to educate the other doctors treating him. In time, Brown recovered enough to go back to work part-time, as a driver. His insurance, Medicaid, covered 80 percent of his brand-name Prograf, and a fund at Cleveland Clinic covered the rest.

Doctors at Cleveland Clinic were not alone in their concerns. In October 2013, a pharmacist at the Loma Linda University Medical Center in California reported to the FDA, through its online complaint database Medwatch, that "multiple patients" who used the Dr. Reddy's tacrolimus had "unpredictable levels leading to inadequate immunosuppression and subsequent transplant failure." The report from Loma Linda noted, "This has only been seen with the Dr. Reddy's brand of Prograf." Tacrolimus was a so-called narrow therapeutic index drug that required precise dosing; minor variations could lead to life-threatening complications. For years, doctors who prescribed other such drugs, for conditions like epilepsy, hypertension, mood, or endocrine disorders, had debated whether certain generics were really interchangeable with the brand, and whether the FDA's bioequivalence standards allowed too much latitude. Medical societies, including the American Academy of Neurology, the Endocrine Society, and the American Heart Association, opposed a switch to generics without a doctor's approval.

Facing medical unease over the interchangeability of certain generics, starting in 2010, the FDA's Office of Generic Drugs began to commission a series of studies of generic narrow-therapeutic index drugs. In 2013, researchers at the University of Cincinnati began a bioequivalence study of generic tacrolimus made by Sandoz and Dr. Reddy's. They tested the lowest dose of the drugs in healthy volunteers and followed two groups of patients who'd gotten kidney and liver transplants. Their results, which they published in 2017, found the drugs to be bioequivalent to the brand and interchangeable with one another.

But at Cleveland Clinic, Starling and his team felt little reassurance. Months after Cedric Brown's admission, it happened again—another heart-transplant patient suffering from organ rejection who had taken the Dr. Reddy's formulation was admitted to the hospital. Several more followed. In studying those cases, the only explanation that Starling's team could find was an inadequate generic. The Dr. Reddy's drug became a variable with an unknown impact, in a treatment plan with no room for error. Starling was a doctor who valued control and was accustomed to solutions, but he was left without one. He had no Cleveland Clinic data to point to. But like his colleague Lever, his "index of suspicion" increased.

"I developed a stance of demanding that my patients take brand-name drugs," Starling explained, "because I didn't want any variable out there." Though he knew this created a financial and logistical burden for Cleveland Clinic, its patients, and their insurers, the stakes could not have been higher. An organ transplant is a "huge investment," said Starling—organs were scarce, and the average cost of transplanting a heart was well over $1 million. "If we're giving a patient inert medication, it's a huge failure of the system. An organ has been wasted potentially."

Both Starling's and Lever's patients had served as sentinels. In their fluctuating heart rhythms, Lever sensed that something was awry. It was a problem he couldn't see and had no way to fix, but he suspected it lay in the distant manufacturing plants where his patients' drugs were made. In the months that followed, as he took his quest to the FDA and to the news media, his instincts were proven true. "In all my innocence," he said, "I stumbled on a mess."

CHAPTER 20

A TEST OF ENDURANCE

SEPTEMBER 2009

Silver Spring, Maryland

Based on the numbers alone, the Ranbaxy case held the promise of being a blockbuster.

Federal agents had emerged from their search warrant at Ranbaxy's headquarters with over 30 million pages of documents. Three top executives—Malvinder Singh, Brian Tempest, and Abha Pant—sat squarely in prosecutors' crosshairs. "I'm inclined to say 'felony or nothing' when it comes to the individuals," one of the case's lead prosecutors, Assistant U.S. Attorney for the District of Maryland Stuart Berman, asserted to colleagues in a September 2009 email. He would have considered "misdemeanors to resolve borderline cases," he noted. But there was nothing borderline about Ranbaxy.

By the spring of 2010, government prosecutors had thrown down a gauntlet. They floated a $3.2 billion settlement to Ranbaxy's lawyers to resolve the company's criminal and civil liability, which would make the judgment the largest against a pharmaceutical company in U.S. Justice Department history.

But over at the FDA, as Debbie Robertson worked alongside other investigators out of a war room stacked high with jackets, well into her fifth year of painstaking work on the case, her feelings of frustration and bitterness were growing by the day. She was not alone. For dozens of investigators, special agents, and attorneys, the seemingly slam-dunk case had turned into a morass and spread recriminations from the U.S. Attorney's Office in Baltimore to the Justice Department's Office of Consumer Litigation to the FDA's Office of Criminal Investigations. Out of a sense of fatalism and superstition, some investigators wouldn't even say the name of the company aloud. They referred to it as the "R Word."

An alphabet soup of government agencies and divisions were involved in the case, including criminal prosecutors, two branches of the Justice Department's civil division, inspectors general from several agencies, and Medicaid fraud control units. At the FDA alone, the agency's Ranbaxy enforcement team now comprised at least thirty people from over a dozen different divisions with bewildering acronyms—ORA, OIP, OCC, CDER, OC, OCI.

Even at its best, this vast machinery could easily become dysfunctional. But the Ranbaxy case posed a unique difficulty: the defendant was headquartered on another continent. Basic investigative tasks—interviewing witnesses, obtaining documents—became major jurisdictional challenges. Could prosecutors get India to extradite anyone, for anything, let alone get an expedited visa for a witness they needed to interview?

A more human problem, however, lay at the center of the case's difficulties: growing animosity and distrust between the FDA's investigators, who had lived and breathed the investigational details, and the Justice Department's prosecutors, who had been tapped to come on board. In theory, the two groups were on the same team. But somehow they'd become opponents as progress on the case slowed to a crawl. In the FDA's war room, a picture of Malvinder Singh was taped to the wall, and someone had drawn in devil's horns. Prosecutors objected when they saw it, and the picture was taken down.

It was a small moment of discord, but one that underscored the tensions that had led Robertson and her colleagues to doubt their very first investigative decision: to take their case to the Justice Department's Maryland district.

The FDA had limited legal authority. For any formal investigation, it needed to partner with prosecutors. The FDA could have taken the case to any Justice Department district able to claim jurisdiction—in this case, New Jersey, where Ranbaxy had its headquarters; or New York, where prosecutors could almost always assert prerogative. But FDA criminal investigators had gone to the federal prosecutors in the agency's backyard, only to discover how difficult it was to get the unbroken attention of a lawyer there. The prosecutors were split between administrative duties and other ongoing trials. They kept getting switched on and off the case, as the wrongdoing being uncovered at Ranbaxy seemed to grow in scope and complexity. Even drafting the affidavit for the search warrant took almost a year. Prosecutors kept returning it to FDA agents for revision, but offered little affirmative guidance.

Once the search warrant was approved, it yielded mountains of incriminating information. But Ranbaxy's defense attorneys immediately claimed the material was covered by attorney-client privilege and managed to freeze review of it for almost eighteen months. Prosecutors did not assign a full-time attorney to review whether the documents were indeed privileged, and so they simply sat in government storage, untouched. "They took every batch record, they took files of emails, they took all the lab notebooks, they took everything . . . and then they had nobody to look through it," said one FDA compliance officer. "They should have had a team of twenty people to go through it. It seemed all for naught."

Robertson and her colleagues conjectured that, had the FDA gone to prosecutors in the Eastern District of New York, with their reputation for speed and aggression, the case would already have been over, with the company shut down and executives in jail. The disarray and lack of follow-through they experienced from the Maryland office led them to joke repeatedly, "If only we had a prosecutor . . ."

But more than anything, a single incident crystallized the distrust. After the search warrant, federal agents left an inventory of seized documents for company officials, as required. Robertson and her colleagues then created a far more sensitive document, which they called the "Princeton Headquarters" spreadsheet. It listed the most important evidence they'd seized, annotated crucial lines of inquiry, and contained potential leads. In late 2008, Ranbaxy's lawyers asked prosecutors for a copy of this spreadsheet, a request that triggered a meeting between Robertson, her colleagues, and Justice Department lawyers, including the senior litigation counsel in the Office of Consumer Litigation, Linda Marks. They decided as a group that the document was too sensitive to share.

But just a few months later, on February 9, 2009, Marks emailed Robertson asking for the spreadsheet, with the intention to turn it over to defense lawyers. Robertson responded, reminding Marks of the group's decision not to turn over the spreadsheet. Marks replied, "Is there still an original copy of the Princeton inventory without highlights or flags?" She added, "I would never disclose to defense anything that indicated what items we're interested in, but a detailed inventory list, albeit agent work product, is routinely provided to defense counsel." Her email triggered a meeting the next day, in which the group again agreed not to share the document.

Two months later, in April, Robertson got a horrible jolt when a Ranbaxy lawyer emailed her, asking for help finding a document listed in the "Princeton seizure log." Robertson was stunned and wrote back several hours later, "I am a little puzzled. Can you tell me what document you are referring to as the Princeton seizure log and where you got it from?" The lawyer responded that the spreadsheet had come from Justice Department lawyer Linda Marks, in late February. With no consultation or warning, Marks had violated their agreement and turned over the annotated spreadsheet, their Rosetta Stone, a map of their thinking about the case, to defense lawyers, a claim that a Justice Department official would later deny.

In cold fury, Robertson followed up with Marks, writing, "Despite our agreement, you took affirmative steps to obtain [the document] and send it to defense counsel." The document, she wrote, gave the defense "a roadmap" of the prosecution's strategy and could even "help them to identify our cooperating witness." She noted, "Actions such as this further diminish the already strained relationship between agents and attorneys, as we find it hard to trust a prosecutor that appears to be less than forthcoming."

Weeks later, Marks pulled Robertson aside and said, "We need to talk."

"Fine," Robertson responded. "But my supervisor isn't here, and I am not talking to you without a witness."

Ranbaxy's lawyers had first asked the government for a global resolution to the case in January 2010. This meant one settlement to resolve every criminal, civil, and regulatory claim that the government had against Ranbaxy. Though a daunting task, it was a simpler and far less costly one than going to trial.

By March, the FDA's associate chief counsel for enforcement, Steven Tave, began to hammer out a possible resolution: a company guilty plea to multiple counts of conspiracy and making false statements; a criminal fine based on sales figures for all of Ranbaxy's U.S. drugs from all its plants; a civil settlement related to the company's false claims; and a consent decree that would require the company to transform its operations before drugs made at its major plants would be allowed into the United States. No foreign company had ever committed to such a decree.

But with a goal of midsummer to wrap up the case, deliberations between the FDA's team and prosecutors bogged down over the size, structure, and even the theory of the penalty. Should it be based on all U.S. sales? If so, gross or net? Should it cover all the company's drugs, or specific ones on specific dates?

As the two groups deliberated, the theory of the case became murkier and the size of the potential settlement began to drop. "Guess we are going with some sort of net sales figures?" Robertson wrote to her colleague Steve Tave in August 2010.

Tave wrote back, "That would require [the U.S. Attorney's Office] to make a decision about something instead of using their canned response that they're still waiting for something from FDA. (Oh, am I sounding a little bitter?)"

As the two sides bickered, and Ranbaxy's lawyers pleaded for a resolution, President Barack Obama signed the Affordable Care Act (ACA) into law that March. It made generic drugs essential to the 20 million Americans who would now have their access to pharmaceutical treatment covered under the law. While the ACA did not enter the tortured deliberations by name, it underscored the importance of fixing the generic drug supply.

As the months stretched on without resolution, prosecutors and the FDA's team traded thinly veiled barbs over who was to blame for the delay. "My management is again asking when this case is going to be resolved—and if we are going to charge any individuals," Assistant U.S. Attorney Berman wrote to Robertson and Tave in July.

Irked, Tave wrote back to Robertson, "I want to ask them about the Consent Decree, which I sent more than a week ago . . . (but FDA is the one holding things up, of course)."

By September 2010, Robertson learned that prosecutors had not even pulled Ranbaxy's sales data, on which a settlement would be based. "Looks to me like DOJ is admitting they have not done anything to bring this case to settlement," Robertson wrote to her supervisor. "What have they been doing for 10 months?"

Finally, as 2010 drew to a close, Tave tried to break through the logjam. In a lengthy email to prosecutors, he exhorted: "This is a strong case. This is an important case. A lot of people have spent a lot of time building this case into what it is now." He detailed why negotiations for a settlement should start at $1.6 billion and not go lower than $817 million. Reflecting the tortured process, he wrote, "At the risk of sounding like a broken record, if anyone believes that the evidence compels a different result, would you please explain what that result is, how you got there, and explain the evidence supporting that view?" He added, "We have a defendant that has been pleading with us for more than a year to come to the table to resolve this case. . . . There is absolutely no reason why we shouldn't be able to do so."

And yet, they couldn't. By March 2011, one of the prosecutors was ready to capitulate and not impose criminal penalties for one of the main manufacturing plants. Robertson sent an exasperated note to her supervisor: "It just keeps getting uglier as we do not have a criminal prosecutor that will get involved."

By August 2011, Ranbaxy's lawyers had whittled down their counteroffer to $260 million. Plans to prosecute individuals—which at one point had included an idea to immunize Pant to testify against Singh—had fallen by the wayside. As Malvinder Singh would remind a journalist years later, "I was never questioned or contacted by investigators." Nor was Abha Pant.

Robertson thought frequently about retiring and longed to quit the case. But one thought kept her going: she felt an obligation to Thakur. "This man came forward and pretty much risked his life," she reflected. The least she could do was help bring the company to justice, even if the effort was putting her own sanity on the line.

In the years since 2005, when Dinesh Thakur had first reported his concerns to the FDA, he built a business called Sciformix that employed medical professionals from India, the Philippines, and elsewhere to help drug companies deal with patient complaints and other regulatory issues. The company had even become profitable. But he spent much of his time at the Stein Mitchell offices in Washington, D.C., helping two dozen lawyers and support staff in Andrew Beato's firm sift through roughly four hundred boxes of documents to help build the government's case.

The twin obligations of Sciformix and the Ranbaxy case kept Thakur working almost eighteen hours a day. He traveled back and forth between India and the United States. Even during his return trips to India, he spent more time in Mumbai on business than he did at home in New Delhi. Though he remained impeccably dressed, with razor-sharp pleats in his khakis and neatly ironed collared shirts, the circles under his eyes grew darker and more pronounced. His temper got shorter. Even when he was with his family, who had remained in India, he seemed to disappear into his laptop and his own doubts about whether it had been smart to bring the fraud at Ranbaxy to the attention of the U.S. government.

Arguably, he was working hard to secure his family's future and Sonal had chosen to remain in India. But day to day she felt isolated with two young children. He was more familiar to them as a face on a computer screen than as a present father and husband. When he was home, he would spend long hours in his basement office, reading all the materials his lawyer sent to him. "Mentally, he was never there," said Sonal, "no matter how much he tried."

Beato had wanted Thakur to tell his wife as little as possible about the government's investigation into Ranbaxy and his role in it. The case and his lawsuit that had started it were both confidential, the related court records under seal, and Beato was worried about the safety of Thakur's family. He was also concerned about Sonal's discretion and how she would react. The resulting silence had strained Thakur and Sonal's relationship to the breaking point.

In February 2009, Thakur returned to India, in time to celebrate Mohavi's third birthday. On that visit, he took Sonal to lunch at a popular Chinese restaurant in Gurgaon, in the Royal Bank of Scotland office building. As their waiter brought the soup course in the crowded restaurant, Thakur said softly, "I need to share something with you." He had decided to defy his own attorney for the sake of his marriage. Gingerly, he began to explain everything: what had actually happened at Ranbaxy; what he had done about it; how he now had a lawyer and was suing the company on behalf of the U.S. government; how his identity would remain secret.

Sonal stopped eating, and her soup grew cold. She was stunned—and frightened. Not only was her husband caught up in the dangerous machinery of an investigation. He had jump-started it. He was a key witness, whose identity could be disclosed at any minute. The family could be in danger. Thakur explained, "Look, they are playing with the lives of people here. I couldn't sit back and not do anything about it."

"Why didn't you tell me about this before?" she asked.

"I wasn't sure if you were going to support me. I thought you might have gotten scared, and then I might not have done it." If he'd consulted her, he said, "that would have made me weaker." On some level, she felt he was right. Had he consulted her, she would have given him ten good reasons not to report the fraud.

"Who's paying for it?" she asked of the lawsuit.

He explained that his lawyer would not charge them, but would be paid out of the government's settlement. That her husband had a lawyer, let alone a free one, was a bewildering concept. Thakur did not mention a possible financial reward for them. The chances of that happening seemed very unlikely and had not motivated him in the first place.

"Are we safe?" she asked. She knew what happened to whistle-blowers in India.

"Nobody knows about it." He explained that the case was filed under seal, another concept that seemed impossible to grasp, given how flimsy and political a tool the law was in India.

"Did you think about all of this before you filed the case?"

"I actually thought that once I reported it," he said, "I was done with my job." Long ago, he had imagined that a single tip to the FDA would be enough to stop Ranbaxy's fraud.

For the first time, it all made sense to Sonal: the parade of bosses, her husband's sudden resignation, the long hours he had spent in his basement office. He was utterly logical and not prone to making rash decisions. For him to leave a job without having another one had made no sense—until now. Though Sonal had a clear explanation for her husband's actual and mental absence, this knowledge did little to salve the family's wounds.

After their discussion in the restaurant, Thakur rarely spoke again about his interactions with the FDA, and Sonal didn't ask. "You are doing whatever you need to do," she would say in frustration, not understanding or feeling like a part of his decision-making. Whenever she did weigh in, her question was: "When is the case going to end?" It was a question he'd asked himself many times and could not answer.

With Thakur's return to the United States, their marriage grew even more strained. Sonal tried to rationalize that he was absent because he was doing something positive to help others. Yet when she felt lonely and overwhelmed with the children, the thought gave her little solace. Sonal toyed with the idea of leaving him, a foreign concept born of desperation.

Couples counseling is not a regular part of Indian culture. So Sonal sought help from an older neighbor to whom she'd become close. Sonal and Thakur met with her as a couple. The neighbor's advice was standard issue: try to make it work; make more of an effort to understand each other's positions. But they remained apart and unhappy, and fought when they were together. "We couldn't find that peaceful ground," said Sonal. Thakur tried to bury himself in work. Sonal told him, "You're just getting crazier and crazier."

In India, "you don't just move out of a marriage," said Sonal. It becomes a family affair. And so their parents began to weigh in. Her mother spoke to his father about Thakur's long absences and the pressure on the family. Thakur's father then spoke to Sonal. The older man, who knew nothing of his son's lonely quest, tried to impart some wisdom: "I know what you're doing is not easy," he advised, "but at the end of it, you have to make it work. The woman of the family can hold everything together or let everything go."

Though hardly an enlightened viewpoint, in this instance it happened to be true. Thakur had chosen his path, and it was up to Sonal to keep the family on it. Her cultural training acted as a counterweight to her feelings of resentment. "In India, we are sort of taught to compromise on things," as she put it. "You don't start looking for options." But with her loneliness intensifying, it was hard not to think about what those options might be.

As the government's case dragged on and the tension between the couple grew, Thakur turned to his lawyer, Andrew Beato, to help keep the peace. Beato spoke to Sonal. He tried to reassure her that her husband's identity would remain secret and that the case *would* end at some point. But he had a somewhat different message to impart, one that Thakur would never think of sharing. "What your husband is doing is unparalleled," Beato told Sonal. There had never been a case of a whistleblower from India, fighting a battle of this magnitude in the United States, he explained. Beato wanted to let her know that, though anxiety about her family's safety was "valid and paramount," as he put it, there were "larger issues about stopping a company that was doing something very bad." Her husband was "a hero," Beato explained.

Thakur didn't feel heroic. Mostly, he felt distraught. As the months—and years—slid by, he felt suspended in a terrible twilight: unable to move forward and caught in the past. Each time a deadline approached and it appeared the government would finally bring charges against the company, a government lawyer instead called Beato to request an extension. Thakur, on whose knowledge the government had built its case, had no choice but to acquiesce if he wanted the case to succeed. "I am someone who plans everything," said Thakur. "I am a very outcome-driven person." But he never planned to have the case hanging over his head for years, with no way to influence the outcome.

In the summer of 2009, Sonal moved the family from their freestanding home on a quiet side street to a gated community behind a guard post off a major Gurgaon thoroughfare. For the ever-social Sonal, the move eased some of her loneliness. Suddenly, she had neighbors with young families and a social life independent of her husband. The complex, named Unitech World Spa, had a gym, a clubhouse, parties, and even a singing group, which she joined. She drafted Thakur to be a part of it, though he was a reluctant and infrequent participant. The perpetually absent Thakur was a mystery to her new friends: "People used to question me, why does he spend so much time in the U.S.? Why is he there all the time?" She didn't tell them. But the move shifted things ever so slightly between them. She became less reliant on him, and he noticed that his wife's new life didn't necessarily include him.

In October 2010, Thakur took his daughter to a Halloween party at the complex's clubhouse, only to come face-to-face with one of his new neighbors, Abha Pant. She was still Ranbaxy's vice president of regulatory affairs. Because of Thakur's disclosures, the FDA's associate chief counsel would soon send prosecutors a lengthy memo recommending that she be criminally prosecuted for her role in Ranbaxy's fraud. Her name, as well as Malvinder Singh's, topped every

Justice Department list of "suspect persons" at Ranbaxy. Thakur's name, too, was on a list, one that Pant may have known about: of possible whistleblowers cooperating with the government compiled by company executives.

The two spoke briefly. She mentioned that Ranbaxy's new CEO had moved into the apartment complex as well, which was not what Thakur wanted to hear.

"I recently ran into Rashmi," she added, referencing Thakur's old boss Barbhaiya, who had recruited him to Ranbaxy. "He hasn't changed much."

"What do you mean by that?" Thakur asked, his discomfort growing.

"He still has a pretty big ego." As Pant tried to make conversation, it seemed to Thakur that she was probing for information. She brought up his other previous boss, Kumar. "He seemed like a fish out of water at Ranbaxy," she said.

"He was a pleasure to work with," Thakur said stiffly, ending the conversation.

As much as he tried to disappear, he felt surrounded by the company and the consequences of his time there. He relayed the encounter to his lawyer, Beato, typing notes as they spoke in his ongoing effort to document the case. In a single sentence, he wrote to himself: "When the news becomes public, leave here."

CHAPTER 21

A DEEP, DARK WELL

1998–2010

Durham, North Carolina

For three decades, Joe and Terry Graedon had been on the side of patients. The husband-and-wife duo, a pharmacologist and a medical anthropologist, respectively, wrote a syndicated newspaper column and hosted a National Public Radio program, *The People's Pharmacy.* Over the years, as they worked to empower and educate patients in their search for cures, the Graedons held a bedrock assumption: the FDA was a competent regulator whose claims could be believed.

Even during the dark days of the generic drug scandal in the late 1980s, the Graedons expressed confidence that the FDA's "exhaustive analyses" had found no problems that would endanger consumers; they remained staunch advocates for generic drugs. "We were steadfast in maintaining they were identical and you would have to be a total idiot to pay for a brand-name drug when a generic was available," Graedon recalled.

A chance encounter at their attorney's office a decade later sparked the Graedons' first real doubts about just how exhaustive the FDA's analyses were. There, an employee whose young son had attention deficit disorder told them, "When he goes off to school on Ritalin, he does very well." But when he was on the generic, the teachers complained about his ability to function for the same length of time. In the late 1990s, the Graedons started to receive letters from readers and radio listeners recounting bad experiences with a range of generic drugs. One patient had suffered anxiety and insomnia after switching to generic Synthroid: "I was sweating more than usual and my heart felt as though it would pound out of my chest." Another patient had become manic after taking generic Fioricet for a migraine: "I was stimulated beyond belief—typing letters and sending faxes at 3:00 a.m."

The Graedons published these cases in a 1998 newspaper column, raising the question of "how well the FDA monitors generics after they have been approved." By 2002, Joe Graedon had contacted the FDA and connected with the director of the Office of Generic Drugs, Gary Buehler. It was the start of a fraught communication that spanned years.

Before long, the People's Pharmacy website became a clearinghouse for patients wrestling with terrible symptoms after a switch to generic drugs. Patients wrote in, desperate for answers. Graedon forwarded the complaints to Buehler at the FDA. Between 2007 and 2009, he relayed complaints about at least twenty different drugs, believing that highly placed officials would want to know what patients were experiencing.

In a January 2008 email to Robert Temple, deputy director for clinical science at the FDA's Center for Drug Evaluation and Research, Graedon forwarded a complaint about generic Dilantin, an epilepsy drug: "After being on 300 mg. of dilantin for more than 20 years, I tried the cost cutting measure of replacing it with the generic. Numerous times I had seizures." Graedon wrote to Buehler about the complaint, "This is most worrisome. We trust you are taking this as seriously as we are."

As Graedon scrutinized the FDA's standards for bioequivalence and the data that companies had to submit, he found that generics were much less equivalent than commonly assumed. The FDA's statistical formula that defined bioequivalence as a range—a generic drug's concentration in the blood could not fall below 80 percent or rise above 125 percent of the brand name's concentration, using a 90 percent confidence interval—still allowed for a potential outside range of 45 percent among generics labeled as being the same. Patients getting switched from one generic to another might be on the low end one day, the high end the next. The FDA allowed drug companies to use different additional ingredients, known as excipients, that could be of lower quality. Those differences could affect a drug's bioavailability, the amount of drug potentially absorbed into the bloodstream.

But there was another problem that really drew Graedon's attention. Generic drug companies submitted the results of patients' blood tests in the form of bioequivalence curves. The graphs consisted of a vertical axis called Cmax, which mapped the maximum concentration of drug in the blood, and a horizontal axis called Tmax, the time to maximum concentration. The resulting curve looked like an upside-down U. The FDA was using the highest point on that curve, peak drug concentration, to assess the rate of absorption into the blood. But peak drug concentration, the point at which the blood had absorbed the largest amount of drug, was a single number at one point in time. The FDA was using that point as a stand-in for "rate of absorption." So long as the generic hit a similar peak of drug concentration in the blood as the brand name, it could be deemed bioequivalent, even if the two curves reflecting the time to that peak looked totally different.

Two different curves indicated two entirely different experiences in the body, Graedon realized. The measurement of time to maximum concentration, the horizontal axis, was crucial for time-release drugs, which had not been widely available when the FDA first created its bioequivalence standard in 1992. That standard had not been meaningfully updated since then. "The time to Tmax can vary all over the place and they don't give a damn," Graedon emailed a reporter. That "seems pretty bizarre to us." Though the FDA asserted that it wouldn't approve generics with "clinically significant" differences in release rates, the agency didn't disclose data filed by the companies, so it was impossible to know how dramatic the differences were.

As more patients wrote in to the People's Pharmacy website about their struggles with generics, the Graedons found that time-release drugs received more complaints than others. One was generic Toprol XL, the drug that had so troubled Harry Lever's patients. Almost as soon as the generic became available, patients taking it began writing in to the Graedons, reporting dramatic increases in blood pressure and heart rate. They were nauseous and dizzy, they had hives and headaches, their hair fell out, and they had trouble sleeping and suffered vivid nightmares. Graedon forwarded some of the accounts to Buehler at the FDA. "What are you doing about this?" he asked. "If there are deaths it could turn into a big scandal." As the Graedons later wrote of the response in one of their columns, "We were told, in essence, 'We'll get back to you.' We heard nothing more."

But it was another drug entirely on which Graedon planted his flag. Wellbutrin XL was a popular antidepressant sold by GlaxoSmithKline. As an "extended-release" formula, the drug dispensed its medicine into the bloodstream over hours, unlike older forms of the drug, which had to be taken several times a day. When the patent on Wellbutrin XL expired in December 2006, the Israeli manufacturer Teva began marketing the first generic version of the pill and contracted with Impax Laboratories to manufacture it. Teva sold the generic under the trade name Budeprion XL. The active ingredient was bupropion. Almost immediately, the People's Pharmacy was flooded with troubling emails from patients complaining of headaches, nausea, dizziness, irritability, sleep problems, and anxiety attacks. Some people said that their generic pills smelled bad. Many cried easily. Some became suicidal. Others reported having tremors and even seizures. "I shake so badly sometimes I have trouble drinking water from a glass or I can't hit my mouth with a forkful of food!" one patient wrote. Nearly all the patients reported that their depression had returned.

The similarity of the reports astounded Graedon. One patient wrote, "I have no history of suicidality, but a day after switching to the generic, I went into a week of steadily rising panic. . . . I was psychotic, self-loathing way WAY beyond anything I have ever experienced. I made it through the worst of it, called a suicide hotline, took two Ativan and didn't take any more of the Budeprion. The next day I felt much better, and today I'm back to my normal self." Another patient reported that after two weeks on generic Wellbutrin, "I ended up getting highly aggressive. I spent all the money in my checking account. I cut off cars in the highway and ran a red stop light. I could have been killed or killed someone else."

One patient suffered a "full-blown panic attack with dizziness at 65 mph. . . . Then I began to cry . . . thinking, what's wrong with me? AM I going crazy? Not getting any relief, I got out of my car and sat on the side of the turnpike as the giant trucks and cars went flying by. I was forced to lie down in a drainage ditch along the road and covered my ears to calm myself down because I had uncontrollable thoughts of running into traffic."

Graedon reported the deluge of messages to Robert Temple at the FDA, who didn't respond. But the FDA had also been amassing patient complaints. Between January and June 2007, more than eighty-five reports of adverse reactions to Budeprion XL rolled into the agency, where officials were largely dismissive, suggesting that the patient reactions could be psychosomatic and might be caused by changes in the tablet's shape and color.

By April 2007, however, the Graedons announced good news

to their readers that seemed to herald cooperation from the FDA: "We have arranged with the FDA to analyze any generic pills that readers of The People's Pharmacy suspect are not equivalent to their branded counterparts," they wrote. "Please describe your experience and send your generic pills with as much information as possible." Medicine from across the country filled the Graedons' mailbox, including "hundreds and hundreds" of Teva's Budeprion XL tablets. Right away, Graedon could tell something was wrong: "They smelled to high heaven." He sent them to the FDA.

At a fund-raiser, Graedon happened to run into a chemist at Burroughs Wellcome, the company that had developed the chemistry behind Wellbutrin. "What's going on with this funny smell?" Graedon asked.

"That's a no-brainer," the man said, explaining that it was a sign that the pills were deteriorating. "It's a problem with the manufacturing process."

As he waited for results from the FDA's tests, Graedon continued to forward complaints about Teva's Budeprion XL to the agency. "What alarms us most," he wrote to Temple on June 21, 2007, "is that many of these people have described suicidal thoughts as a result of this switch."

Graedon was sick of waiting for the FDA's test results. He spoke with experts about what could produce the symptoms that patients were reporting. He even reached out for help to independent laboratories. Tod Cooperman, the president of ConsumerLab in White Plains, New York, was quick to join his cause. ConsumerLab tested the 300-milligram dose of Teva's Budeprion XL against that of GSK's Wellbutrin XL. The results revealed the likely source of patient distress: the generic dumped four times as much active ingredient during the first two hours as the brand name did. Graedon compared the effect to guzzling alcohol. "If you sip a glass of wine over the course of two or three hours, you're not going to feel drunk," he explained. "But if you drink the whole thing in fifteen minutes, you're getting too much too fast."

The Graedons believed that this "dose dumping" explained why many patients were experiencing signs of overdose, such as headaches and anxiety, followed by symptoms of withdrawal, including renewed depression and suicidal thoughts. Teva flatly rejected the ConsumerLab report and claimed that the independent laboratory's testing method was "inappropriate." The FDA was silent.

In December 2007, Graedon and Temple of the FDA were both asked to be guest speakers on a Los Angeles radio show to discuss the FDA's requirements for generic drug approvals. During the show, the host asked Temple about the differences between generic and brand-name Wellbutrin uncovered by ConsumerLab. For the first time, Temple admitted that the generic and brand-name drugs released their active ingredient at different rates. "The main point here is that yes, the generic releases a little sooner," Temple said. "That could be described as an advantage." He added that the early release was "unlikely" to make any real difference in treating patients' depression. Graedon was astonished to hear Temple describe the early release as an advantage, as he wrote to Tod Cooperman after the show. "One wonders what Alice-in-Wonderland world the FDA is living in where up is down and a rapid release of 300 mg. bupropion is desirable."

Temple later told a journalist, "Most of the excitement from people who worry about this come from isolated reports. Then you look at the results of bioequivalence studies, and there is nothing to worry about."

Graedon wanted to speak to Temple personally. But in January, after getting no reply from him, he tried reaching Buehler: "We are at our wits end about how to proceed with this Budeprion XL situation. . . . You cannot seriously believe that the hundreds of similar reports are all coincidence," he wrote. He continued to pepper the FDA with messages, sometimes on a daily basis. As he wrote to a colleague, "We have been nagging, cajoling, badgering, whining, complaining and otherwise making asses of ourselves with regard to the FDA."

By March 2008, Graedon was furious. It had now been almost a year since the FDA's promise to investigate Teva's drug. But seemingly nothing had happened. The issue was devouring his time, and he'd have preferred to drop it. But complaints from patients continued to stream in: they were hospitalized, underwent invasive workups, lost jobs and homes to depression. As one patient wrote, "I have a long road ahead of me—just to get back to where I was—but then I'll never get back. Because my experience with the generic destroyed everything in my life that was important to me."

On March 17, 2008, unable to restrain himself any longer, Graedon wrote Buehler about the messages, "They JUST KEEP COMING! . . . These people are YOUR boss! They pay your salary. You are accountable to them. They are not nutcases, crackpots or idiots. They are real people with real problems and you must be responsible." The next day, Buehler sent a curt reply to say that the agency was still working on the report. Graedon wasn't pacified. The next month he followed up again: "I keep saying this . . . lives are at stake here. We must have your report sooner rather than later."

On April 16, 2008, more than a year after Graedon first alerted the agency to patients' troubles with the drug, the FDA released its report and assured consumers that it had been right to approve the 300-milligram dose of Budeprion XL. Despite "small differences" in how it dissolved, the FDA said that Teva's drug met the agency's standards and therefore was "therapeutically equivalent" to Wellbutrin XL. It concluded that the "recurrent nature" of depression—not drug failure—was the likely cause of patients' problems.

As Graedon read through the report, he was appalled. Rather than test samples in a laboratory, which Graedon had been promised in 2007, the FDA simply reviewed the bioequivalence data that had been submitted by Teva with the drug's application in 2003. Even worse, the data that the FDA had reviewed was from the 150-milligram dose, not the 300-milligram dose. Typically, generic manufacturers test only the highest dose, known as the "reference listed drug." The FDA assumes that the lesser doses are proportional and will behave similarly in the body. But in the case of Budeprion XL, the higher 300-milligram dose was never tested "because of the potential risk of seizures" in volunteers, the FDA explained. Graedon was astonished. How could the FDA know the 300-milligram dose was bioequivalent without any data? Yet the agency had approved it for millions of people to take, based on test data from a lower dose.

What really floored Graedon were the test results for the generic and brand versions of the 150-milligram dose, which the FDA had included in its report. These were reflected in two curves that mapped the subjects' blood test results. The curves were completely different. With one glance, Graedon knew that they couldn't possibly have the same effect on a patient. The brand had a gradual slope leading to a high peak at around five and a half hours. The generic had a sharp rise that peaked at around two hours. The difference was "so incredibly obvious," he said, that "any school kid" could have figured it out.

If that was what the FDA was relying on to call *the same*—a drug the agency had no data on, which they didn't even test for their report, which was presumed the same based on results for a lower dose, and which were obviously different to the naked eye—things were even worse than the Graedons had suspected. "We were like, 'Holy cow, we've got something big here,'" Graedon recalled. "This is the card that brings down the house of cards. In some respects, it changed everything. I just never imagined the approval process was so screwed up."

The Graedons wrote to Temple and Buehler: "After reviewing the report posted today, we believe you have misled the American public." Graedon had hoped that the FDA would come to the rescue of patients, but the agency seemed to be operating instead to their detriment, and from behind a closed door. Except for those two curves, the agency did not publicly release any bioequivalence data submitted by the companies. Temple later told a journalist that while the FDA study was "not misleading, it may have been inadequate. . . ." Uncertain what else to do, Graedon cast a wide net throughout the scientific community. He saw the issue as a public health emergency, but he felt stumped. Should he go to Congress? To medical lobbyists? To journalists? What was the magic pressure point to make the agency act?

By now, Graedon had become email companions and confidants with Harry Lever at Cleveland Clinic, both men swimming against the tide of accepted medical and political wisdom. Both were asking similar questions. In a sympathetic email to Graedon, Dr. Steven Nissen, chairman of Cleveland Clinic's Cardiovascular Medicine Department, who had initially put the two men in touch, called the agency's stonewalling "appalling." As a high-profile patient advocate, Nissen was familiar with FDA inaction. He'd led numerous inquiries into drug safety that raised questions about the agency's review process. He wrote to Graedon, "Don't expect any action from the FDA. In these cases, they invariably operate in denial mode because, to acknowledge the problem, it makes it look like they screwed up."

Graedon started hearing from a former high-ranking FDA official, who had offered Graedon guidance and information on the condition of anonymity. Graedon referred to his source as Deep Throat. "He basically reinforced our position that there is a problem and that we are on the right track and that we should not give up or give in," Graedon wrote to Cooperman. In July 2008, Graedon met with FDA officials. They agreed to collaborate with him on a study evaluating the bioequivalence of Budeprion XL in patients who had problems with the generic. He figured the agency had agreed to the study to "shut us up once and for all," Graedon told a reporter. Yet six months later, in January 2009, the FDA had yet to make any headway. As Graedon wrote to an FDA official, "It frequently feels as if we are dropping pebbles down a very deep, dark well. We never hear a splash."

He continued to send complaints, including one in which a man wrote that the Teva version of Wellbutrin XL "almost caused my wife to commit suicide." Graedon forwarded the report to Buehler and Temple on a Friday evening in February 2009. "Guys, this has to affect you, right?" he wrote. "I mean seriously, after all this time you are not going to tell us that this is psychosomatic. . . . When are we going to do something about this?"

In December 2009, the *Wall Street Journal* reported that Teva and Impax planned to conduct a bioequivalence study of the 300-milligram dose of Budeprion XL under the agency's guidance. After a year, that study faltered, and the FDA decided to undertake its own.

Finally, in October 2012, the agency issued a press release about its findings, which confirmed what the Graedons had known all along: the 300-milligram Budeprion XL dose was *not* therapeutically equivalent to Wellbutrin XL, as the drug did not deliver enough active ingredient. The generic "is not absorbed into the bloodstream at the same rate and to the same extent" as the original drug, the FDA reported. Tod Cooperman of ConsumerLab was pleased but unimpressed. "We're proud we were able to help uncover this problem, but it's unfortunate that it's taken the FDA five years to get the product removed," he told the Associated Press. By then, the FDA had asked the other four manufacturers of generic Wellbutrin XL to conduct bioequivalence studies of their own 300-milligram pills. Of the four generic brands, one made by Watson Laboratories also failed, and the company withdrew its product.

Graedon wanted to contact Gary Buehler, director of the FDA's Office of Generic Drugs, and ask: Why did it take the agency five years after he'd first reported patient problems with it to test Teva's drug? Why had the drug been approved without specific bioequivalence data? But he couldn't, because in October 2010, about eighteen months after plaintiff lawyers geared up to sue Teva on behalf of patients, Buehler had left the FDA. He'd taken a job as vice president of global regulatory intelligence and policy—at Teva.

Had the FDA wanted to tackle the problem of bioequivalence head-on, it needed to look no further than its own adverse event database, which collected reports from the public. It was also chock-full of complaints about generic drugs: patients had found their medicine covered in mold or smelling intensely of "rotten fish" or "cat urine"; some reported no therapeutic effect from their drugs, "as if the drug was not working at all," as one patient wrote in; others had been directed by the FDA to contact the drug makers directly, only to hear nothing in return when they sent in their drugs for testing; and there were reports of foreign objects in pills, from eyelashes to insects.

While the FDA viewed these complaints as "an important source of information about potential safety concerns," an agency spokesperson later explained that they required "careful review and interpretation." The number of complaints could be affected by anything from legal proceedings to press coverage.

In Mount Laurel, New Jersey, in January 2016, Carla Stouffer, a seventy-one-year-old retiree, would have simply swallowed her daily capsule for high blood pressure, amlodipine/benazepril, were it not for a flash of movement. On closer inspection, she noticed a small, centipede-like bug stuck halfway inside the capsule, alive and wiggling. Horrified, she watched as the bug tried to free itself from the pill casing. She'd never thought before about who made her medicine. But she learned that the maintenance drug, which she'd gotten in a three-month supply from the prescription benefit manager Express Scripts, was made by the Indian company Dr. Reddy's

Laboratories. Her complaint became one of over a million related to pharmaceuticals filed with the FDA that year.

Much like the doubtful patients, Graedon had lost his trust in the FDA. "I had always believed the FDA had its stuff together, and just accepted that the approval process was perfect, because that's what everybody believed," said Graedon. But the more he learned, the less confident in the FDA he felt. One night at a party, he spoke with a GlaxoSmithKline executive who'd had too much wine. The man shared with him that many companies were moving their manufacturing operations to China in order to reduce operating costs—a new problem for Graedon to consider. He suspected that while the FDA was busy with its decades-old regulations, the companies making America's medicine were playing hide-and-seek with the agency all over the world.

This flipped Graedon's thinking on its head. He had assumed all along that the drugs the FDA approved met the agency's standards, but the standards, he realized, were flawed. What if the drugs plaguing America's patients were so bad that they did not meet even the FDA's flawed standards and were being wrongly approved? That would mean that both the FDA's standards and its approval process were broken. Even if the FDA fixed both, however, that would do nothing to solve the larger manufacturing problem: America no longer made its own drugs.

CHAPTER 22

THE $600 MILLION JACKET

2011

Silver Spring, Maryland

Scattered across the desk of Karen Takahashi, a meticulous FDA compliance officer, lay Ranbaxy's most valuable asset: its application to launch the first generic version of Lipitor. The controversial jacket was stamped ANDA 76-477, and it would fall to the soft-spoken Takahashi and her colleagues in the FDA's international compliance branch to review it. The clock was ticking.

In the generic drug world, nothing was more lucrative than atorvastatin. Each year, U.S. government programs alone spent $2.5 billion on brand-name Lipitor. Any delay in launching a generic version could cost Americans up to $18 million a day, a group of U.S. senators reminded the FDA commissioner in a March 2011 letter. Ranbaxy was first in the FDA's queue to make atorvastatin and, owing to a settlement with Pfizer, could legally start selling the drug by November 30, 2011. All the company needed was final approval from the FDA. But Ranbaxy's mounting problems had turned the once seemingly surefire launch into a nail-biter. "For Ranbaxy, this is the fight of their lives," a lawyer for a pharma company told *Fortune* magazine. "This is the biggest generic opportunity in history. None of us know where this is going to come out."

Inside the FDA, despite murky and chaotic deliberations over Ranbaxy's fate, it seemed clear to some that the agency would never allow a company so saturated in fraud to keep exclusive rights to launch the nation's most important generic drug. Ranbaxy's first-to-file application seemed "unsalvageable," a word that had emerged from government huddles. "This application will never be approved," Debbie Robertson had told a prosecutor in mid-2010. As the FDA's deputy director of the Office of Pharmaceutical Science wrote in an internal memo, "To point out the obvious . . . an approval of Ranbaxy's ANDA 76-477 may not seem in harmony with the agency's regulatory actions against Ranbaxy." One way or another, the company would be forced to forfeit its application, and some other company in the queue would step in. "Unfortunately for Ranbaxy, choices are very bleak," Dinesh Thakur jotted down during a discussion with his lawyer, Andrew Beato, in early 2010. "Even if they go to court, they don't have much choice. FDA is not going to sign off on Atorva."

Publicly, the agency said nothing about its deliberations. But as Takahashi and her colleagues delved into the application, fully expecting to stop it, the agency found itself in a quagmire of evidentiary, regulatory, procedural, and bureaucratic problems, a number of the FDA's own making. The battle with Ranbaxy had become so convoluted that many of the regulators waging it barely even understood it themselves. They joked bitterly about needing RANBAXY—BE STRONG bracelets. But they soon realized that they faced a serious checkmate: if the FDA did not approve Ranbaxy's atorvastatin, it was not clear when—or even if—Americans would get access to a low-cost version.

Ranbaxy had filed its atorvastatin application in August 2002.

Under rules in place then (which changed months later), unless the FDA revoked Ranbaxy's application owing to a specific finding of fault, the company could simply park its exclusive rights. Competitors would pile up, unable to launch, and the public would get no price relief. As one FDA lawyer explained the rules to his colleagues, "As long as Ranbaxy is not able to sell generic Lipitor, no one can sell generic Lipitor."

Ostensibly, the FDA had plenty of cause to shut down Ranbaxy's bid for six months of exclusive atorvastatin sales. The company had committed extensive fraud and needed to be punished. But in the topsy-turvy world of the case, an argument had begun seeping into the government's internal deliberations. Because the generic drug company operated on razor-thin margins, it needed atorvastatin profits in order to pay the record-breaking fine it deserved. Unless Takahashi found irrefutable proof of fraud in the application, Ranbaxy might just get away with a blockbuster launch of atorvastatin.

Takahashi systematically reviewed Ranbaxy's application from the beginning. Since 2002, individual reviewers had flagged anomalies: missing electronic data, discrepancies in resubmitted data, impurity data that made no sense. As the application crept through review, Ranbaxy had blamed rounding errors, copying or calculation blunders, and even its own lax laboratory protocols for the glitches. Company executives had claimed that test dates were inconsistent because the company had been inconsistent in what it taught its analysts.

But agency reviewers had found numerous "unexplainable" discrepancies between the original application and amendments that Ranbaxy had filed in 2007. Some test results changed significantly. Impurities suspiciously decreased. Pills originally described as "white" became "off-white." In several instances, tests originally reported as "out of specification" now fell "within specification." The changes suggested that portions of data had been entirely false, in total disarray—or both.

These were only the small problems. There were more ominous signs that data had been compromised. The mysterious 4 degree Celsius refrigerators, which the FDA suspected were being used to artificially slow the degradation of test samples, had held bottles of atorvastatin. Takahashi learned that in October 2007 an informant notified the FDA that a Ranbaxy vice president had faked atorvastatin records just before the FDA inspected Paonta Sahib earlier that year. In the summer of 2008, while inspecting documents from Paonta Sahib, the FDA learned that Ranbaxy had been throwing out failed stability tests for atorvastatin and other drugs, then retesting the medicines until they passed and reporting only the successful tests. An inspection of Ranbaxy's research and development laboratory in Gurgaon, India, in the spring of 2009 uncovered similar deceptions. Takahashi hadn't found proof yet that the atorvastatin application itself had been compromised, but the chances of it being uncompromised were almost zero.

The agency had had a surefire way to shut down the application—and failed to. In February 2009, it had leveled a rare Application Integrity Policy against the company, which compelled Ranbaxy to prove its products weren't fraudulent in order to get them approved. Takahashi and her colleague down the hall, Doug Campbell, had written memo after memo arguing that the punishment should be imposed on the entire company.

Remarkably, the FDA cracked down on Ranbaxy with one hand and opened an escape hatch with the other. That month, the agency announced that it would impose the AIP on only one plant, Paonta Sahib. Although this put the brakes on eighty-five drug applications, including atorvastatin, it also left Ranbaxy free to shift its most lucrative applications out of the embargoed plant to other manufacturing sites, so long as it submitted fresh data.

In December 2009, ten months after the FDA imposed the AIP on Paonta Sahib, Ranbaxy filed new amendments to its atorvastatin application. After settling its patent lawsuit and making a deal with Pfizer, it proposed using Pfizer's active ingredient and moving production to the Ohm Laboratories plant in New Jersey, which had a comparatively clean regulatory record. Even high-level bureaucrats struggled to understand why the agency would suddenly permit the company to shift manufacturing sites. "From what I know Compliance did not want us to permit this because Ranbaxy could use this tactic as a simple way to circumvent AIP," the chief of the regulatory support branch for the Office of Generic Drugs wrote to his colleagues. Yet somehow the FDA had authorized the shift. The move left Takahashi and her colleagues with yet more work: they had to sift through all of Ranbaxy's amendments to ensure that fraudulent data from the embargoed plant wasn't being reused.

In April 2010, Takahashi began to follow another important lead. The AIP required Ranbaxy to hire an outside auditor to verify that the data in its applications was accurate. Ranbaxy had chosen Quintiles Consulting, which was obligated to report its findings directly to the FDA. In November 2009, Quintiles auditors went to India to examine the atorvastatin application and ended up notifying the agency of a "very hot" finding, as Campbell relayed to his FDA colleagues. The original records for the dissolution data submitted in 2002 were nowhere to be found, and the raw data on hand didn't match the data submitted to FDA. The situation threatened to upend Ranbaxy's application. The FDA required that applications be "substantially complete" when submitted. This was to prevent companies from submitting sham applications simply to be first, before even finishing their studies or generating complete data. But the absence of original records meant that the data had either been lost or never existed in the first place. If the tests had been faked, it could mean that the original application had been incomplete, a disqualifying lapse. At the FDA's Office of Criminal Investigations, Debbie Robertson flagged the missing data for prosecutors. "Sound familiar?" she quipped.

Quintiles's report made clear that unless the original data could be located, Ranbaxy might have to repeat tests with new ingredients and a new clinical study. And this would certainly raise the question as to whether Ranbaxy really was the first generic company to have submitted a completed application. Should the company keep its place at the front of the line or move to the back?

In late May 2010, the FDA's criminal investigators learned that Ranbaxy's newest CEO, Atul Sobti, had summoned a top Quintiles official, Bob Rhoades, to India the previous November over the Thanksgiving weekend and demanded to know why he would report such egregious failings to the FDA without first consulting the company. "He was not happy," Rhoades recalled of the meeting, where Sobti dressed him down. Sobti later expressed surprise "that an interaction that was just part of a long process is being given not only an exaggerated, but an inaccurate, view."

But inside the FDA, officials learning about the meeting viewed it as an apparent effort at obstruction by Sobti. It "should be the absolute last straw," Debbie Robertson's supervisor declared to an FDA lawyer. "The CEO no less. Not only is the culture of fraud and deceit still alive and well at Ranbaxy—it exists all the way to the top. All bets should be off."

Few at the FDA disagreed. Meanwhile, Takahashi returned to the question of the missing data. What did it mean? And what could the FDA do about it? "Is it written anywhere that the raw data for a drug application must be kept forever?" the director of the Office of Compliance's Division of Manufacturing and Product Quality emailed his staff. Unfortunately, it wasn't. As the regulatory counsel for the Office of Compliance noted, "We have not, as of yet, been able to point to a regulation that says they have to keep the data forever."

On May 5, 2010, a group of almost a dozen FDA lawyers, regulators, and criminal investigators met at the agency's headquarters with prosecutors and Justice Department officials to discuss the sprawling effort to settle the case against Ranbaxy. With every imaginable punishment on the table—from prosecuting individual executives to debarring the company to imposing a record-breaking fine—the conversation kept returning to atorvastatin.

"What's in the best interest of the government?" Douglas Stearn, assistant director of the FDA's Office of Compliance, asked those gathered. He noted that Ranbaxy had filed suspicious amendments for its atorvastatin ANDA, and that the FDA should not be rewarding the company with approval but imposing a heavy fine instead. "Big numbers talk."

Stuart Berman, the assistant U.S. attorney for Maryland, pointed to the remarkable statement made to Quintiles auditors by a vice president at Ranbaxy's research headquarters in Gurgaon: "Recording and verifying data contemporaneously with data generation is not a cultural trait in India. We need to learn this as a habit." It seemed irrefutable, said Berman, that "every bit of data received [from Ranbaxy] is junk."

"Why are we working so hard to keep Ranbaxy in business?" Steven Tave, the FDA's associate chief counsel for enforcement, chimed in. "Why not take a broad sweep—take everything on the market from 2006 on and pull it off?"

Or, said Berman, "put the settlement aside and start indicting people." The question of a settlement was simply "wagging the dog," he said, unless they knew how much Ranbaxy could pay.

But as they debated their options, the discussion turned to the implications if the company was denied permission to make atorvastatin. Tave acknowledged that "if we take strong action, we may not have a huge settlement because the company can't fund it."

This concerned Stearn. "If we let them have the first-to-file and they get a huge settlement, what does that say?"

As they mulled how to stop Ranbaxy from making atorvastatin and still get a settlement that would send a strong message to the drug industry, Roann Nichols, an assistant U.S. attorney, asked, "Is there a way we can *not* approve the first-to-file and let them sell the rights?" That way, there would still be money for a settlement.

Berman pointed out that the Justice Department's position should not be to fund the fine through "ill-gotten gains."

Without atorvastatin, "we won't get any money," said Linda Marks, the senior litigation counsel for the Justice Department's consumer protection branch, who had skirmished with Robertson over disclosing critical leads to Ranbaxy's defense team.

"The company has to relinquish the exclusivity totally," an FDA lawyer, Paige Taylor, interjected. "If the application is tainted, it's tainted."

There seemed to be no perfect path to justice. As Nichols concluded, "Someone will have to make a decision. Atorva will hit the fan either way." That was almost certainly true. Officials at the meeting left with the view that the application was unsupportable, and therefore almost certainly doomed.

At the FDA, the essential compromises of regulating overseas plants were already whittling away the officials' tough stance. Just three weeks after the May meeting, regulators got a request from Ranbaxy to inspect a plant in Mohali, India, that the agency had not yet approved. The compliance staff understood this to be part of Ranbaxy's "creative tactics," an effort to get more plants approved to circumvent the crackdown on Paonta Sahib. By now, Ranbaxy had been under criminal investigation for almost five years. But the agency responded to the company's request using the same approach it took for all foreign facilities—by asking for permission to come inspect the plant, weeks in advance.

On August 19, 2010, an FDA investigator wrote to the firm

requesting "your availability and the firm readiness for our next visit Oct. 4–8, 2010," and asked for "your availability to assist in or information on round trip transportation airport/hotel/firm for the Investigators." A Ranbaxy official wrote back providing more details about the plant, promising to work out hotel reservations, and saying that the company would "arrange the invitation letter as well." In other words, the company would once again function as a host and travel agent and get six weeks of advance notice to transform the plant for the FDA's arrival.

Meanwhile, at the international compliance branch, Takahashi continued reaching out to different scientific experts at the agency to see if, somewhere, ANDA 76-477 had crossed some obvious bright line. Each inquiry seemed to loop back to the same place: the application could not be verified as legitimate but couldn't be proven obviously false. On the morning of March 16, 2011, however, Takahashi had an idea: to stop studying the application and to study the company's modus operandi instead. She asked Debbie Robertson for a copy of the notes from her division's interview with Raj Kumar back in 2007. As a witness with impeccable credentials and unshakable integrity, Kumar had been a vital source of information. Now, as Takahashi scoured the notes, one particular falsification that Kumar described grabbed her attention. Kumar told investigators that for two U.S. drug products, the company had taken the chemical profiles of brand-name drugs, and duplicated them to make them appear as though they were for Ranbaxy drug products.

This got Takahashi thinking. The dissolution data that Ranbaxy had submitted in its original 2002 application looked suspiciously similar to the dissolution data for brand-name Lipitor. The company also claimed that it couldn't find the corresponding raw data for those tests. Perhaps Ranbaxy had cribbed all of Pfizer's data and passed it off as its own. This possibility had to be pursued. Though something short of a "smoking gun," it became known inside the agency as a "smoking gun theory." If true, it would explain why some of the initial data was missing—perhaps because it had never existed in the first place.

The best way to figure out whether Ranbaxy had passed off the innovator data as its own would be to comb through Pfizer's Lipitor application, patents, and other public sources in hopes of finding chromatograms or other results that perfectly matched the data in Ranbaxy's application. It was perhaps a mark of Takahashi's desperation that she thought this was even a doable idea. The task turned out to be monumental. Pfizer's Lipitor application, which had been approved in 1996, was stored in 220 paper volumes in a government storehouse in West Virginia. Takahashi learned that it would take four to six weeks to get the documents shipped to CDER headquarters in Maryland.

Takahashi was forced to limit her search to public records. Thousands had been available when Ranbaxy filed its application. It was a true needle-in-a-haystack effort that turned up little of significance.

As her options dwindled, pressure was mounting on the FDA to make a decision. Not only had U.S. senators demanded quick approval, but the six other drug companies that had submitted atorvastatin applications were also demanding action. No one knew when—or even if—the FDA would approve Ranbaxy's application, and competitors worried that the agency's hesitation would delay their own approvals.

On March 18, Mylan filed a lawsuit against the FDA in federal district court in Washington, D.C., alleging that Ranbaxy should be forced to surrender its exclusivity, owing to its use of false data, and that the "FDA's indecision is permitting Ranbaxy to maintain a benefit to which it otherwise is not entitled." As news articles circulated about the suit, an FDA official based in India emailed his colleagues: "These stories are not good in this environment."

The FDA responded on April 4 with a hard-hitting motion

to dismiss Mylan's suit, arguing that it was under no obligation to disclose confidential deliberations or to help Mylan with its business planning. On May 2, a judge agreed, ruling that one company couldn't intervene in another company's application. Investors interpreted the FDA's stance in the Mylan case as a hint that the agency was planning to allow Ranbaxy to sell atorvastatin in November. Ranbaxy's shares rose on the news. But no one, including Ranbaxy, knew where the agency's deliberations would lead. By then, Wall Street analysts had mapped out complex flow charts of potential scenarios and their market impact.

But the compliance staff knew where it should come out. One month earlier, they circulated a memo with questions and possible answers. One was, "Should the government insist that Ranbaxy forfeit exclusivity for Atorvastatin, even if it means a substantial decrease in the settlement's potentially large recovery?" The answer was, "Yes, it is in the best interest of public health."

Yet the machinery of the agency kept moving toward approving the drug. In May 2011, Deb Autor's division, the Office of Compliance, sent a memo to the director of the Center for Drug Evaluation and Research, with a fateful recommendation. The memo urged that the Office of Generic Drugs formally review Ranbaxy's atorvastatin application and expedite the process under an exception to the Application Integrity Policy imposed on Ranbaxy. It argued that if there was fraud, a review could detect it, an argument that could be easily controverted, if anyone was inclined to argue. But with the U.S. government spending $2.5 billion on brand-name Lipitor each year, there were larger forces at work. The logic, as Doug Campbell later replayed it, was: "'All we can do is block it, but we're not going to block it, because we're all going to save millions of dollars.'"

As the Office of Generic Drugs started formally reviewing Ranbaxy's revamped atorvastatin application, objections that once seemed insurmountable fell away. By mid-October, the necessary review divisions had signed off on the application. The FDA prepared to announce its approval. By early November, the agency finalized a press release and drafted responses to anticipated questions. It seemed that November 7, just three weeks before the final deadline for approval, would be the big day for the agency to greenlight Ranbaxy's drug.

But that morning agency officials panicked. They realized—in a major oversight—that the FDA had not conducted a preapproval inspection of a plant that Ranbaxy had now listed as a main source for the active ingredients. Previously, Ranbaxy had told the agency that it would assemble and bottle its atorvastatin at its plant in New Jersey and use active ingredients from Pfizer's plant in Cork, Ireland. But in late July 2011, Ranbaxy changed this and told the FDA that it would also use active ingredients made at its own plant Toansa, in Punjab, a decision that had bewildered the agency. "Seems ridiculous to me that Ranbaxy decided to use its own API," the FDA lawyer Marci Norton wrote to Doug Campbell. "Given how high profile this drug is, you'd have thought the smarter course would've been to use Pfizer's API."

By its own standards, the agency was supposed to inspect every drug manufacturing facility once every two years. The Toansa plant was due for another visit. At an emergency meeting that afternoon, FDA officers decided to delay an announcement and rush an inspection there.

On November 21, nine days before Ranbaxy could potentially start shipping generic Lipitor, two FDA investigators arrived at Toansa to determine two things: whether the plant followed good manufacturing practices, and whether it was capable of safely making the active ingredient for atorvastatin. One of the investigators, Regina Brown, had originally found the unregistered refrigerators full of samples at Paonta Sahib.

As the inspection of Toansa progressed, the regulators back

in Maryland waited anxiously for the outcome. On Thanksgiving, Brown wrote back to Campbell, summarizing problems she'd observed so far. She found workers poorly trained and openly using scrap paper while making a batch of atorvastatin, a breach that pointed to a lack of control and potential data manipulation. Fourteen hours later, when she sent him a draft of her findings, it was clear the inspection had deteriorated further. She'd found a shredder in the middle of the production floor, and evidence that records related to batches were being destroyed. "Each scrap paper incident led us into deeper doo-doo," she wrote to Campbell.

On Friday, November 25, with five days to go until Ranbaxy could start manufacturing atorvastatin and with the FDA's decision still unmade, Campbell was troubled not just by Brown's findings but by what he felt to be the agency's inexorable drift toward approval. He was also running out of time. The Office of Generic Drugs had been pressuring him to submit his team's official recommendation either for or against approval by the end of the week—and it was now Friday morning. The lawyer Marci Norton, who'd been immersed in settlement negotiations with Ranbaxy's lawyers, told Campbell to wait until Monday, when everyone would be back in the office. There was no room for mistakes. "This decision likely will undergo heavy scrutiny, perhaps in court and likely by Congress," she wrote.

Still, Norton worried that Brown might classify the inspection as Official Action Indicated (OAI), the worst rating a drug facility can get. In the face of such a finding, it would make no sense for the FDA to approve Ranbaxy's atorvastatin application. Campbell had reassured Norton: "I will touch bases with you before anything goes OAI."

Norton knew that the agency was staring down the barrel of a no-win situation. If it didn't approve Ranbaxy's application, criticism would rain down from Congress and patient groups that were demanding an affordable Lipitor alternative. Also, Ranbaxy would likely refuse to settle the larger case if the company didn't get its atorvastatin approved. But if the FDA did approve the application, it could knowingly be greenlighting a drug made in a subpar plant.

The white-knuckled regulators wanted a clear path to say yes. But less than an hour later, Regina Brown sent in a copy of the final inspection report. Norton's biggest fear had come true. Brown recommended giving Toansa a flunking grade, or OAI, for failing to use good manufacturing practices. As she summed up in an email, "We saw too much scrap paper in use." She noted: "It was suspect to even have a shredder there for general use."

Presumably, since the plant couldn't function safely, this should have been enough to disqualify its bid to make atorvastatin. But Brown essentially split her vote. In a remarkable decision, she deemed the plant "acceptable" to make the drug ingredients and recommended that the application be approved. She concluded that for the purposes of generic Lipitor, the deficiencies could be overlooked. "The firm has got a lot of experience now" making atorvastatin active ingredients, she wrote. It was, Brown admitted, "an unusual recommendation." Under pressure, she had essentially rolled the dice: the plant was bad, but it could likely make a good drug.

Regulators could see the outlines of a disaster looming. The plant had failed a review of its good manufacturing practices. It didn't meet the FDA's standards. But there were only twenty-four hours left to either green-light Ranbaxy's generic Lipitor or leave millions of Americans in the lurch.

In an emergency call with the FDA's India office, regulators weighed the essential conundrum: The agency had found unacceptable procedures but still had faith in the active ingredients. What was the explanation for that? Several regulators wanted more time before they made their decision. Another asked a more fateful question: "Why should the American people want this [active ingredient] from this company in India?" The lawyer Norton chimed in, explaining that unless the agency approved the atorvastatin, Ranbaxy wouldn't sign a consent decree, a settlement agreement that had been in the works for years. Ranbaxy, despite having committed fraud on a vast scale, seemed to have backed the agency into a corner. She concluded, "The commissioner may have to weigh in."

November 30, 2011, dawned cold and dry in Washington, D.C. Doug Campbell awoke with a feeling of dread. He did not relish the day ahead. By the time he got to work, media calls about atorvastatin were pouring in. The Office of Compliance had yet to issue its final recommendation for Ranbaxy's application. Though the inspection's facts were the facts and presumably could not be changed, Regina Brown's recommendation had left the FDA with a "narrative" problem: how could the agency justify approving Ranbaxy's drug if the plant making the principal ingredient couldn't meet basic manufacturing standards?

The facts could be altered. The Office of Compliance prepared a draft memo for approval in which it changed investigator Regina Brown's dim assessment of Toansa's good manufacturing practices from "Official Action Indicated" to "acceptable CGMP status." It stated that after interviewing the inspection team "at length" and reviewing a draft report, "every subject matter expert consulted during the review process agreed that the facts surrounding the deficiencies do not warrant regulatory action." To get from *never* to *yes,* the agency had changed an inconvenient fact and overlooked obvious warning signs.

By day's end, the FDA informed Ranbaxy that its ANDA was approved. The scandal-ravaged company had gotten its prize: the lucrative rights to make generic Lipitor exclusively for six months, then alongside competitors after that. At 8:12 p.m., the agency issued a press release announcing the news to the public. In India, Ranbaxy's latest CEO and managing director, Arun Sawhney, addressed his staff. The U.S. launch of generic Lipitor, he said, was a "historic moment for every Indian and the entire generic industry of the world." He exulted, "It's a spirit of never say die," adding that the story of generic Lipitor had been "nothing short of a thriller" and that, in a competitive world, "Ranbaxy will turn out on top in the Atorvastatin market." Ranbaxy started shipping the drug the next morning. Based on preorders alone, it generated $100 million of revenue in the first twenty-four hours, and nearly $600 million within six months.

At the FDA, the compliance staff congratulated one another and convinced themselves that, ultimately, they had done the right thing. But a deeper unease about the decision, and the process by which it was reached, settled over the group. And it didn't take long for the regulators to realize just how badly they'd been manipulated. Ranbaxy had initially reassured the FDA that the drug would be made in the United States under the highest manufacturing quality it could muster. But on December 1, one day after the agency had given the company the green light, Ranbaxy filed a request with the FDA to use a different manufacturing site to make the finished doses of atorvastatin: its plant in Mohali, India. With its approval in place, the company planned to shift manufacturing back to a lower-cost plant with less oversight. It was the same plant where the FDA had asked for an "invitation" to inspect and gave the company six weeks advance notice of its arrival.

Even the regulators well versed in the company's "creative tactics" were stunned. "What do we know about the Mohali site?" the deputy director of the pharmaceutical science office asked a group of colleagues.

"Well, it's not in New Jersey!" a regulatory lawyer shot back.

From its seemingly unwinnable position, Ranbaxy had played

its losing hand perfectly—and won. "They worked the Agency like a Hammond organ, a fine Stradivarius," Campbell later said.

Ranbaxy continued to offer the FDA the veneer of compliance. In a March 2012 email, the head of Ranbaxy global quality wrote to Campbell: "In the spirit of transparency & cooperation with the FDA, I confirm that the shredder at Toansa is now confirmed as relocated out of the manufacturing area and that it now resides in the Toansa QA block with continuing controls for its use." However, in the coming months, the concerns of Takahashi and her colleagues—that Ranbaxy was *still* concealing fraud and Americans would be getting an inferior drug—would prove dismayingly well founded.

PART VI

THE WATERSHED

CHAPTER 23

THE LIGHT SWITCH

DECEMBER 7, 2012

Toansa, Punjab India

As the FDA looked for investigators willing to relocate and join its sparsely staffed offices in India, the agency did not have many takers. But Peter Baker, the young consumer safety officer, raised his hand. He had a love of travel and adventure and took yearly trips with his motorcycle group to far-flung locations, from Mongolia to the Philippines. He also had a more pragmatic reason to volunteer. By reputation, India was the world leader in aseptic manufacturing, the exacting science of producing sterile drugs. Baker figured that after a stint in India reviewing best practices, he'd return to the United States with expert knowledge that would advance his career.

Just three months after arriving in New Delhi, he got a critical assignment: to inspect Ranbaxy's Toansa plant in a remote area of northern Punjab, where the company made the active ingredient for its generic Lipitor, atorvastatin.

It was the same troubled plant that had squeaked through a last-minute inspection, largely owing to the intervention of FDA bureaucrats. But only ten months after Ranbaxy began its lucrative launch of atorvastatin, the company made a staggering admission to the FDA. It had discovered that some of the millions of pills already dispensed to patients across the United States were suffused with tiny shards of blue glass. The manufacturing failure confirmed the worst fears of the FDA's compliance staff. Ranbaxy should never have been approved to make the drug. Now, as the company recalled millions of implicated pills, the FDA dispatched Baker to the Toansa plant, where the active ingredient had been made, to figure out what went wrong.

For the high-stakes inspection, Baker was paired with another veteran FDA investigator, Atul Agrawal, who was stationed in Mumbai. The two men were opposites: Baker was young, handsome, and athletic, with close-cut, brownish-blond hair; Agrawal was short, balding, and walked with a limp. But the two men shared one trait: they disliked being fooled. This time, because the inspection was so vital, the FDA's India office had given Ranbaxy only a few hours of advance notice—less time to conceal evidence. But Baker and Agrawal wound up in the wrong town and had to backtrack, which took hours and bought Ranbaxy additional time.

The investigators finally arrived at the vast plant complex, with dozens of buildings stretched over acres. A corporate executive from Ranbaxy's headquarters had beaten them there. The executive had taken the first flight from New Delhi that morning, which meant the company must have gotten word of the inspection the night before. Someone from the FDA's India office had almost certainly leaked it.

Plant officials proposed that the investigators first sit for a presentation. Agrawal insisted that they be taken immediately to the building where the glass contamination had occurred. The officials took them to the MP-11 plant. There they explained that the contamination had come from a blue-glass protection ring that encircled the top hatch of a reactor, a big rounded vessel into which they shoveled ingredients, and mixed them to cause a chemical reaction. They claimed that they could not show the investigators the protection ring because they had since removed it; instead, they showed the men a photograph of it with a small chip in the glass. Baker studied the image of the small round imperfection, thinking that it looked too perfect. It was almost as though someone had carefully taken a hammer to it. How could that have caused enough broken glass to contaminate millions of pills?

As Agrawal and Baker inspected more of the reactors, they peered down through the hatches on top at the glass-lined walls. They even had plant employees hold their legs, as they dangled upside-down inside the reactors. As a knot of Ranbaxy officials waited silently, Baker cast a flashlight beam across the cylindrical walls of an enormous steel silo. He caught sight of a spiderweb-like crack along one of the glass linings and followed it with his flashlight. He doubted that the crack was the cause of the catastrophe. But he was certain that the hovering officials were doing everything in their power to keep him in the dark.

Inside another reactor he saw heaps of powder residue, as though it had never been cleaned. He asked to see the inspection forms. All the results were documented simply as "OK," with no notes next to them. The cleaning checklist had been signed by two different technicians. As Baker questioned plant officials, they first spoke among themselves in Hindi, as though to get their facts straight. Amid pointed questioning, they admitted that those who'd signed the logs had not actually done the cleaning.

As the inspection continued, a plant official mentioned that the company planned to grind, refilter, and resell the glass-infused ingredients. The investigators explained that the material was too compromised, so plant officials pledged to destroy it. In the United States, it was standard protocol for investigators to witness the destruction of products deemed too dangerous to use. Baker himself had once visited a dump and watched as a company poured bleach over mountains of fish that had gone bad, before running over the piles with a bulldozer, as the fish exploded from the built-up toxic gases. Baker and Agrawal didn't trust Ranbaxy to destroy the compromised ingredients. They later asked the director of the FDA's India office for permission to witness the destruction. He told them it wasn't necessary.

The investigators spent eight days at the plant. They found significant violations, but left with the sinking feeling that they'd been played—that the perfectly round hole in the glass had been faked; that they'd been diverted from a major glass failure in a reactor they hadn't been shown. They feared that Ranbaxy would almost certainly resell the contaminated ingredients instead of destroying them, as promised.

The experience led Baker to a resolution: he vowed to get smarter and more aggressive, to pierce the coordinated defenses and denials of plant officials. His determination, and what he would soon uncover, would make him an infamous figure among Indian drug makers and raise questions around the world about their practices.

Baker grew up about as far from a world of global deceit as could be, on a grass seed farm in Lebanon, Oregon, where he was raised by Mennonite parents. His family's church preached service to others, and the community's missionary work took his family to remote destinations—indigenous villages down the Amazon River in Colombia and orphanages in the Dominican Republic. Baker spent summers on his neighbor's farm, driving tractors and combines beneath an unbroken blue sky.

After attending a Christian college in San Diego, where he discovered his love of chemistry, he met a Swedish woman during a European backpacking trip. Together, they settled back in San Diego, where he got a job at a contract research laboratory that tested drugs for manufacturers. He spent hours each day running tests on high-performance liquid chromatography machines. The company seemed more interested in billing hours to its clients than in scrupulous adherence to good manufacturing practices, an attitude that trickled down to the lab. There, minimally supervised chemists used shortcuts to nudge data up or down in order to finish their work quickly, particularly on Friday afternoons when the company brought in a keg and the employees played basketball.

After a year, Baker and his girlfriend moved to China, where he taught English and Western culture to engineering students. On their return to San Diego, they married. Baker wound up at Abbott Laboratories, working as an engineer at the company's quality assurance laboratory. His real learning began there. His boss assigned him the task of taking a lab with no formal controls and bringing it up to code. This meant creating an entire infrastructure of transparent and verifiable standards. Baker learned GMP from the ground up.

In 2008, while completing a master's degree in analytical chemistry at San Francisco State University, he joined the FDA. Within a year, he was inspecting plants in the United States. The facilities were largely compliant, and he'd spend weeks at each one, urging companies to perfect the manufacturing process. Only two of his fifty-five U.S. inspections resulted in a warning letter. Afterward, one of the companies ended up moving manufacturing plants to Taiwan, in part to evade so much scrutiny. By mid-2011, with the FDA facing a backlog overseas, Baker volunteered to do some inspections in China, planning to make use of the primitive Mandarin he'd studied in college.

Months later, Baker was asleep in a hotel in Pucheng, a remote town in Fujian province, where he had traveled to inspect a veterinary medicine plant. He awoke at 1:00 a.m. to pounding on his hotel door. Through the peephole, he saw more than a dozen men in street clothes, who turned out to be part of a government security team. Before he could unlatch the door, they busted it open, swarmed in, briefly detained him, and searched his room. The hotel later claimed to have confused him with someone else—though he was almost certainly the only foreigner at the hotel.

The FDA had given him only cursory training for the world of foreign inspections he had entered, which included the admonishment not to accept any gift worth more than $20. Topics like government surveillance, offers of gifts that included gold coins or call girls who materialized in hotel rooms late at night, and how to stay safe in remote towns controlled by the companies whose plants he would be inspecting never came up. As Baker learned to navigate the dicey situations, he found that the long hours, forlorn towns, and unpredictable nature of the inspections energized rather than daunted him. Within months, he agreed to relocate full-time to India.

In late 2008, the FDA opened several foreign outposts, including in India. The move was a response to the adulterated heparin crisis and an acknowledgment of what had long been obvious: the FDA couldn't hope to regulate plants in a country where it had no permanent presence. But the FDA's India office had gotten off to a rickety start, with a skeletal crew of eight employees—four in Mumbai and four in New Delhi—tasked with inspecting hundreds of plants across the entire country. Investigator Mike Gavini, who'd been so lenient with Indian companies on his visits there, was among the first to volunteer. He joined the Mumbai group in June 2009.

In Mumbai, the investigators worked out of the building that housed the U.S. Consulate General, where mice chewed through the power cables on their computers. Gavini was housed miles away and commuted on a halting local train. He traveled to manufacturing plants that had never seen an FDA investigator before. He found untrained and uneducated workers, some of them illiterate, traipsing through factories in street shoes. But in his effort not to "indiscriminately [shame]" the country people, as he would later put it, he nudged, cajoled, and guided India's reluctant factory owners, often giving them green lights to proceed based on their promises to do better. As the companies thrived under his lenient inspections, the number of Indian plants approved to sell drug products to the United States soared. By the time Baker arrived in September 2012, India had become a pharmaceutical-exporting juggernaut.

As Baker started his inspections, he saw what had given India its vaunted reputation: many of the manufacturing plants were new and appeared to be in immaculate condition. The equipment was pristine. But he noticed something else. From the minute the inspections began to the minute they ended, the managers led the FDA investigators around the plants like dogs on a leash. First, a welcome reception. Then, an opening slide show. Guided tours of the facilities. The elaborate show of hospitality amounted to running down the clock.

On his fifth inspection in India, at RPG Life Sciences in Mumbai, Baker took a detour and visited the company's quality control laboratory, an out-of-the way spot where the plant aggregated its testing data. There, he asked to see all the records related to failing drug products. A nervous manager admitted that the records had been destroyed. This troubled Baker. But as he scrolled through the company's computer systems, he found some documentation about the products that had not been officially logged. When everything was supposed to be transparently documented, why did the company have an unofficial set of records? He gave the plant a flunking grade of Official Action Indicated. In safe, well-run plants like those he had inspected in the United States, there were not supposed to be any hidden or destroyed records, back doors, or *unofficial* anything.

In January 2013, his fourth month in India, Baker arrived at a manufacturing plant in Kalyani that made active ingredients for injectable cancer drugs. It was owned by Fresenius Kabi, a German brand-name drug company. The last FDA inspection there four years earlier had only one finding: the company had not correctly documented the chain of custody for the drug samples it was testing. Since his experience at the Toansa plant a month earlier, Baker vowed that he would no longer allow himself to be steered through a manufacturing site. Instead, he would choose where to go and what to inspect. As at RPG, he headed directly to the quality control laboratory—the ultimate back room.

The ostensible purpose of a quality control lab was to audit the test results coming in from the factory floor to ensure that they were unaltered and preserved and that any irregular data was investigated. It was also a last line of defense for a plant interested in detecting, and eliminating, failing drug products.

But the labs could also play a nefarious role, as hubs for manipulating or discarding data that might expose those failing drugs. At the Kalyani lab, instead of asking for data, Baker simply sat down at a computer and began to scrutinize the results from tests done on high-performance liquid chromatography (HPLC) machines. The bulky machines separated and measured the impurities in a drug sample and displayed them as a series of peaks in a record called a chromatogram.

Toggling back and forth between computer files, Baker discovered something strange. He found the official test, with the result stored in the correct data folder. But in a file called "MISC.," he saw what looked like earlier tests of the same drug sample, some a day apart, some a month apart. Some of the earlier tests were in ancillary folders, like "MISC.," or within the proper folder but labeled "DEMO." This unofficial pretesting and the subsequent retesting were not explained. Nor were there any procedures that permitted it. He figured out that some of the HPLC machines were not officially registered and had not been connected to the plant's main server. The plant was conducting *offline* tests.

Baker had uncovered the outlines of a secondary manufacturing operation hidden beneath the surface of the first. The technicians were using the initial hidden tests to get preliminary results, which they used as a guide to tinker with the test settings—adjusting the parameters, the amount of solvent. Then they retested in the plant's official system to get the desired results. As Baker showed the plant managers point-blank evidence of this malfeasance, they denied his findings. *All test results were stored on a central server,* one manager insisted, despite all the data Baker put in front of him. As Baker's inspection progressed, he moved deeper into the plant. He opened an unmarked binder and found scratch paper. A manager snatched it from him and tried to hide it in his pocket. Baker demanded it back. The paper contained notes about manufacturing errors and other problems that needed to be fixed. The hidden notes clearly sidestepped the plant's obligation to document all quality issues in official records. Baker also noticed foreign particles in a closed vessel used to filter solvents, which the manager denied, claiming it was due to the reflection of light on the glass. When he returned later, the particles had been removed. The manager denied this, asserting that the technicians would not have undertaken "cover-up activities."

The next day Baker unearthed a report that confirmed his suspicions: it stated that operators had removed the "particulate matter" during his one-hour absence from the processing room. The fragments had come from a deteriorating internal gasket. A statement from the manager was attached: "During the USFDA inspection on 14th January 2013, I panicked and advised my junior to open the equipment and clean it."

As Baker asked questions, plant managers in India were relaying these to quality managers in Germany, sparking alarm. On the last day of the inspection, Baker faced a stunned and exhausted executive vice president, who'd flown all night from Germany with an entire team to be there when Baker rendered his final verdict. It was evident that the Germans hadn't known what was happening in their own plant. Baker's questions had caused them to grill their employees there.

The Germans were forthcoming with Baker and shared what they'd learned in the prior seventy-two hours. The plant's managing director had overseen a scheme to pretest drug ingredients, preview the results, make secret adjustments to the test settings, and then retest the drug samples until they passed. The director had ordered the HPLC machines on which they'd engineered the pretesting shipped offsite before Baker's arrival. The plant had reblended failing drug ingredients that had high impurities with higher-quality ingredients until they passed testing. Baker was particularly stunned to learn that the most senior person at the plant had directed these activities.

In the months that followed, Fresenius Kabi launched a probe into its own plant. The findings were grim. The company later admitted to the FDA that it had found unofficial testing, unauthorized blending, false manufacturing records, and the deletion of thousands of records, and that it could not trust any data from the plant, or the compliance of any batches, prior to Baker's inspection. It was a debacle. To its credit, Fresenius Kabi halted production at the plant, fired the entire management, and recalled all the medicine made with reblended ingredients. As shocking as the situation was, Fresenius Kabi seemed to be a victim of its own poorly managed outsourcing. Had Baker not uncovered the malfeasance, the company might never have discovered what was happening in its own plant.

But Fresenius Kabi wasn't alone. Brand-name and generic drug companies the world over were snapping up plants in India to make their own active ingredients and drugs at a fraction of the cost. They could rapidly increase their profits, all from saving on the cost of labor and supplies. But Baker suspected that company owners had little insight into what really went on in their own money-saving plants.

Before Baker arrived in India, the FDA's investigators had long suspected that something was not right. Typically, a compliant manufacturing plant will reject a certain percentage of drug batches for many different reasons. But in India, investigators rarely saw a rejected batch. Somehow, almost all of them passed. Plants were also missing documentation. "I've always known [Indian companies] had a hard time keeping paperwork," one FDA investigator explained. "Their habit and practice is not to keep documentation. They have a *chalta-hai* attitude," he said, using a popular phrase in India usually accompanied by a shrug, which means a willingness to accept a less-than-ideal outcome. "In my mind, I knew something had to be *not right*."

But with advance notice and only a week's time at a sprawling plant complex, it was hard to know precisely what was *not right*. Baker changed that, in part by looking in places where no one else had. As he scoured data, he followed clues: audit trails within the supposedly impermeable software systems had been disabled; tests had been repeated, and then deleted from the official network servers. Through a painstaking forensic effort, he matched up the metadata from deleted tests with the later official tests of the same drug samples. Among hundreds of thousands of tests at any given plant, he followed his instincts: he zeroed in on drug tests he suspected the facility might be faking, and the equipment they might be using to fake them.

Baker became good at finding the HPLC machines that were used for the secret tests and not linked up to a plant's central software system. In some cases, the company hid them in concealed labs, or even amid other networked machines. But Baker began finding them—largely because he began looking for them. Previously, investigators walked into a laboratory with dozens of HPLC machines and allowed company officials to guide them to a machine, to show them how it worked and what data was being generated. With only five days at a facility and so much to inspect, finding the one or two non-networked machines was nearly impossible. During one inspection, he simply asked lab workers, "Are all these units networked?" One of the lab technicians piped up, "That one is not." A regional manager tried to cut him off: "Yes, it is." "No, it is not," said the employee. Back and forth they went. But Baker had uncovered a stand-alone machine simply by asking for it.

As he worked, other FDA investigators learned from his techniques. The difference was dramatic. "It was like 'holy shit,'" one FDA employee recalled. "Like you walk into a dark room and suddenly someone just turns on the light. It was shocking." But it was a fateful inspection two months later—when he discovered the torn batch records in a garbage bag that an employee tried to slip from the plant—that darkened Baker's view considerably and focused his thinking about the dangers of faked data.

On March 18, 2013, Baker arrived at the main plant of Wockhardt Ltd. in the Waluj area of Aurangabad, two hundred miles east of Mumbai. At the inspection's opening meeting, the company's vice president of manufacturing insisted repeatedly to Baker that the plant had only one manufacturing line that it used to make products for the U.S. market. But on the inspection's second day, Baker recovered a garbage bag that an employee had hurled beneath a stairwell. Inside were torn batch records for the company's insulin products, which showed that numerous vials contained black particles and had failed visual inspection. As Baker soon discovered, the test results had not been logged into the company's official system. The manufacturing line on which the company had made the drugs didn't exist in official records but operated secretly within the plant. In official records, the insulin products passed inspection at a much higher rate than in the deleted batch records. The drugs had been released into Indian and Middle Eastern markets.

Though Baker's mandate only extended to drugs for the U.S. market, he and his colleague, a microbiologist, continued to follow the clues in the torn records. The next day they found the secret formulation area and unearthed an unofficial "investigation report" into the black particles. The report was not dated or signed but was written to the head of the plant. It made clear that the black particles were "metallic" and had come from damaged heating coils inside a machine used to sterilize vials and cartridges before they were filled with medicine. The blistering heat inside the machine had broken down the coils. They had been repaired, but hadn't been replaced due to cost. The medicine with the particles had been released to patients.

Baker was staggered. It would have been one thing for the company not to know about the particles, or to accidentally order the drugs released. That would have been shoddy or negligent. But the plant director had *ordered* the insulin released, with full knowledge that the drugs were potentially lethal to patients. The metallic particles could easily cause an immune-compromised patient to suffer anaphylactic shock and die. Worse yet, Baker discovered that an injectable drug for the U.S. market, adenosine, used to treat an irregular heartbeat, was manufactured on the same secret line, and with the same perilous equipment. Though the unofficial report noted nothing about that drug, Baker was sure the results were no better.

If he'd found such deceit in the United States, it would have triggered a law enforcement raid and, almost certainly, prosecution. Someone would be going to jail. But in India, Baker had no such authority. The only remedy was a regulatory one. He had enough evidence to restrict all the drugs made at Wockhardt's Waluj plant from entering the U.S. market.

The inspection lasted five days. Baker's findings—seven major ones—were devastating. They painted a picture of elaborate fraud, extreme hazard, and filth. Plant officials had repeatedly refused to cooperate. One official dumped vials into a drainage sink after

Baker inquired about their contents. In the bathroom, about twenty feet from the gowning area for the sterile formulation lab, the urinals lacked drainage piping. As Baker noted in his report, "Urine was found to fall directly onto the floor, where it was collected in an open drain." The room had an "overwhelming sewer stench," he wrote. By mid-inspection, both Baker and his colleague had fallen ill, after plant officials gave them unsealed water bottles. They suspected that plant officials had tried to sicken them, to shorten their inspection time.

Word of Baker's findings, let alone his published inspection report, could wipe millions off the company's stock. At the inspection closeout meeting, a part of agency protocol, Baker presented his findings to the company's vice president of manufacturing. The executive demanded, with a menacing glare, that Baker remove the first observation, about the discrepancy between unofficial and official insulin batch records, from his findings. It was a threat.

"I can't do that, sorry," Baker responded, getting increasingly nervous about safe passage from the plant as the man stared him down. "Let's get out of here," he told his colleague. Neither wanted to get into a company car. The firm was in the middle of nowhere, and a deadly traffic accident would surprise no one, given the number of trucks in the area and the craziness of the roads.

His colleague wanted to mail back the evidence she'd collected rather than travel with it. The company volunteered to call DHL. The next thing they knew, a man dressed in what appeared to be a fake DHL uniform wandered in. A plant employee clearly intended to take her exhibits and disrupt their inspection efforts by any means possible. When Baker demanded to see the man's DHL van, he left and didn't return. The FDA investigators were spooked and headed out to the road with all their gear to hail a *tuk-tuk,* a three-wheeled makeshift cab.

Baker noted in his inspection report: "Due to the threatening behavior and personal safety concerns encountered during this inspection, it is suggested that an inspectional team perform the follow-up inspection with a clear emergency plan in place prior to arrival."

For Baker, the Wockhardt inspection of March 2013 was a watershed. Amid the hidden laboratories, secretly repeated tests, and altered results, Baker realized that he'd unearthed far more than individual acts of fraud. He had uncovered the larger game being played by India's generic drug industry: making third-world drugs and selling them at first-world prices. The companies exploited their technological savvy, captive employees and a corporate culture inured to fraud. They were also aided by the FDA's archaic inspection methods and the West's dependence on cheap drugs.

Americans had embraced the promise of cheap generics: identical drugs for a fraction of the cost. They assumed that faraway companies were making them with no compromise to quality, in large part because the FDA claimed it was so. But the Indian companies Baker had caught were aiming to make the lowest-quality drugs they could get away with, to make the biggest profits. Without a doubt, the companies could have made perfect medicine. There was no knowledge gap. Their equipment was first-rate. The difference was simply cost. Exacting controls cost about 25 percent more, according to some industry estimates.

To avoid significant up-front costs without a guarantee that their drugs would be approved, the companies pretested everything in secret labs tucked within their manufacturing plants. There, they screened in advance for failing results in order to make secret adjustments to tests. They worked to make their formulations look perfect on paper, regardless of their actual quality. They jury-rigged the results by fiddling with the tests, retesting already proven batches, or even testing the brand-name products instead. Only then would they move the data to the computer system the FDA was going to examine. Until Baker showed up in India, only Ranbaxy had gotten caught—because of the whistleblower Dinesh Thakur.

Baker uncovered all this by breaking the mold of traditional inspections. Instead of roaming the factory floor and checking equipment maintenance records, he focused instead on forensic analysis of the companies' computer systems, an effort in which he had no formal training. He'd taught himself. It was a risky approach. If he'd spent an entire week combing through computer files and emerged empty-handed, he'd have little time left for the more conventional inspections he was required to do. But Baker had figured out what to look for: in files like "MISC," "CHRON," and "DEFAULT" he found tens of thousands of secret tests not entered into the quality systems.

This bogus manufacturing system required the knowledge and participation of hundreds of people, which would be nearly impossible in the United States, where plants faced the constant threat of unannounced inspections and employees had access to more responsive regulators and a chance of becoming legally protected whistleblowers. But in India, if employees challenged the practices, they could face banishment from the industry, if not worse. In order to get a new job, they needed a letter of recommendation from their previous employer. Survival dictated that they leave on good terms. Becoming a whistleblower could be lethal. It was a perfect system for fraudulent manufacturers.

Many of the employees at generic pharmaceutical plants were contract laborers, and at remote plants, some were even poor farmers. They were poorly trained, if at all, and often illiterate, despite the requirement that they document and sign off on activity logs during their shifts. By regulation, they had to be regularly tested on the rules they had to follow. One plant kept all the answers to a test up on the wall, so that the workers just had to look and copy.

Most of the employees lived on one meal a day. In their daily lives, many lacked access to a toilet or running water. To Baker, it seemed ludicrous to expect them to walk into a sterile manufacturing plant and suddenly follow all the rules.

Over time Baker refined his inspectional methods, getting better and sharper and quicker at finding hidden fraud, whether concealed in ancillary files, deleted from computer systems, tossed in the trash—or even shipped out of the factory before his arrival. Fear spurred him on. If there was fraud and he failed to find it, no one else would. The finished doses manufactured at Wockhardt and other Indian plants were shipped straight from those factories to American wholesalers and drugstores. People had the right to know what they were taking and to choose what they didn't want to take. But American patients had no idea of the subterfuge that went into the manufacturing of their low-cost medicine, and the FDA had no plans to tell them. He felt that he was the last person standing between the American patient and companies determined to cheat when given the chance.

CHAPTER 24

WE ARE THE CHAMPIONS

AUGUST 2011

Silver Spring, Maryland

As the government's case against Ranbaxy dragged into its sixth year, the FDA's criminal investigator, Debbie Robertson, decided to retire at age fifty-five. She had never intended to leave the Ranbaxy case, or turn her back on Dinesh Thakur, until a clear resolution was in sight. But life was too short to battle prosecutors anymore.

Dinesh Thakur felt the loss keenly. He knew that without her, the investigation would probably still be where it had started—with him hiding behind a pseudonym, beseeching the agency to put a stop to Ranbaxy's crimes. From Gurgaon, Thakur sent Robertson a heartfelt note. "It's hard for me to put in words the comfort that your steadying hand gave to me during the early years of the investigation," he wrote. "Whether you realize it or not, our emails and calls helped me through one of the most difficult periods of my life." He thanked her for the vital role she'd played in making medicine "a little better and safer for all of us." He also noted that in India, where public health systems lacked regulation and law enforcement was "all but corrupt," people like her were a rarity: "There are no Special Agents here as conscientious as you."

The long years of the case had also taken their toll on Thakur's lawyer, Andrew Beato. Night after night at his home in Bethesda, Maryland, he lay awake, worrying about Thakur's safety and the solvency of his firm. The case was costing the firm millions of dollars, with two dozen employees toiling on it full-time. If it didn't settle, Stein Mitchell might not survive the financial hit. Meanwhile, with the case under seal, Beato was the only one Thakur could talk to about it. The nine-and-a-half-hour time difference had led to regular middle-of-the-night talks between the two men. When Beato had trouble returning to sleep, he took late-night walks with his dog, Ziggy, who remained a faithful companion.

Year after year, the two men had waited for a global resolution to the case, one that would bundle Ranbaxy's criminal, civil, and regulatory liabilities into a single settlement. There was one clear indicator of an actual deal: a commitment from Ranbaxy to put down money. By December 2011, Beato was waiting daily for word from the Justice Department. Would Ranbaxy make that commitment? If so, what would the magic number be? What percentage would his client get? How much would his firm recoup? After all the years of work, could Stein Mitchell do more than break even?

On the expected day, Beato waited for hours to hear from the Justice Department. Just as he was leaving the office, already late for a scheduled dinner with his wife, his phone finally rang. Ranbaxy, through Daiichi Sankyo, had agreed to set aside $500 million, an amount that would resolve all outstanding liabilities. This meant the Japanese company was willing to avoid a trial and take legal responsibility for Ranbaxy's past acts. Some portion of the settlement, still to be negotiated, would go to Thakur and to Beato's firm. On December 21, just three weeks after its blockbuster launch of atorvastatin, Ranbaxy issued a press release, confirming it would pay $500 million to settle the case.

"Merry Christmas," Beato texted Thakur, who was with his family in a cabin in northern India, in the foothills of the Himalayas. Beato then drove home and sideswiped his entire car, from bumper to bumper, against a concrete pillar in his garage, overwhelmed by the cumulative stress. Though the agreement was made in principle, he still wasn't sure whether the company would sign it.

Weeks later, when the company did sign, officials at the FDA rejoiced. "What a long road!!!" one high-ranking FDA official emailed her colleagues. "I hope they signed in pen so they can't erase it!" Carmelo Rosa, who led the FDA's Division of International Drug Quality, sent kind words to Karen Takahashi and Doug Campbell, acknowledging the "countless hours" they'd spent on the case. "You both consistently had only one goal in mind, to protect consumers."

Even after news of the expected settlement, government and Ranbaxy lawyers continued to haggle over details of the deal, and company executives continued to exasperate regulators. Almost immediately, Ranbaxy executives angled for a meeting with FDA officials. The reason, as an agency lawyer told her colleagues, was that "Ranbaxy management lacks trust that the agency will give the company a fair shake." This perspective stunned some of the FDA's higher-ups. Rosa wrote back to the FDA lawyer, "I am just puzzled by the nerve" of Ranbaxy officials "to even suggest the below comments."

In May 2012, after clashes with his partner, Thakur stepped down from his position as CEO of his business, Sciformix. Meanwhile, his savings had dwindled and Sonal had given up hope for a resolution to the case. He expected that as soon as his role in the Ranbaxy case became public, he would become a pariah in the pharmaceutical industry.

With a final deal not yet inked, Justice Department lawyers continued to ask Beato and Thakur for extensions, which only they had power to grant under the rules of a whistleblower case. By January 3, 2013, Beato responded angrily to a government lawyer who requested another two-month extension. "Why do you need sixty days? You are killing me with my clients and my partners. Dinesh is in a bad financial situation. His family situation is not good either. He is running out of patience. I too have challenges at my firm. I am on a very short leash. If you give Ranbaxy sixty days, they will use all that time." Though Thakur was willing to agree to the extension, having no other choice if he wanted the case to succeed, Beato urged the government not to tell Ranbaxy that it had sixty more days, since it would only run out the clock again. An assistant U.S. attorney tried to reassure him: "We are at the inches line here."

Meanwhile, Beato was battling Ranbaxy's lawyers, who had known for some time that Thakur was the whistleblower. They not only wanted Thakur to agree not to disparage the company publicly; they wanted all his documents back, especially the explosive Self-Assessment Report, which they hoped to bury permanently. Ranbaxy's lawyers threatened to withhold the funds it would owe to Beato's firm to cover expenses, as part of the anticipated resolution, unless Thakur returned the documents, which he did. But finally, a settlement hearing appeared in the calendar of Maryland's U.S. District Court for Monday, May 13, 2013. A real end was in sight.

On the day before the hearing, Thakur woke up early at the Washington Guest Suites Hotel, made his coffee, watched the news, and then FaceTimed with his children, Mohavi and Ishan. This had become his Sunday morning ritual, and he was a man of habit, though there was nothing ordinary about the day.

He headed out into the morning, which was still cool. Steam rose from the manhole covers. Though the blossoms had long fallen from the Japanese cherry trees along the Potomac River, everything else was in bloom and vibrant green as he walked past the rectilinear State Department and toward the Washington Mall. The reflecting pool before the Lincoln Monument barely wavered in the early-morning calm. When he was growing up, the monuments and temples of Hyderabad—like the grand Falaknuma Palace—had filled his imagination. He could not have fathomed that the biggest fight of his life would lead him to Washington, D.C., with the might of the U.S. government as a backdrop.

Thakur had turned to the United States to intervene on behalf of patients around the world, a calculation that he'd made at a dark moment, knowing nothing about the process he might end up triggering. He'd had no idea that so many branches and offices of government—from inspectors general to Medicaid fraud control units to chief counsel and litigation branches—would be needed to make Ranbaxy pay, in some measure, for what it had done. For eight long years, he'd witnessed the creaking and imperfect legal machinery of the U.S. government up close.

At the Lincoln Memorial, he jogged up the steps and stood for some time beneath the towering figure of the seated president. Lincoln, to some extent, had helped to craft Thakur's journey. The False Claims Act, which his lawyers had used to sue Ranbaxy, was also known as "Lincoln's Law." Lincoln had introduced it during the Civil War, to allow whistleblowers to sue on behalf of the government to prevent profiteers from selling defective goods to the Union army. The Gettysburg Address was carved onto one wall of the monument, and as Thakur stood there, he read it several times. It was perhaps no surprise that the nation dedicated to the idea that all men are created equal would be willing to fight for the uniform quality of its medicine. The laws and customs of the United States had offered him an avenue to pursue justice, and by one measurement—with a legal resolution scheduled for tomorrow—he'd almost succeeded.

He took a seat on a bench crossing the Arlington Memorial Bridge, which connected Washington, D.C., to Virginia. He called one of his project managers at Ranbaxy who'd helped him investigate the fraud years earlier. He told him to look out for some news that would break the next day. Then he called Raj Kumar, his former boss who'd set all the events into motion. They had remained in contact through the years. Neither had fared well professionally, though Kumar—who had seemed like CEO material, with his thoughtful demeanor and professional stature—had suffered particularly.

After Ranbaxy, Kumar had headed to Dr. Reddy's, one of India's largest generic drug companies, as president of research and development. He hadn't stayed long. After two years with the company, he returned to Cambridge, where he spoke little about his experience. He was a physician and scientist lost in a world he could never have imagined: one that valued profits above patients. Thakur told him as well to look out for news that was going to break the next day. He wanted to say more but couldn't.

After making his calls, he returned to the Guest Suites Hotel to do something he'd become far too good at: wait.

On Monday morning, a throng of people stood outside Courtroom 5A at the U.S. Federal Courthouse in Baltimore, Maryland. With the courtroom doors still locked, government prosecutors mingled uneasily in the hallway with Ranbaxy executives and their attorneys. Andrew Beato and his colleagues helped steer Thakur past the crowd to stake out an empty spot. Heads turned as they walked past. Ranbaxy executives already knew who had caused them so much misery. But seeing him and sharing a hallway was a difficult exercise. Thakur, who had never been inside a courthouse before, waited nervously.

Once the doors opened and the crowd filed into the courtroom, a rare smile crept onto Beato's face. Inside, everyone took their seats. Though flanked by lawyers, Thakur continued to look toward the door, hoping to spot Robertson. And at the last minute, there she was. He smiled and rose to his feet, but Beato gently restrained him.

Judge J. Frederick Motz called the court to order. It was a dark morning for Ranbaxy, and a rare one for any corporation. Though companies paid fines all the time, they were seldom criminally convicted.

The company had agreed to plead guilty to seven federal criminal counts of selling adulterated drugs with intent to defraud, failing to report that its drugs didn't meet specifications, and making intentionally false statements to the government. To make its case, the government had zeroed in on three drugs that offered the most egregious examples of Ranbaxy's fraud: the anti-acne drug Sotret, the epilepsy drug gabapentin, and the antibiotic ciprofloxacin. Ranbaxy had agreed to pay $500 million in fines, forfeitures, and penalties. Though it was a far smaller amount than the $3 billion the government started with, it was the most ever levied against a generic drug company.

Judge Motz accidentally called the firm Ranksberry and misread the settlement amount as $500,000. Assistant U.S. Attorney Stuart Berman corrected the judge on the penalty amount, stating that it was $500 million.

Judge Motz cracked, "All the numbers look similar to me."

The Ranbaxy executive who'd agreed to plead guilty on behalf of the company stood and said that he'd prefer the previous amount, prompting laughter in the courtroom. But when he said, "On behalf of the corporation, I wish to plead guilty," a silence settled over the courtroom. Motz approved the settlement. No individual executives were held criminally responsible.

And just like that, it was over. Outside of the courtroom, Thakur gave Robertson a hug. The events of the morning seemed to stand for "more than just a court hearing with a company representative," Beato later recalled. He felt proud of his work on the case, but also hopeful that his firm was helping to clean up an industry on which millions of people depended. *This might be a turning point,* he thought.

As Beato and Thakur drove back to D.C. in Beato's Honda Pilot, the Justice Department's press release hit the wires. Beato's colleague read aloud from the backseat, "In the largest drug safety settlement to date with a generic drug manufacturer . . ."

"Great!" Beato called out.

As the news rocketed from D.C. to New Delhi, Beato turned on the CD player and blasted "We Are the Champions" by Queen. The men sang along, their windows down. By the time they arrived at Beato's office, inquiries from the news media were piling up.

It was late that night by the time Thakur called Sonal in India, and the two were able to compare notes. She'd kept the children home from school and stationed a security guard outside the family's front door. Ranbaxy's U.S. guilty plea was the lead story in the Indian media, alongside news that Thakur had been awarded $48 million, a percentage of the settlement, for his role in the case. His picture was on television. Sonal's phone rang continuously with astonished and congratulatory calls from friends and family. Scared and stressed, she was exhausted from trying to explain his role in all of it. The following day, though still afraid, she let the children return to school.

Two days after Ranbaxy's guilty plea, a 10,000-word article about the case and Thakur's role in it appeared on *Fortune* magazine's U.S. website. The article made public the document that Ranbaxy, and Malvinder Singh, had spent years trying to suppress, both from the public and from Daiichi Sankyo: the Self-Assessment Report that Raj Kumar had shown to the science committee of the board of directors. One question posed by the article was whether Daiichi Sankyo knew the extent and depth of the fraud at Ranbaxy. Interviewed for the article in 2010, Daiichi Sankyo's head of global strategy, Tsutomu Une, had told *Fortune,* "I never thought that we were fooled."

But a week after the article came out, Daiichi Sankyo issued a press release that took clear aim at Malvinder Singh. The company essentially admitted that it *had* been fooled and stated that "certain former shareholders of Ranbaxy concealed and misrepresented critical information concerning the U.S. DOJ and FDA investigations." The company announced that it was pursuing "available legal remedies." In fact, the company had already begun proceedings against Malvinder in the International Court of Arbitration in Singapore.

Thakur spent the following months in a blur of activity. He celebrated at a festive dinner with Robertson, Beato, and others who'd worked on the case. He received a number of whistleblower awards, including the Joe A. Callaway Award for Civic Courage, for which he attended a banquet in Washington, D.C. The award noted his "commitment to drug safety globally, at considerable professional and personal peril to challenge fraudulent pharmaceutical industry practices, beginning with his former employer."

He returned, partially, to his family. The Thakurs bought a condominium in Tampa with a stunning view of the ocean. Sonal and the children spent a month there over the summer, and the family went to Disney World. "We finally have a life," Sonal reflected, marveling at the fact that her husband finally seemed relaxed. "Now we can enjoy being together."

But nothing was that simple. What, exactly, had the Ranbaxy case accomplished? The day Ranbaxy pleaded guilty, Rod Rosenstein, then–U.S. attorney for the District of Maryland, all but acknowledged to a journalist the limits of the case: overseas pharmaceutical manufacturers "self-certify, to a large extent," that they are following regulations, he said. "If a manufacturer decides to violate [the regulations], it's very difficult to prove what's going on." Given this problem, he acknowledged, "prosecuting individuals would be a more effective deterrent." But that hadn't happened.

At the FDA, as far as Doug Campbell could see, the FDA had lost by winning. On the one hand, securing an international consent decree was unprecedented. "Nothing will ever be as big as this," Campbell acknowledged. But the size of the decree and its strict provisions meant that there were now "three people at the FDA who do nothing but" ensure that Ranbaxy complied. Was that the best use of the FDA's resources?

The FDA investigator Mike Gavini was scornful of the laborious process that led to Ranbaxy's guilty plea. By never holding any individuals accountable, the agency "dug up the mountain to kill the mouse," he said, yet "the mouse got away." That was true. Many of Ranbaxy's executives had become experts at data fraud. They had spent years immersed in the intricacies of altering test data, from the research-and-development phase through to commercial manufacturing, all while managing questions from skeptical regulators. They had learned a system that aimed to get drug applications approved at record speed, even before the company had mastered how to make the drug in question.

Now these executives were leaving Ranbaxy in droves, forced out by Daiichi Sankyo and the upheaval of Ranbaxy's guilty plea, and getting jobs throughout the industry, taking their colleagues and their skill sets with them. To the regulators, investigators, and investigators who'd worked on the Ranbaxy case, this company diaspora meant one thing: the best way to figure out where they'd find the next fraud was to follow former Ranbaxy executives and see where they landed.

CHAPTER 25

CRASHING FILES

JANUARY 2013

Canonsburg, Pennsylvania

At the glass-walled headquarters of Mylan Laboratories, the brash former Ranbaxy chemist Rajiv Malik became executive director of the board, in addition to his role as president of the company. His ascent surprised former colleagues. It was rare enough for a chemist with a background in research and development to reach a U.S. executive suite. To do so from Malik's background—as a scientist trained in Punjab who had worked only at Indian companies—was extraordinary. But Malik, with his perspicacious mind and buoyant manner, hand always outstretched in greeting, had never been an ordinary bench scientist. He was known for "amazing vision and the willpower to achieve the impossible," as a former associate put it, and as someone who "never fails the mission provided to him by the management." But now, as the same colleague noted, "he is capable of setting his own missions—as he himself is the management."

His latest mission was to oversee Mylan's biggest foreign acquisition yet: an upcoming $1.6 billion purchase of Agila Specialties, a sterile injectable drug maker based in India with nine manufacturing plants worldwide. As Mylan grew, Malik spoke often of a larger, and more complex, mission: to "raise the bar" at every Mylan plant globally and ensure that the company made the same quality of medicine for every world market. This was easier said than done. Malik needed to ensure that India's "low bar," as he put it, did not lower the quality at Mylan or remake the company in the image of his former employer, Ranbaxy, which he called a "beautiful story gone sad."

For decades, Mylan had enjoyed a reputation as a standard-bearer, a company "on the right side of the story," as Heather Bresch, who became the company's CEO in 2012, put it. But as companies looked to buy factories where labor was cheap and oversight less onerous, the story—and being on the right side of it—became more complicated. Publicly, Mylan played a leading role in dragging a reluctant industry toward improvement. When Bresch returned from a global trip by way of Australia, where she visited a company plant, she discovered that the FDA had not been to inspect the plant in over a decade. Though the number of foreign facilities making drugs for the United States was "going through the roof," she said, the inspections lagged far behind those at plants on American soil.

So Bresch—the glamorous, stiletto-heel-wearing daughter of U.S. senator Joe Manchin—began an unlikely campaign to tackle the inspection disparity. She sought to convince her colleagues and competitors to pay fees to the FDA in order to be inspected more. That seemed like a tall order. Why would any company part with money to place themselves under greater scrutiny? But Bresch had a convincing argument. The fees could go not just toward increasing inspections overseas, but also toward speeding up application reviews, thus reducing the notorious backlogs that slowed down approval.

The result, the Generic Drug User Fee Amendment (GDUFA), was signed into law in January 2012. The achievement, largely credited to Bresch, enhanced Mylan's reputation for being on the right side of the story. Ideally, GDUFA would allow the FDA to more effectively regulate a global industry, while also leveling the playing field for disadvantaged American companies, which faced far more scrutiny at their plants in the United States. The result could be higher-quality drugs everywhere, said Bresch: "I still am very hopeful and optimistic that we're raising the bar across the world."

But raising the bar was not simply a matter of law and regulation. It often required a transformation of company culture, something that Mylan soon found itself confronting, from both within and without.

In June 2013, the FDA scheduled an inspection at a sterile injectable plant in Bangalore, Karnataka, which Mylan had announced it would purchase from Agila just months earlier. There are few inspections more complex than those of a sterile injectable plant. Ideally, the FDA should have sent a team to the Bangalore plant that included a microbiologist expert in aseptic techniques. But the FDA—seeking to find someone, anyone, to travel to the far-flung site and looking to spread the pain among its employees—tapped an investigator from the Buffalo, New York, resident post.

The investigator had spent his time principally in upstate New York, where he'd inspected a dairy farm and a cow veterinarian, among other assignments. Those inspections didn't have life-or-death implications. The Bangalore plant was a different matter entirely: Americans could die if he overlooked something. To his credit, he was terrified. In a stroke of luck, a more experienced investigator was asked to accompany him: Peter Baker. The two men arrived on June 17 and stayed for ten days.

At an aseptic manufacturing plant, every motion and action must be considered and controlled for its impact on the sterile environment. In Bangalore, the investigators found a dangerously sloppy plant. They discovered a used mop left haphazardly near a conveyor belt with open vials. Untrained workers moved rapidly rather than in the slow and deliberate manner required, thereby "creating the potential for disruptions of the unidirectional air flow," Baker noted in his report. Key pieces of equipment were stored in nonsterile areas, then never resanitized before use. In the bathroom, several employees failed to wash their hands after using the toilet.

But the gloves really told the story of the plant. The men saw technicians wearing gloves that were flaking and had pinholes, exposing the medicine they'd been working on to contamination. Inside a storage closet, the investigators found "crushed insects" in a shipping box for the gloves. Additional gloves stored there were cracking and discolored. Though Baker had flagged the problem on his fourth day in the plant, technicians were still using the corroded gloves at inspection's end.

It was a disaster, one that grew as the FDA found serious problems in two more Agila facilities. In a little over two years, the FDA had censured three of Mylan's plants, two originally owned by Agila, for failing to ensure a sterile environment. The problems at the plants—which also made active ingredients for Pfizer and GlaxoSmithKline—resonated across the globe. But they landed with the heaviest thud in Canonsburg, Pennsylvania, at Mylan's corporate headquarters, where Rajiv Malik fumed over the problems he'd inherited.

"When we bought Agila, there were six sites [in India] approved by FDA, approved by ANVISA [the Brazilian regulator], approved by every agency in the world," as Malik would later explain. "Pfizer. GSK. It was state-of-the-art, everything robotic, video cameras. . . . Six months there, we got slapped with a warning letter." The topic then turned, perhaps inevitably, to Peter Baker and his aggressive methods. "He created an atmosphere of panic," Malik contended, one in which even the workers' fear and silence were held against them. Nonetheless, Mylan's response was thorough, Malik recounted. The company took 119 of the 199 batches of drugs potentially impacted by the flaking gloves off the market and tested them. They were free from particulates, and the company turned this data over to the FDA, he said.

By then, the company had hired one of the FDA's top officials, Deb Autor, as its senior vice president of strategic global quality and regulatory policy. "If I hadn't been 100 percent satisfied that Mylan was going to do the right thing, I would have walked out the door," Autor later said. In the Agila case, said Malik, "We shut down three plants for almost three years." The actions showed Mylan at its best, he stated: transparent and committed to quality. It was the spirit of the white glove, the dust-free machine, and the "do it right" ethos, all rolled into one.

But in fact, Mylan was changing, and not for the better, some of its employees believed. Internally, as Malik moved with laser-like focus to bring drugs to market, employees in both India and the United States began to experience a shift in the company. Malik and his deputies seemed to prize speed above all else, said several former employees. Those who insisted on adhering to the well-articulated rules of good manufacturing practices felt sidelined, said one senior executive who resigned. "When you're rigid," he said, "you're tagged as being slow."

Under Malik's leadership, Mylan-India became a hothouse of productivity. Malik's own compensation was based, in part, on the number of applications Mylan filed with global regulators. Year after year, he and his team exceeded targets. With their development pipeline full and their laboratories humming with favorable data, they often filed dozens more applications than expected by the company. But employees—some of whom allegedly quit after being asked to tamper with data—were left to wonder: had Malik's handpicked team left behind their Ranbaxy training—or brought it to Mylan instead?

Mylan's reputation as a standard-bearer would soon take a spectacular hit. In August 2016, at the height of the U.S. presidential election and just before the nation's children returned to school, Mylan dove headfirst into the wrong side of a different story—and suddenly became notorious as a company doing the wrong thing. It hiked the price of its EpiPen—an injectable form of epinephrine often used by children with life-threatening allergies—by more than 400 percent.

Mylan came to own the EpiPen in 2007, when it bought the generic drug division of Merck KGaA. After making some innovations in the auto-injectable device, Mylan began selling the drug at $100 for a pack of two. After the FDA rejected a competitor's generic version, owing to design flaws, Mylan had the field to itself and hiked the price. By 2016, EpiPen's listed price was $600. Suddenly parents who needed to buy their children enough EpiPen packs to cover home, school, and their backpacks—found themselves with a prohibitively high out-of-pocket expense.

Outrage built quickly on social media. The hashtag #Epigate gained traction, as did a public narrative about Bresch and her soaring salary. Bresch had made $2.4 million in 2007. But by 2015, she had made almost $19 million. And owing to the company's 2014 decision to incorporate in Ireland to lower its taxes, she and the other executives had made a great deal more: in 2014, both she and Malik earned over $25 million each in total compensation. By then, the EpiPen provided roughly 10 percent of the company's revenue.

Overnight, Bresch became the face of pharmaceutical greed. She was compared in the media to Martin Shkreli, the former hedge-fund manager turned Big Pharma CEO who'd raised the price of a decades-old drug to treat AIDS infections by 5,000 percent. As a wave of public condemnation washed over Bresch, she did herself few favors. In a disastrous CNBC interview, she declared of the price hikes, "No one's more frustrated than me." She then went on to blame others in the supply chain and proposed a national conversation about the broken health care system. The effect was more Marie Antoinette than Florence Nightingale.

Mylan struggled to contain the PR debacle. It offered discount cards. It explained that it would soon market a generic version of the EpiPen at half price. It gave lengthy explanations of byzantine drug pricing and all the middlemen who took a cut. But to the public, which couldn't follow why a generic drug company was alone in selling an overpriced drug in the first place, none of this made any sense.

Suddenly, a history of the company's past scandals spooled out before the press. There was Bresch's MBA. In December 2007, the *Pittsburgh Post-Gazette* uncovered that Bresch had not completed necessary coursework for the degree, but West Virginia University had altered her transcripts and awarded it retroactively once Bresch's father, Joe Manchin, became governor. An uproar ensued, and in 2008 the university revoked her degree. There were allegations that Executive Chairman Robert Coury frequently misused the company jet to fly his son, a musician, to gigs around the country. There had been a shady land deal, involving a company vice chairman, for the property where Mylan had built its new headquarters.

But the EpiPen scandal was in a league of its own. Before long, there were congressional inquiries, class-action lawsuits, and investigations by multiple attorneys general into antitrust violations. On September 21, 2016, an unhappy-looking Bresch found herself grilled under oath in a nationally televised inquiry by the House Committee on Oversight and Government Reform. The lawmakers demanded to know why she'd turned her back on families who could no longer afford the drug.

But even as all this played out, a far more consequential set of developments—ones that raised questions about the company's integrity and the quality of its medicine—had been unfolding far from public view.

About a year earlier, a former Mylan employee arrived at the FDA's headquarters in Silver Spring, Maryland, and sat down with a group of senior FDA officials. Confidentially, he made specific allegations: that under Rajiv Malik's leadership, Mylan's research and development center in Hyderabad had become a hub for data fraud that had disseminated its methods of falsification throughout Mylan's Indian operations. The whistleblower alleged that people who now held key leadership positions at Mylan, among them former Ranbaxy employees, were using their skills at data manipulation.

The Mylan whistleblower identified specific applications for drugs that were due to be launched into the American market. He claimed that in order to generate passing results for some drug products, Mylan had switched samples from commercial batches, which were less stable, with samples from smaller pilot batches that were easier to control. But perhaps the most surprising allegation he made was that the Mylan team had evolved its fraudulent methods to evade detection. Instead of deleting manipulated data from the plant's software systems—which would have left a trail of metadata that FDA investigators like Peter Baker could uncover—plant managers were deliberately corrupting the data they wanted to hide. This was considered a better way to evade investigators.

Though the FDA officials found the whistleblower's claims credible, remarkably they did nothing for about a year. Mylan seemed to exist within a charmed circle of protection at the FDA. Not only was its chief executive officer the daughter of a U.S. senator, but now one of its top regulatory executives, in charge of overseeing the company's relationship with the FDA, was the former FDA official Deb Autor.

In July 2016, the whistleblower jolted the FDA officials by sending an email that expressed his dismay over their inaction. He made clear that they were accountable for what happened to American patients. He suggested that the company's political connections and the revolving door between the FDA and Mylan had been factors in the FDA's passivity.

"Honestly—I had supreme faith & trust in the agency's approach—towards bringing those to justice who commit fraud," he wrote. "However, I learn that Mylan's strategy of providing employment to FDA members has been working very well." He went on: "Perhaps, the agency awaits a definitive tragedy to occur on U.S. soil due to substandard generic drug products not meeting the safety & efficacy standards" (as it has been the case in Africa—where the anti-retroviral drugs aren't showing adequate efficacy)."

He speculated that something or someone was clearly blocking the FDA from inspecting Mylan: "This kind of bureaucratic scenario can be considered as 'common practice' in a country like India—but I certainly had much higher expectations from U.S. government agency—which are known for higher ethical standards & higher moral values."

His unusually pointed note of admonishment set off a desperate scramble inside the agency. Two months later, on September 5—about two weeks before Bresch took her seat on the witness stand in front of Congress—an FDA investigator turned off the main thoroughfare in Nashik, India, onto a dusty side road amid wandering goats and chickens and arrived at Mylan's flagship Indian plant. This time, the investigator had come unannounced.

Mylan's Nashik plant is a five-hour drive from Mumbai, past burning farm fields and desolate road stops. Despite the forlorn spot, the plant there is both massive and cutting-edge. It sprawls over twenty-two acres and has the capacity to make 8 billion doses of drugs a year for every world market, from Australia and Africa to the United States.

Over the course of nine days, the FDA investigator who arrived there found evidence that the plant's software system was riddled with error messages showing "instrument malfunction," "power loss," and "connection to chromatography system lost." Plant managers had apparently conducted no investigation into the repeated crashes. They had simply retested the drugs after receiving the error messages, leading the FDA to suspect that the crashes had been intentional, just as the whistleblower had alleged. It appeared that, instead of deleting unwelcome data, which would have left a trail of clues, Mylan had crashed its system, as though technicians had literally pulled the computer plug from the wall. The technique was so notable that FDA officials gave it a name: "crashing files."

Within two months, three investigators arrived unannounced at Mylan's plant in Morgantown, West Virginia. There, they were stunned to find what looked like suspect data practices. The technicians had been pre-injecting samples into the HPLC machines, prior to the official tests, as though to get a forecast of the results. There were also instances where drug batches had yielded failing or aberrant results and the analysts did not investigate, as required. Instead, they retested the drugs and got passing results, raising questions about what manipulations they may have made.

To the FDA's compliance officials, the crashed software, the pretesting, and the failure to investigate aberrant results all smacked of deception. In correspondence, they demanded answers from Mylan, noting that the error messages at Nashik "raise questions regarding the integrity and reliability of data generated." That perception posed a grave risk to the West Virginia company. If the FDA made a final determination that quality problems were systemwide, not just isolated to one plant, the penalties and sanctions could escalate dramatically. The FDA's suspicions also threatened Mylan's image as a company of integrity, a reputation it had cultivated assiduously, down to its glass-walled headquarters and partly transparent business cards. Despite its political clout, Mylan risked being lumped in with other global generic drug companies that could not be trusted to run a clean laboratory.

In a later meeting with a journalist, Mylan officials played down the FDA's findings, explaining that the ominous term of "data integrity" lapses actually encompassed any number of simple regulatory shortcomings. Said R. Derek Glover, Mylan's head of global quality systems and compliance, "We have found no evidence that any of this was associated with data fraud."

Mylan responded forcefully to the Nashik and Morgantown inspections. In a series of meetings, calls, and correspondence with the FDA, the company's top lawyers and well-connected executives pledged full cooperation and transparency. They flooded the agency with data and analyses intended to prove that the company had thorough quality systems and was prepared to investigate itself.

In January 2017, Mylan sent a lengthy, confidential letter to the agency, attempting to explain the reason for the high number of error messages at the Nashik plant—forty-two over a seven-day period. The company offered no single explanation. The messages were "not related to the disconnection of the Ethernet cable or power cord." It then added: "It is not evident through retrospective review whether these disconnection events were caused by manual intervention of cables (accidental knocking of cables), or through an electronic loss of signal." For a different error that appeared 150 times over seven days, it gave a partial explanation: some of its software settings had led to "unintended consequence of a number of repetitive error messages." In a confidential letter the following month, Mylan assured the agency that "there was no resultant impact to the integrity and appropriateness of the results considered for batch release decisions."

The FDA didn't buy it. On April 3, 2017, it gave the Nashik plant a warning letter, effectively freezing agency review of the site's applications until the company made corrections. The letter noted that Mylan's quality system "does not adequately ensure the accuracy and integrity of data." It made clear that the agency harbored ongoing suspicions about the error messages: "Your quality unit did not comprehensively address the error signals or determine the scope or impact of lost or deleted data until after these problems were reviewed during our inspection." The company's share price fell 2 percent on the news.

Less than three weeks after the warning letter at Nashik, Malik and six other Mylan officials sat down with nineteen displeased FDA bureaucrats at the agency's headquarters to try to fend off regulatory action against its Morgantown, West Virginia, plant. As agency officials grilled them about why laboratory technicians had failed to investigate anomalous results and instead had retested the drugs and recorded passing results, Malik's team found itself facing a larger question from agency officials: what had happened to Mylan? The regulators said that they were "stunned" by the lapses at Morgantown, found the practices "egregious," and questioned whether the company was being "transparent at all of its sites." One FDA bureaucrat put it bluntly: "The FDA is questioning how such violations could have ever happened at the Morgantown facility in light of Mylan's broader quality culture."

It fell to Malik, who both reflected the transformation of Mylan and had helped to lead it, to argue that the underlying values of Mylan and Morgantown had not changed. "Mylan's philosophy of quality," he explained to the officials, "is that there is no compromise on data integrity or patient safety." Aiming to disentangle the Morgantown site from the agency's scrutiny, he explained that the plant was unique "as the business started there and from day one the facility was founded upon the principle of integrity." Ultimately, the firm blamed the retesting without investigation on an old standard operating procedure that had needed updating.

This time the company's approach appeared to work. In May

2017, the FDA's director of the Office of Manufacturing Quality, Tom Cosgrove, made a controversial decision over the strenuous objections of staff in two separate FDA divisions: he downgraded the investigators' negative findings at Morgantown, from Official Action Indicated to Voluntary Action Indicated. He also chose to send an untitled letter to the company that was not visible to the public—the second time in two years that Cosgrove had downgraded findings against Mylan and concealed the agency's response.

In an email to FDA colleagues, Cosgrove acknowledged their view that the company's retesting practices were "more widespread and that Mylan's investigation was insufficient." But he defended his decision: "Mylan has been responsive and forthcoming, and I have no reason to believe they will not remediate voluntarily."

The move briefly got Mylan's Morgantown plant off the hook of intensifying agency scrutiny. But it did little to resolve the storm brewing there. In early 2018, a whistleblower from inside the plant reached out to the FDA to report deteriorating conditions, from understaffing to cleaning lapses. The whistleblower claimed that Mylan's management, instead of working proactively to remedy problems, was more focused on creating a "facade of documents" to fend off the FDA, according to an agency memo that detailed the allegations. The whistleblower described how a team of employees from India was brought in to rapidly close a backlog of company investigations at Morgantown, and employees there were instructed not to question their work. Mylan had developed an "embedded culture" that permitted fraud, the whistleblower claimed, an observation shared by some former employees.

From the Sea Lounge in Mumbai's most famous hotel, the Taj Mahal Palace, a former Mylan chemist looked out to Mumbai Harbor and the 1920 triumphal arch known as the Gateway of India. But he took no pleasure in the view. Even surrounded by silk pillows and attentive waiters, he was distraught. He was there in secrecy, to describe to a journalist how Mylan had moved dozens of drug applications swiftly through the system, using "cooked" data at each step of the manufacturing process. This data manipulation occurred under the leadership of Rajiv Malik and a group of his associates, the chemist explained. His team built up the company's India operations into a powerhouse central to the company's success, along the way transforming—and corrupting—the West Virginia company, said the chemist.

Malik's team used an array of deceptive methods to hasten approval of critical products, he said. They did "what's needed" to make the development data pass and "managed" the manufacturing of the submission batches. They generated the bioequivalence data by switching the samples, if necessary. "Wise people" prepared the submission packages to regulatory agencies. Post-approval manufacturing was "taken care" of by specialists. Global experts, held in esteem by regulatory agencies, were consulted to "bless" the packages, but given only partial information. All of these interventions served to "short-circuit" the timeline typically required to develop and manufacture a generic drug.

At the Sea Lounge, the chemist described the well-oiled machinery of industrial-scale data manipulation, with teams from research and development deployed at each manufacturing step to manage the failing data. Malik had bracketed every manufacturing system with his people, who moved entirely in sync. "One person starts a sentence, the other person can complete it," he explained, adding that Malik's people could execute his directions with very few instructions. "The command does not need to be very specific." The goal is to get the drugs to market as quickly as possible, said the chemist, and those working under Malik do everything they can to make that happen.

At each step, he said, they used workarounds, from hidden equipment to tinker with tests to secret substitutions. When the process got transferred to the plant and was scaled up, the larger batches would start to fail. "Then the phone call goes, no email," said the chemist. "You send a guy from analytical to QC [quality control], they manage the data. The data is clean."

Then comes commercialization, or large-scale manufacturing, which is far harder to control. "Commercial batches will fail on stability," he said, and again, the solution was data manipulation. "You play with parameters so impurities don't show up." At each step, "people come from R&D to show how to fix the issue."

It was under this system that the data for a number of products—submitted to the FDA for approval—were manipulated, which the chemist said hastened his departure from the company.

Mylan's general counsel Brian Roman said that the company "absolutely and vehemently" denied allegations of data manipulation. He pointed out that the FDA had not confirmed any such activity at the company. "If someone is telling you they have evidence that we switched samples," he told a journalist, "my belief is, they're lying to you." He added that anyone leveling such a claim was obligated to "make a report that's capable of being investigated." But in fact, the chemist had. After his resignation, he detailed his allegations in writing to senior managers.

At the Sea Lounge, the chemist explained that the FDA investigator Peter Baker "got the exact pulse of India," by uncovering the subterranean system of testing that paralleled the official one. "It always comes from the top down." He cried silently, tears rolling down his face. "What's going on in the industry is very, very, very dirty."

CHAPTER 26

THE ULTIMATE TESTING LABORATORY

FEBRUARY 7, 2013

Kampala, Uganda

At the Mulago National Referral Hospital, Dr. Brian Westerberg, a Canadian surgeon on a volunteer mission, examined a very sick thirteen-year-old boy. He had a fever and chills and was vomiting. Fluid oozed from his ear canal, the likely source of infection. Westerberg suspected bacterial meningitis, though he couldn't confirm his diagnosis because the CT scanner had broken down again. The boy was given intravenous ceftriaxone, a broad-spectrum antibiotic that Westerberg believed would kill the bacteria and reduce the swelling around his brain. He was confident that the boy would be cured.

For the sixteen years that Westerberg had worked volunteer missions at the Mulago hospital, scarcity had been the norm. The throng of patients usually exceeded the 1,500 allotted beds. Running water was once cut off when the debt-ridden hospital was unable to pay its bills. For years, medicine had been scarce, forcing

Westerberg to bring his own drug supplies from Canada. But more recently, low-cost generics had become widely available through the local government and international aid agencies, a hopeful development that Westerberg saw as a step in the right direction.

But after four days of treatment, the boy was still sick. His headache was even worse, and the "draining ear" had turned into a pussy, boil-like mass. Westerberg prepped for surgery, assuming that he would have to slice into the child's ear and scoop out the infected tissue. Right before the procedure, the boy had a seizure. The hospital's CT scanner had started to work again, so Westerberg sent the boy for an urgent scan, which revealed small abscesses in his brain, likely caused by the infection.

A hospital neurosurgeon looked at the images and confidently told Westerberg, "We're not going to operate." He was sure that the swelling and abscesses would abate with effective antibiotic treatment. This confused Westerberg. They had already treated the boy with antibiotics—the intravenous ceftriaxone—and it had failed to beat back his infection. Westerberg's confusion deepened when his colleague suggested that they switch the boy to a more expensive version of the same drug. "Why swap one ceftriaxone for another?" Westerberg wondered. But he would soon learn that the hospital's drug supply was plagued by a phenomenon that Ugandan doctors knew all too well.

In a world of scarcity, Africa was saturated with Indian- and Chinese-made generic drugs that too often didn't work. Doctors throughout the continent had adjusted their medical treatment in response, sometimes doubling or tripling recommended doses to produce a therapeutic effect. Many hospitals kept a stash of what they called "fancy drugs"—either brand-name drugs or higher-quality generics—to treat patients who should have recovered after a round of treatment but didn't.

Westerberg's colleagues were prepared with a backup stash of ceftriaxone purchased outside the hospital. They swapped in the more expensive version and added two more drugs to the boy's treatment plan. They did not operate fully, but Westerberg did clean out the draining ear. The treatment might have been effective, but was administered too late. The boy never recovered. On the eleventh day of his treatment, he was declared brain-dead.

The Ugandan doctors were not surprised by the death of the thirteen-year-old boy. Their patients frequently died when treated with drugs that should have saved them. Even though doctors turned to backup reserves of "fancy" drugs, there were not enough to go around, making every day an exercise in pharmaceutical triage. "We are tired, honestly," as one doctor in western Uganda put it. She found it hard to keep track of which generics were safe and which were not to be trusted. "It's anesthesia today, ceftriaxone tomorrow, amoxicillin the next day."

Each time Peter Baker inspected a new manufacturing plant, his sole job was to focus on the drugs headed for the U.S. market and protect Americans who would be taking them. But as he documented fraud and manufacturing failures in India's plants, he noticed with increasing alarm that the drugs bound for developing markets were even worse than those headed for the United States.

In May 2013, he arrived at a manufacturing plant south of Hyderabad, run by an Indian generic company. There, he saw vials of a chemotherapy drug, gemcitabine, which had improperly sealed caps and therefore weren't sterile. "What the heck did you do with this?" he asked officials. The answer: "We sent it to Africa." At another plant, he saw drug ingredients that the FDA had restricted from import into the United States and asked where those were headed. The Ukraine, he was told. He passed that information to the Ukrainian government but didn't hear back.

Baker slowly began to see a map of the world similar to the

one that had so shocked Dinesh Thakur when he uncovered it at Ranbaxy. Companies claimed that they produced the same high-quality drugs for every world market. At Ranbaxy, Thakur exposed this as a lie by uncovering fraudulent data. Baker was looking at the actual medicine. At plant after plant, he found the most blatant fraud and the most egregious quality lapses in the manufacture of drugs bound for the least-regulated markets: Africa, eastern Europe, Asia, and South America. One reason he rarely saw a rejected batch in India was that, no matter how evidently defective the drug, there was always some world market where it could be sent.

What Thakur and Baker had stumbled into was not a glitch or exception but rather common practice in the generic drug industry. The unequal production standards had interchangeable names: "dual-track," "multi-tier," or "row A/row B production." Companies routinely adjusted their manufacturing quality depending on the country buying their drugs. They sent their highest-quality drugs to markets with the most vigilant regulators and their worst drugs to countries with the weakest review. In an industry with slim margins, companies slashed costs by using lower-quality ingredients, fewer manufacturing steps, and lower standards, then sold those drugs in poorly regulated countries.

Racism undoubtedly played a role, as it had at Ranbaxy when the medical director said of the poor-quality AIDS drugs bound for Africa, "Who cares? It's just blacks dying." But at root, a cold calculation drove the disparate standards: companies could make their cheapest drugs for markets where they would be least likely to get caught. When asked, the companies insisted that they made drugs of different standards because quality standards differed from market to market. But Patrick H. Lukulay, former vice president of global health impact programs for the USP, the world's top pharmaceutical standard-setting organization, called that argument "totally garbage." For any given drug, he said, "There's only one standard, and that standard was set by the originator," meaning the brand-name company that developed the product.

Baker had no control or authority over the drugs sold to patients anywhere outside the United States. But he wrote up his findings in his inspection reports nonetheless. Those reports became a road-map of warnings for regulators in other countries, some of whom followed up on his findings. Baker was documenting the tip of a much larger public health crisis. He couldn't be sure what was happening to patients in other countries, but the generics pouring into the developing world were of such low quality that he assumed the consequences had to be dire, nothing short of a "ticking time bomb."

In Uganda, Dr. Westerberg was shaken by his newfound knowledge that patients were dying after taking substandard drugs. He flew back to Canada and teamed up with a Canadian respiratory therapist, Jason Nickerson, who'd had similar experiences with bad medicine in Ghana. They decided to test the chemical properties of the generic ceftriaxone that had been implicated in the Ugandan boy's death.

Westerberg's colleague brought him a vial of the suspect ceftriaxone from the pharmacy at the Mulago hospital. The drug had been made in China's Hebei province by an international drug company, CSPC Zhongnuo Pharmaceutical, which also exported to the United States and other developed markets. But when they tested the ceftriaxone at Nickerson's lab, it contained less than half the active drug ingredient stated on the label.

Both Westerberg and Nickerson were shocked. At such low concentration, the drug was basically useless and would not have cured a single patient, said Nickerson. He and Westerberg published a case report in the CDC's *Morbidity and Mortality Weekly Report*. Although they couldn't say with certainty that substandard ceftriaxone had killed the boy, their report offered compelling evidence that it had.

CSPC Zhongnuo Pharmaceutical had a track record of quality failures. In 2009, Jackson Lauwo, a professor at the University of Papua New Guinea's School of Medicine and Health Sciences, grew concerned about the quality of drugs in the country and contacted a pharmaceutical scientist in Frankfurt, Germany, Jennifer Dressman, to ask if she would be willing to test drug samples in her lab. She agreed to help, so Lauwo collected samples of the anti-infective drugs amodiaquine and amoxicillin made by different manufacturers from five registered pharmacies in the capital city of Port Moresby. He brought them to Dressman's lab. After several months the results were in: every single sample failed quality evaluations.

Three of the fourteen samples were counterfeit: entirely fake pills made by criminals. The rest were substandard, meaning that they appeared to have been manufactured by legitimate firms intentionally making a low-quality product. One such sample came from CSPC Zhongnuo Pharmaceutical.

Even the company's drug products for the U.S. market were of poor quality. Since 2013, the FDA has cited the company four times for quality violations. Peter Baker inspected the company's plant northeast of Beijing and found blatant data manipulation. He discovered that failed tests were routinely deleted and samples were retested until they yielded passing results.

The United States has the resources to send investigators abroad. Poorer countries like Uganda and Papua New Guinea typically import drugs from the lowest bidder and don't have the regulatory apparatus to check quality. Minimal oversight by countries buying the drugs, coupled with weak laws and poor regulation in countries manufacturing them, have allowed dual-track production to flourish.

India's regulators, for example, won't take legal action unless a company is making a drug with less than 70 percent of the required active ingredient, which is far below the acceptable standards set by the FDA, the World Health Organization, and other major drug regulators. Dual-track production is not illegal in India, said Anant Phadke of the All India Drug Action Network, an Indian organization that works to increase access to essential medicines. "Whether morally this is correct or not is open for interpretation."

The same year Westerberg treated the boy in Uganda, Dr. Sean Runnels, an American anesthesiologist from Utah, arrived in Rwanda to work in the country's national health care system. One of the first things he noticed was that many generic drugs supplied by Rwanda's health care program "simply didn't work." Back then, the country did not have an official drug agency to test the medicine it was buying and could not verify its quality.

At the University Hospital of Kigali, in the nation's capital, Runnels realized that he could no longer count on anesthesia to put patients to sleep and on antibiotics to fend off infections. He watched as new mothers, after giving birth by cesarean section, succumbed to bacterial infections despite taking a full course of antibiotics. In lieu of effective medication, Runnels and his colleagues resorted to surgery, flushing out women's abdomens and cutting out infected tissue in a last-ditch effort to save their lives. "Very few of them survived," he said.

Runnels was initially flabbergasted to discover such variable drug quality, but like the doctors in Uganda, his Rwandan colleagues were accustomed to the problem and had a system to address it. If a generic wasn't working, they tried to find another version produced by a different manufacturer, or they switched to another type of drug. If neither option was available, they increased the dose of the substandard generic in an effort to achieve a therapeutic effect.

Wealthy patients were lucky. They could escape the quagmire by purchasing brand-name medicine from private pharmacies. The difference was astonishing. "As soon as you saw a brand-name drug, those are the patients that would suddenly get better," Runnels said. It was a phenomenon so pronounced that he started calling it "the Lazarus Effect," after the biblical character who rose from the dead.

Over the past decade, Africa's pharmaceutical problems changed dramatically. Previously, the continent's drugs came largely from more-developed countries, through donations and small purchases. The biggest problems were the high costs and resulting scarcity. By 2004, Indian drug reps started arriving throughout Africa, offering cheap generics. In Ghana, though the initial feeling was positive, the result was not, recalled Dr. Anita Appiah, coordinator of community and institutional care for the National Catholic Health Service. Africa became an avenue "to send anything at all," said Professor Kwabena Ofori-Kwakya, head of the Pharmaceutics Department at the Kwame Nkrumah University of Science and Technology in Kumasi, Ghana. The adverse impact on health has been "astronomical," he said. The poor quality has affected every type of medication.

Dr. Gordon Donnir, a psychiatrist who heads the Psychiatry Department at the Komfo Anoyke Teaching Hospital in Kumasi and treats middle-class Ghanaians in his private practice, said that he and his colleagues are bedeviled by low-quality medicine in all sectors. Almost all of the generic drugs he prescribes—olanzapine, risperidone, diazepam—are substandard. The situation has forced him to increase dosages. To treat psychosis, his colleagues from Europe typically prescribe 2.5 milligrams of haloperidol, the generic version of Haldol, several times a day. They are shocked to learn that he'll prescribe 10 milligrams, three times a day, because he knows that the 2.5 milligram dose "won't do anything." Those initially shocked colleagues become converts once they realize they have to increase doses to get an effect. Donnir once gave ten times the typical dose of diazepam, an anti-anxiety drug, to a fifteen-year old boy, an amount that should have knocked him out. The patient was "still smiling," said Donnir.

Though data is scarce, in 2012 the Ghana Food and Drugs Authority, with help from the USP and USAID, tested the quality of maternal health drugs on the market. They zeroed in on oxytocin and ergometrine, both essential medicines used to prevent postpartum hemorrhaging. The results were devastating. The report, published in 2013, found that over half of all the oxytocin samples, and almost three-quarters of the ergometrine injection samples, made by various generic companies, failed. All of the ergometrine tablets failed—that is, they produced no effect at all. Some samples contained no active ingredient, and others failed sterility tests, results that pointed to only one conclusion: substandard manufacturing. The drugs, almost all of which came from India and China, were a death sentence for women bleeding due to childbirth.

Most of the time, Ghana's patients have no idea what kind of medicine they take or who manufactures it. The culture in Ghana is faith-based, explained Bright Simons, a technology innovator: "People actually pray [their] medicine will work."

In 2008, Alexandra Graham, an African scientist, opened LaGray Chemical Co., a drug manufacturing company outside Ghana's capital, Accra, with the goal of making high-quality drugs that would follow international standards for good manufacturing practices. Graham, who is Nigerian, and her Ghanaian husband, Paul Lartey, had ideal backgrounds for the effort. They met while working at Abbott Laboratories in Chicago, she as a chemist and manager of the Specialty Products Division, and he, also a chemist, as director of the company's infectious disease drug discovery unit. Initially, they looked to India's generic drug industry as a model for manufacturing high-quality drugs in a low-cost environment. It was the model that Dr. Yusuf Hamied at Cipla had set for the world.

Graham traveled to India to learn how the country had accomplished such a feat. Instead of being inspired, she was shocked by what she saw. Graham toured a manufacturing facility so decrepit that it was essentially a house with "little bedrooms that were the manufacturing sites." The site lacked the quality controls needed to prevent cross-contamination. "There was no air conditioning, no [ventilation] system, dust everywhere," she recalled. As she left, she noticed the security guard packaging the drugs, which were destined for Nigeria and Kenya. They were not even approved for sale in India.

Graham resolved to run her business differently, but doing so was a formidable challenge. With unreliable electricity, it was difficult to keep the lights on, much less run complex chemical reactions with sensitive equipment. But the bigger challenge came from competition abroad. With Indian and Chinese drug representatives "aggressively coming into the country," selling cheap medicine, and "offering all sorts of incentives and bribes," Graham said, her company struggled to compete. Corrupt local wholesalers even negotiated with the drug companies to sell drugs with reduced amounts of active ingredient, making them cheaper.

Seeking help, Graham turned to a respected colleague, the CEO and chairman of a major Indian drug maker. His advice perturbed her. He suggested that she set up one "good factory" capable of producing a "showpiece product" and a second factory that met "local standards." It became clear to Graham that the idea of high quality at low cost was a myth. Companies that flaunted their factories as "US-FDA inspected" or "WHO-certified" happily sold a cheaper, second-tier product to Africans. Graham found herself caught in a vise. Even if she insisted on making drugs of one high standard only, she couldn't guarantee that the companies supplying her with active ingredients were doing the same.

In 2014, to buy the active ingredients for HIV drugs, Graham turned to the Chinese firm Shanghai Desano, whose factory in Binhai had been inspected and approved by the WHO. Early in negotiations, a Desano sales rep offered LaGray a discounted product reserved for "African customers." The ingredients were cheaper, the sales rep explained, because they were not produced at the WHO-approved Binhai plant, but at a clandestine plant in Puxin, China. Graham was furious. She dashed out an email to the company's vice president, exclaiming, "Whatever is good for any other market is appropriate for Ghana!"

Graham's standards were expensive, and her refusal to prioritize profit over quality did not sit well with her investors, who began to sell their shares. By 2016, her company had closed and the manufacturing plant outside Accra was "overgrown with grasses," she said. Without strong regulatory reform in Africa, India, and China, she cannot imagine a way to rein in substandard drug production. "The landscape is such that there are no incentives for quality," she said. "When you're playing by the rules, you just cannot survive."

Public health experts first began to expose the problem of dual-track manufacturing years earlier. In 2003, an international aid worker and pharmacist, Jean-Michel Caudron, decided to visit the international generic drug companies to observe the production process. Gaining access to the manufacturing sites was "quite simple," said Caudron. "At the time, I was working with Doctors Without Borders and UNICEF, and we were purchasing a lot of medicines . . . so I was welcome [to] visit." Caudron was treated like a potential customer, and he and his colleagues were granted extraordinary access to manufacturing facilities in India and other parts of the world, where he witnessed multi-tiered production. "I heard the managers of a famous company in India explaining that they were very proud to be validated by the US FDA and Europe, [then] they said very clearly they were producing different quality for non-regulated countries," Caudron recalled. When he asked a manager why the company would produce different standards for different markets, the manager replied, "Oh, those markets in Africa and Asia? They don't ask for proof of safety."

After cataloging dual-track production at 180 facilities over a four-year period, Caudron and his colleagues published a landmark paper in the *European Journal of Tropical Medicine and International Health*. It described the alarmingly common business tactic of undercutting quality for less-regulated markets, effectively creating "one standard for the rich, another for the poor."

As international researchers grapple with the problem of dual-track production, an increasing number are linking the phenomenon to another global catastrophe: drug resistance, in which bacterial and other infections evolve to resist the very medicine designed to treat them. It is exactly the sort of ticking time bomb that Peter Baker feared. In 2014, the British government commissioned an ambitious effort to map the harm of drug resistance and propose possible solutions. The first of a series of reports estimated that if current trends continue, 10 million people will die every year from drug-resistant infections by 2050. "If we fail to act," said the former British prime minister David Cameron, "we are looking at an almost unthinkable scenario where antibiotics no longer work and we are cast back into the dark ages of medicine."

Ultimately, the project culminated in nine major reports, most of which blamed three factors typically associated with drug resistance: pollution from drug manufacturing in developing countries, which spews medicine into lakes and rivers; the widespread overuse of antibiotics in livestock; and medication misuse, when patients don't take their medicine exactly as prescribed. But one of those reports blamed a fourth factor: substandard generics that are systematically underdosing wide populations in the developing world. Researchers working in developing countries, where poor-quality generic drugs and drug resistance are both prevalent, have increasingly examined a link between the two.

Low-income countries are plagued by poor-quality medicine: counterfeit pills made by criminal gangs and substandard generics made by second-rate drug companies. Though counterfeit drugs look like genuine products, they often contain no active ingredients. By contrast, substandard drugs typically contain active ingredients, but they do not contain enough, or are not formulated well enough, to be effective. Though counterfeit medicine attracts the lion's share of political outrage and media attention, some experts now argue that substandard drugs are actually a greater threat to public health. Often, substandard drugs do not contain enough active ingredient to effectively treat sick patients. But they do contain enough to kill off the weakest microbes while leaving the strongest intact. These surviving microbes go on to reproduce, creating a new generation of pathogens capable of resisting even fully potent, properly made medicine.

In 2011, during an outbreak of drug-resistant malaria on the Thailand-Cambodia border, American public health expert Christopher Raymond fingered substandard drugs as a culprit. Treating patients with drugs that contain a little bit of active ingredient is like "putting out fire with gasoline," said Raymond, former chief of party for the USP in Indonesia. From his vantage point in Southeast Asia, Raymond could see a clear correlation between regions with high volumes of substandard drugs and "hotspots" of drug resistance.

Paul Newton, a British malaria expert who works in Southeast Asia, has observed the overlapping problems of substandard antimalarial drugs and emerging resistance for nearly two decades. In 2016, he coauthored an editorial that explained that drugs containing "sub-lethal" concentrations of antimalarial medicine create a "survival advantage" for the parasites that can withstand a subpotent dose. Still, he cautioned that the connection, although logical and highly likely, was not yet backed by robust scientific evidence.

By 2017, with circumstantial evidence piling up, the nonprofit group USP launched a new center called the Quality Institute, which funds research into the link between drug quality and resistance. In late 2018, that funding bore fruit. Dr. Muhammad Zaman at Boston University coauthored the first study to link substandard drugs to antimicrobial resistance. Working in a laboratory, Zaman studied a commonly used antibiotic called rifampicin. If it's not manufactured properly, it can yield an impurity as it degrades called rifampicin quinone. When Zaman subjected bacteria to the impurity, it developed mutations that helped it resist rifampicin and other similar drugs. Zaman hopes his laboratory work will help convince policymakers that substandard drugs are an "independent pillar" in the global menace of drug resistance, potentially on par with poor adherence and prescription misuse.

Elizabeth Pisani, an epidemiologist who has studied drug quality in Indonesia, wrote in a 2015 report, "Antimicrobial Resistance: What Does Medicine Quality Have to Do with It," that subpotent drugs are fueling a crisis of drug resistance in lower-income countries that rich countries soon won't be able to ignore: "The fact is, pathogens know no borders."

When a pathogen evolves to resist every known treatment, every patient in the world becomes a potential victim. In August 2016, a Nevada woman in her seventies returned home from an extended trip to India, where she had broken her femur. An infection that had started in her thigh bone soon spread to her hip. She checked into a Reno hospital, where doctors immediately tested her for multi-drug-resistant bacteria. The CDC confirmed that she had carbapenem-resistant Enterobacteriaceae (CRE), a "nightmare bacteria" with no known cure, as Thomas Frieden, the CDC's former director, described it at a news conference.

In Nevada, there wasn't much doctors could do to save the woman. Their bigger concern was saving other patients from the same fate. The hospital immediately set up an isolation room so that the infection wouldn't spread, and the staff donned masks, gloves, and gowns while caring for the woman. In less than a month, she was dead.

What started in Mahatma Gandhi's ashram as a campaign of Indian self-reliance had morphed into a pharmaceutical rescue mission for the world's most unfortunate patients. Dr. Hamied's revolution, as it thundered along, offered generic drug companies the chance to act as global equalizers and make the same cures available to the wealthy and impoverished alike. But what Thakur had first documented on a spreadsheet at Ranbaxy, and what Baker observed in manufacturing plants throughout India, was not the fulfillment of that ideal. It was the ultimate subversion and exploitation of it—making the worst medicine for the poorest patients, with life-and-death consequences for us all.

In Ghana, the technology innovator Bright Simons summed up the stark reality: "All medicines are poisonous. It's only under the most controlled conditions that they do good." Only a drug accompanied by data that traces a minute-by-minute path through the manufacturing process can be trusted to work properly. How could anyone guarantee that standard in an underpoliced global marketplace? Only one-tenth of African countries had effective regulators, and two-fifths lacked laboratories capable of routinely testing medicine quality.

Those deficits brought Dr. Patrick H. Lukulay to Ghana's capital, where he set out to train a generation of African Peter Bakers. In Accra, he ran the Center for Pharmaceutical Advancement and Training (CePAT), an outpost of the USP that opened in 2013. Located down an unpaved dirt lane and set behind a long metal gate, the facility appears unremarkable from the outside. Inside, however, it's a marvel: door locks are controlled by biometric fingerprint readers, a state-of-the-art microbiology lab is full of costly HPLC machines, and a stability room has specialized refrigerators that test how quickly drugs degrade.

The center runs the continent's only dedicated program for training African drug regulators and also operates a fully accredited laboratory to test drug quality. With these resources, Lukulay hoped to lift the quality of medicine across Africa. It did little good for one country to up its game, said Lukulay, when "its neighbors are all screwed up."

A highly trained scientist with a commanding though kindly air, Lukulay was uniquely suited for his mission in Ghana. He grew up in an impoverished village in Sierra Leone, where his father, the village chief, had twenty-five children. Lukulay studied by kerosene lantern and slept on a bed of palm leaves. Through stellar academic performance and sheer grit, he propelled himself to a school in Sierra Leone's capital, Freetown, and ultimately to graduate studies at Michigan State University. He went on to work at Pfizer and Wyeth in the United States before moving to the USP. His childhood experiences left him with an indelible sense of the obstacles in Africa.

In March 2016, several dozen young regulators who worked for various African governments, including Mozambique, Swaziland, Rwanda, Zambia and Liberia, arrived at CePAT for a training workshop. They spent two weeks learning how to more rigorously vet the dossiers submitted by drug companies and think critically about their claims, as one of their trainers put it, instead of running through a perfunctory checklist. The training program's graduation ceremony was solemn, to emphasize to the regulators their critical role as the first line of defense against poor-quality medicine. Lukulay's colleague urged them to "go forth like living water" into Africa, the "ultimate testing laboratory."

Lukulay then stepped forward, wearing a ceremonial outfit of a long tunic, flowing pants, and embroidered cap. "You are soldiers of the state," he told the graduates. "You have been given ammunition, you have been equipped to do battle with those who want to kill our people." Expecting them to encounter corruption or political interference at some point, he emphasized that their jobs were "noble" and "ethical." But most importantly, he emphasized, when they had a dossier before them, they had to ask: "Am I going to check the boxes, or am I going to be a real assessor?"

PART VII

RECKONINGS

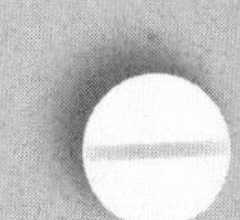

CHAPTER 27

FLIES TOO NUMEROUS TO COUNT

JUNE 2013

New Delhi, India

One month after the U.S. government settled its case against Ranbaxy, the FDA's relations with Indian regulators were in tatters. The agency needed a public health diplomat to head its India office and turned to an American of Indian origin with a warm and elegant manner, thinning silver hair, and sterling public health credentials: Altaf Lal, who had spent six years as the U.S. health attaché in New Delhi.

The FDA rolled him out with about as much fanfare as a health agency can muster. It published a blog post from Lal on the FDA website, in which he outlined his three goals: to work closely with Indian regulators to establish a trusting relationship; to conduct "prompt and thorough inspections"; and to help Indian "industry and regulators understand that protecting the quality, safety and effectiveness of every product is essential." In the blog post, Lal wrote: "A colleague recently likened my new role at FDA to scaling Mount

Everest. But you know, I'm fond of trekking and climbing, and view this next challenge . . . [as] an adventure."

When the FDA tapped him to run its India office, he seemed to be exactly what the agency needed: someone who understood U.S. government expectations, but also understood how India worked. He had grown up in Kashmir, the son of a high-ranking accountant in the central government. He went on to become a scientist with a PhD in chemistry. After a postdoctoral fellowship at the National Institutes of Health, he was recruited by the CDC to research malaria parasites. He stayed at the CDC for fourteen years and then worked in New Delhi for the U.S. Department of Health and Human Services.

For the FDA, an essential part of Lal's job was to boost *mutual reliance,* the idea that the U.S. and Indian governments could work together to improve product safety. The idea seemed obvious enough: the United States was India's biggest pharmaceutical customer, and India was one of its biggest suppliers. Of course they would want to cooperate. But in India, Peter Baker had felt like a lone policeman in an essentially lawless environment. The Indian regulators didn't behave as counterparts—in fact, they often reacted with indifference or downright hostility to his findings.

In the wake of the Ranbaxy debacle, Lal was effectively the agency's reset button in India. He was there to build a sense of camaraderie with Indian regulators, but also to communicate to Indian companies that nothing less than strict adherence to good manufacturing practices would suffice.

One of Lal's first efforts was to plan a series of workshops for the companies on how to comply with FDA standards. As he immersed himself in the issues, he cultivated sources within the Indian government and the industry. One senior pharmaceutical executive summed up for him the FDA's performance in India to date: "You guys must be in La La Land." The FDA seemed to have little idea what was happening on the ground. But Lal recognized swiftly that he did have one transformational tool at his disposal: Peter Baker.

In July 2013, one month after Lal arrived, Baker set out to inspect another Wockhardt plant in Aurangabad, in the area of Chikalthana. This was the facility that manufactured the company's biggest seller, a generic version of the beta blocker Toprol XL, an essential drug for treating heart disease and high blood pressure. Wockhardt's version, metoprolol succinate, reached about a third of the American market. Though the FDA had not yet acknowledged that something was wrong with the drug, America's patients had. *The People's Pharmacy* radio program had been flooded with complaints about this generic. This time, the company got only three days' advance notice that FDA investigators would be arriving.

Seven months earlier, Harry Lever, the Cleveland Clinic cardiologist, had sent a detailed letter of concern, specifically flagging Wockhardt's metoprolol succinate formulation, directly to Dr. Janet Woodcock, director of the FDA's Center for Drug Evaluation and Research. In his letter, he noted that when his patients took that drug, he could not properly control their chest pain, heart rate, or blood pressure. Their symptoms abated if he switched them back to the brand-name drug. "While I have no data to back up my concerns, I have a significant amount of experience with this disease and see a large number of patients," he wrote. "It is absolutely clear to me that there are differences that are clinically significant."

Within two days, Lever received a detailed response from the FDA's Office of Pharmaceutical Quality, which promised a comparative study of the brand and generic metoprolol products. At least in its correspondence, the agency seemed to be running like a well-oiled machine. As Lever waited for these results, he had little idea that an FDA investigator in India, Peter Baker, would be miles ahead of agency officials in detecting what was wrong with Wockhardt's medicine.

On July 24, Baker landed at the Aurangabad airport on his way to Wockhardt's Chikalthana plant. After he'd gotten into a hired car, a strange man wrenched open the door and jumped in, took a long look at his face, asked him where he was going and what facility he was there to inspect, then leapt out again at a red light. Baker assumed that he was being surveilled by the company. He was Wockhardt's worst nightmare, and the feeling was mutual. The prospect of another encounter with executives there filled him with dread. He knew that whatever evidence he unearthed would be met with either a surly response or one completely divorced from reality.

The other FDA investigators, Dipesh Shah and Atul Agrawal, had arrived two days earlier. They lost no time. They started the inspection at the plant's quality control laboratory. There, Agrawal scanned audit trails for each of the HPLC machines. He sat for hours and found that for ten of the instruments, folders labeled "Trial Injections" had been deleted from the hard drives, just after Baker's last inspection of Wockhardt in March. The company was erasing evidence of the unofficial pretests it had done.

As Agrawal sifted through the computer drives, he found data folders for each of the machines, labeled "Default May 2013." Inside, he found records of hundreds of trial injections, clear evidence that the practice of pretesting medicine was continuing. But Wockhardt was getting smarter. At the last inspection in Waluj, it had been relatively easy for the FDA investigators to correlate the pretests to the official tests, as they bore the same lot numbers. This time Wockhardt had tried to conceal the connection, by stripping any identifying numbers from the pretests.

Agrawal was undeterred. Over the course of the week, by overlaying the two sets of tests and comparing them, he correlated the hidden tests to the official ones. But because Wockhardt had tried to mask the connection, the investigators wanted someone at the company to confess to the pretesting, so Wockhardt would not be able to deny the connection.

Each night, the investigators gathered in Agrawal's hotel room, discussing how to best isolate the lower-level chemists and plant officials and encourage them to confess that the two sets of tests were linked and part of a company effort to manipulate results. Eventually, Agrawal got one official alone and questioned him intently. The man begged him, "Sir, please understand what I am saying. If I ever admitted to doing trials," another word for pretesting, "I would lose my job."

After eight grueling days, the investigators emerged with two partial admissions and more evidence of ghastly conditions at the plant. They found bathrooms that lacked drainage piping, with urine puddled on the floor. A manager tried to argue that the stickers above the urinals meant that they were being repaired. The FDA investigators pointed out that there were no stickers and warned the man not to lie. Elsewhere, they found a worker weighing drug samples but not recording the results. The worker told the investigators that he remembered the results and would enter them at a later time.

This was a plant manifestly out of control, governed by the principles of expedience and cost-cutting. The investigators' observations ran ninety pages. In the middle of the inspection, the managing director—the son of Wockhardt's chairman—announced that he was departing for Switzerland. The investigators wondered whether he'd pulled money out of the company and was going to conceal it in a Swiss bank account before the stock price crashed once the FDA published its findings. Months later, India's Securities and Exchange Board inquired into significant shares of Wockhardt that had been sold just before the FDA disclosed its previous enforcement action at Waluj. The company's stock ultimately lost 70 percent of its value in the wake of the FDA issuing warning letters and import restrictions to the company.

By the inspection's end, the managing director was back. At one point, he brushed by Agrawal and asked angrily, "Do you think we have quality issues?" Agrawal responded, "It's pretty apparent."

There were two other reasons why the experience at Wockhardt ended up shaking the inspectorate. In the middle of the inspection, Agrawal fell ill. The investigators suspected that the company had slipped tap water into his food. Baker was also particularly troubled by the managers' through-the-looking-glass response to the findings of trial injections. Their coordinated denials infuriated him, since other companies, once caught, had admitted their guilt. But the investigators would later learn through an Indian government official that night after night, as the investigators met in Agrawal's hotel room to plan how to build their case, company officials were listening in. They had bugged the room.

As Lal settled into his job, he recognized a bigger problem than figuring out how to engage Indian regulators in a dialogue over shared regulatory goals. The system of inspections he was expected to supervise had become so thoroughly corrupted that it was practically useless. Giving advance notice had not only encouraged rampant fakery at Indian companies but corrupted the FDA inspectorate as well. With only a skeletal staff working in the country, the FDA still had to deploy investigators from the United States for the majority of inspections. On these junkets, the investigators' standard hotel rooms, the cost of which fell inside the per diem rates permitted by the U.S. government, would suddenly be upgraded, and the inspectors would never see a bill. Some FDA investigators brought along their spouses or partners, who would then go on shopping trips, which the companies subsidized. There were golf outings, massages, and trips to the Taj Mahal, excursions that Lal called "regulatory tourism." The system, or lack of it, left the FDA investigators "captive and compromised," he observed.

Lal also scrutinized the too-close relationship that had developed between the companies and FDA investigators like Mike Gavini. His investigators had witnessed Gavini in conference rooms, inviting plant officials to bring him documents, a practice that had earned him the moniker "Conference-Room Inspector." His approach gave the companies the opportunity to fabricate documents. Investigators also heard him speak openly to plant officials on his cell phone, before and after inspections, and he sent his inspection reports to the companies for review before he submitted them officially. Gavini said that in sharing information with the companies, he achieved as much as the investigators who "keep it to themselves."

Gavini had spent years traveling alone to Hyderabad. There, as he green-lighted plants—approving roughly 85 percent of the ones he entered that had never been exported to the U.S.—the manufacturing sector grew. Now that his colleagues were going back and uncovering fraudulent practices at those plants, they wondered if he'd done any inspecting at all. In June 2011, a whistleblower from inside Ranbaxy had e-mailed an FDA official and alleged that years earlier Gavini and, more recently, another investigator had taken favors from Ranbaxy to minimize negative findings, an allegation that Gavini denies. The whistleblower went on to suggest, "When you send inspectorss to India there should be at leaast 2 per facility and preferably with different back ground and enthicity." Group inspections did offer a bulwark against corruption, as it was harder to bribe or co-opt multiple people at once. In one notable instance, three FDA investigators were each offered a gold coin at one manufacturing plant and all refused it. But at headquarters, some FDA officials responded to the whistleblower's allegations by insisting that his email no longer be forwarded internally.

An FDA spokesperson later insisted, "Any allegations of improper conduct by FDA personnel are investigated."

Some of the plants the inspectors went to were in remote towns dominated by the companies, where the hotels acted as welcoming committees and surveillance operations. The hotel staff knew who the investigators were in advance and were often familiar with the purpose of their visits. The investigators' itineraries were shared quickly throughout the tightly wired manufacturing industry, with executives across companies communicating secretly through a chat group on WhatsApp. On one occasion, Lal got a call from Agrawal, the supervisory consumer safety officer who had been sickened during the inspection at Wockhardt's Chikalthana plant: "'The hotel wants to know, where is the inspector going next?'" But Lal responded, "The hotel doesn't need to know where an officer of the U.S. government is going."

To clean up this swamp, Lal alighted on a long-overdue solution, which he pitched to officials at FDA headquarters: eliminate the months-long advance notice of inspections and company-arranged travel plans and give only short notice—or no notice—of investigators' arrival for all inspections in India. His proposal addressed the most glaring discrepancy between domestic and overseas inspections: the former were always unannounced, while the latter were announced weeks and even months in advance, with few exceptions.

By December 2013, the FDA signed off on Lal's proposal, and Lal started what came to be known as the India pilot program. He directed Agrawal to take over all communications for the FDA investigators coming from the United States. This meant that the companies wouldn't know who was about to walk through their doors or when. Lal even directed Agrawal not to tell *him* who he was sending where, so he couldn't intervene, even if he'd wanted to. Agrawal went a step further. He arranged the investigators' travel through the U.S. embassy instead of the FDA's India office, to bypass the office staff. Even with short notice, he switched up inspection dates to give the companies less time to prepare. The India pilot program, as conceived by Lal and implemented by Agrawal, would give the FDA the most candid look yet at what was happening inside Indian plants. It had no parallel anywhere in the world: India would be the only country, other than the United States, where FDA investigators would arrive without notice.

On January 2, 2014, a Thursday, the FDA's India office gave Ranbaxy officials short notice that investigators would be coming the following Monday to reinspect the Toansa plant, which was responsible for the glass fragments found in millions of tablets of generic Lipitor. The contamination—already the target of class-action lawsuits in the United States on behalf of consumers—had never been fully explained. Agrawal wanted the investigators to get the real picture of what was happening at the facility. Without telling the company, he decided to move the inspection to Sunday morning instead. He made plane reservations outside of the official FDA travel system. Investigators arriving unannounced on a Sunday was something no Indian drug executive would plan for or even fathom.

In the early hours of a Sunday morning, FDA investigators Peter Baker and Dipesh Shah stood outside Ranbaxy's huge Toansa plant and showed their IDs to the guard at the security gate. From the outside, the plant looked quiet and empty, which was exactly what Baker wanted.

The men proceeded swiftly to the quality control laboratory with the hope that, at least for a while, they could move unnoticed. In the lab, they were stunned to see a hive of activity. They found dozens of workers hunched over documents, backdating them, in preparation for the investigators' anticipated arrival the following day. On one desk, Baker found a notebook listing all the documents the workers needed to forge in anticipation of his arrival. There were Post-It notes stuck to every surface, noting what data to change. Workers were backdating stacks of partially completed forms—for employee training, laboratory analysis, cleaning records—that were supposed to have been filled out contemporaneously.

As Baker and Shah took in the operation unfolding around them, the analysts largely ignored them, figuring that they were company consultants. But as higher-ups arrived and word got out that the men were from the FDA, workers frantically stuffed documents into desk drawers. By showing up unannounced, the men saw things they never would have otherwise as they pushed through the facility: vials stuck in the back of drawers; a sample preparation room swarmed by flies because windows were stuck open with piles of trash directly outside—or, as Baker noted in his final report, "flies TNTC" (too numerous to count). The inspection resulted in a warning letter and an embargo of the plant's drugs.

In theory, it shouldn't have mattered whether U.S. regulators announced their inspections in advance or not. Drug manufacturers are supposed to follow good manufacturing practices *all the time.* Well-run plants operate in a state of continuous regulatory readiness. Compliance isn't a part-time thing. As Lal said, "The regulations are non-negotiable. You can't have the month of January as your good manufacturing month."

But in India, the FDA's new program of short- or no-notice inspections exposed widespread malfeasance that had previously been hidden. By showing up unannounced, the investigators uncovered an entire machinery that had existed for years: one dedicated not to producing perfect drugs but to producing perfect results. With advance notice and low-cost labor, the plants could make *anything* look like *anything.* "You give them a weekend, they'll put up a building," as one FDA investigator put it.

The investigators found a bird infestation at one sterile manufacturing site, a snake coiled next to laboratory equipment at another. At one plant, Baker went straight to the microbiology laboratory and found the facility's paperwork for its sterility environment testing in perfect order: microbial limits testing, bacterial endotoxins, all the samples with perfect results. Yet the samples didn't exist. They were testing nothing. The entire laboratory was a fake. Before long, the investigators uncovered a second plant that had also faked all its data proving that the plant was sterile, a "shocking" finding, as one FDA official later described it.

Under Agrawal's direction, the investigators flagged violations at company after company, which resulted in a growing number of 483 findings and warning letters. Before long, drugs from forty-one plants in India had been restricted from the U.S. market. Outrage grew in the tight-knit industry, and U.S. embassy officials increased their scrutiny of the agency's activities in India. One U.S. State Department official admonished Agrawal that the investigators were "acting like cowboys." On one level, the conflict made sense. The State Department's role was to promote American economic interests in India. The FDA investigators' mission was to guarantee that consumers in the United States were safe. And yet the displeasure of both the U.S. and Indian governments hung over Lal's investigators.

India's drug manufacturers—who were adept at getting their own regulators to follow along meekly—fought back. Not only did some of their representatives publicly disparage the new inspections and claim that they sprang from anti-Indian bias. But they engaged in a dark battle of wits. Though they no longer received advance notice of inspections and who the investigators would be, they worked to find out who was coming, and when. The companies surveilled the airports and hotels. They tried to stay a step ahead of investigators at the plants. Once they got wise to the investigators' practice of rifling through trash cans looking for records, they removed all trash from the plants.

Like most battles, this one escalated. Baker and some other investigators took to searching through areas where the plants aggregated their trash. This often involved scaling walls and leaping into dumpsters. In one dumpster, investigators unearthed heaps of complaints from patients, which had been tossed with no follow-up. As his investigators worked, Lal blocked out growing skepticism from FDA headquarters and the U.S. Embassy demands that they ease up—especially before the Indian elections, so they didn't generate any negative press for the Indian government. Lal's response was firm. "The inspections are the inspections. I am not going to change the inspections."

The unannounced inspections clearly marked a new era, one that put the FDA's investigators on a collision course not only with India's powerful generic drug executives and the regulators most often seen as their protectors, but with their own agency as well.

India's top drug regulator, Dr. G. N. Singh, ran the Central Drugs Standard Control Organization (CDSCO) from a run-down building fringed by sparse bushes on Kotla Road in New Delhi. Inside at the reception desk, one "E" in the RECEPTION sign dangled, despite an effort to keep it up with Scotch tape. On the second floor, the international collaboration office was frequently padlocked shut during office hours.

The CDSCO was the Indian equivalent of the FDA, and the downtrodden air of its headquarters seemed to reflect the accusation that had surrounded it for decades: that the CDSCO had done far more to protect India's drug companies from regulation than it had to protect Indian consumers from bad drugs. Forty years of national reports had excoriated India's drug regulators as inept, understaffed, and corrupt and called for the overhaul of the CDSCO. With the Indian market awash in poor-quality drugs, the reports had noted the agency's paralytic inertia, yawning staff vacancies, and missing files.

The most common accusation against Singh's agency was that his bureaucrats colluded with both drug makers and supposedly impartial medical experts. Drugs that had been banned the world over were approved by the CDSCO for Indian consumers after experts in different parts of the country submitted identically worded evaluations apparently drafted by drug companies. "The nexus between the regulators and the industry is so tight, you won't be able to break it," as one prominent Indian pharmaceutical journalist put it.

In January 2015, Singh agreed to a rare interview with an American journalist. In his outer office, a receptionist kept a paper logbook of visitors, where the CEOs of India's top drug companies scribbled their names. In his small inner office, Singh remained polite and animated, if a little unhappy to be questioned. Over the next half hour, he spoke of "joining hands with other counterparts" and explained that the CDSCO would "never compromise" in the effort to protect patients. He also declared that there was "no nexus" between his agency and the drug companies he regulated. But in the next breath, he made clear why the companies had so little to fear from CDSCO regulators: "We always give them the chance to improve," he said. As for whistleblowers, most of the time their information is "fake," he said.

About a year earlier, he had given a more candid interview to an Indian newspaper, the *Business Standard,* explaining that even findings of flies in the plants and hair in pills was not reason enough for him to shut down Indian manufacturing plants. "If I have to follow U.S. standards in inspecting facilities supplying to the Indian market, we will have to shut almost all of those," he acknowledged.

For decades, FDA officials had been struggling to negotiate agreements with foreign governments over what was called "mutual recognition"—the notion that U.S. regulators could collaborate with, and rely on, foreign regulators to make judgments about what plants to approve. But even when dealing with developed nations with similarly high regulatory standards, FDA officials battled each other over whether any country's standards matched those of the United States. One camp within the FDA argued that the agency should be able to use the inspections of trusted foreign regulators in lieu of their own. Another camp vehemently opposed mutual recognition on the grounds that U.S. standards were superior to those of any other nation.

With the FDA struggling to enforce its own rules around the globe and unable to agree on whether any country's regulators were as good as its own, the FDA rebranded its diplomatic strategy as "mutual reliance." Under this strategy, regulators from different countries could engage in a discussion about health and safety, but without using each other's inspections interchangeably. But even an agreement on mutual reliance with India seemed beyond the agency's reach. By the time Lal started working at the FDA, Indian and U.S. regulators had spent three years trying to negotiate the most basic mutual reliance agreement. The process was so fraught that FDA officials had refused requests by their own staff to conduct certain unannounced inspections in India—with the claim that doing so would disrupt these larger negotiations.

But finally, the moment had come. Lal helped oversee the completion of the "Statement of Intent," a four-page document stipulating that the United States and India would share information, collaborate, and include each other's regulators at inspections, along with a disclaimer that this did not "create rights or obligations." In February 2014, the FDA commissioner, Dr. Margaret Hamburg, traveled to India for a ten-day trip and signing ceremony in New Delhi.

The trip underscored the importance of India's role as one of the largest pharmaceutical exporters to the United States. But the staged show of diplomacy could not conceal the ongoing battle over regulatory standards. At a private meeting, the FDA's press officer quietly expressed anguish over a mistaken seating arrangement, when Hamburg found herself sandwiched between a Ranbaxy executive and a Wockhardt executive. The Ranbaxy executive took that moment to lobby Hamburg. The company needed money to fix its quality problems, the executive asserted, which it could get only if Hamburg lifted the import restriction on some of its products. Hamburg politely refused his plea.

No amount of photo ops could obscure the high stakes of the visit—or the gulf between U.S. and Indian standards. At the closing event of the trip, G. N. Singh shot back: "We don't recognize and are not bound by what the U.S. is doing and is inspecting. The FDA may regulate its country, but it can't regulate India on how India has to behave or how to deliver." This wasn't quite true. So long as Indian companies wanted to export drugs to the U.S. market, they were required to follow U.S. rules—and to face sanctions when they didn't. After her trip, at Hamburg's request, Lal drafted a proposal to rework how the FDA staffed foreign offices. He envisioned a "specialized, highly trained and qualified" workforce that would make a years-long commitment to serve overseas and become a "go-to" group for emergency assignments.

The proposal would remedy the problems behind the FDA's anemic recruitment to foreign posts: a lack of clear "career progression and promotion opportunities." Those who served overseas often returned to the FDA's U.S. headquarters without a guaranteed job, and sometimes had to accept demotions. Instead, the superior training, pay, and professional pathway that Lal proposed, similar to that for State Department officers, would help cultivate more elite investigators like Peter Baker, a plan the FDA seemed certain to embrace.

As Baker's extraordinary findings rippled through the FDA, the reaction was decidedly mixed. In March 2014, Lal and Agrawal nominated Baker for the agencywide FDA Investigator of the Year Award, which he won. One official referred to Baker and his colleagues as Lal's "Navy Seals." The agency's most ardent public health advocates called Baker "noble" and "one of a kind." But other FDA officials grew concerned that his methods were unsustainable and impossible to replicate. The agency couldn't require its investigators to leap into dumpsters.

The mixed reaction to Baker's findings was bigger than a dispute over methods. His findings required the FDA to punish companies and restrict the importation of their products. This put the agency in a bind. It wanted to approve more generic drug applications, reduce drug shortages, and show those numbers to Congress. Restricting companies and their drugs would have the opposite effect. Once Baker's reports arrived at FDA headquarters, they entered a murky deliberative process clouded by politics and horse-trading that allowed companies to dodge the worst sanctions.

For example, Baker's findings in October 2012 of serious sterility lapses at a plant run by Hospira Healthcare India in Chennai led the Division of International Drug Quality to recommend that the company receive a warning letter and have the importation of its products restricted. Instead, agency higher-ups reversed that decision, rescinded the "import alert," and decided to send a private or "untitled" letter to the firm that would be invisible to public scrutiny. Furious, Carmelo Rosa, the head of the Division of International Drug Quality, wrote to the deputy director of CDER's Office of Compliance: "Everyone will do as told, but unfortunately some are very upset as the entire compliance review process has been undermined." He added that others in the division "plan to look for opportunities elsewhere as decisions are no longer based on science, policies and regulations, but politically motivated."

The more violations Baker and his colleagues found, the more agency officials intervened. Lal, too, was amassing enemies. At the FDA India office, he viewed a group of policy analysts and some investigators as unproductive. Each one cost U.S. taxpayers roughly $500,000 a year, once salary and various living expenses were calculated. In Lal's view, some of these analysts did little to advance the interests of public health. Funds were being wasted. He discovered that almost $300,000 in furniture had been ordered for an FDA outpost in Hyderabad that was never slated to open, a purchase approved by bureaucrats in Maryland. It was routine for the entire office to decamp for conferences, which were effectively paid vacations. As Lal demanded more productivity from some employees in the FDA-India office, they, in turn, filed Equal Employment Opportunity Commission (EEOC) complaints against him, alleging discrimination.

Then there was Mike Gavini. Though he was a welcome sight to most Indian pharmaceutical executives, he was disliked by his colleagues, who continued to find multiple violations at plants that he'd previously inspected and cleared. He, in turn, had nothing but scorn for the revolution that Lal was leading. He thought Lal's investigators were unfairly treating India's drug makers like criminals, from isolating and questioning the lowest-level workers, who couldn't communicate well, to showing up unannounced on Sundays. "Investigators cannot become god," he later said, adding that to investigate fraud, they "have to look at 3,000 chromatograms. Who has the energy to do that—aside from Peter Baker?" Gavini had a theory about what was happening, and it had little to do with company conduct. "People [at FDA] were getting greedy for awards," he said. They were cracking down to advance their careers, with poor results. Ranbaxy had been the "most respected company," as he put it. But the FDA "killed the company by drumming all the good people out of it."

In December 2013, just as Gavini planned to relocate back to the United States, Lal sent a confidential letter to the FDA's Office of Criminal Investigations that detailed long-held concerns about Gavini. Lal alleged that the investigator had closed-door meetings with pharmaceutical executives, sent them drafts of his inspection reports before they were officially released, and was lenient toward companies based only on a promise of improvement. He noted rumors, though there was lack of evidence, of Gavini accepting gifts.

Indian firms had clearly gotten wind of information before it had been made public, Lal said. He urged the agency not to assign Gavini to a new position that allowed him access to any "inspection-related materials" until a review of his conduct was completed. In response to these allegations, Gavini later told a journalist that he never "took any money, any payoffs, to clear a company," adding that he "would be rich by now" if he had.

No review of Gavini took place that he is aware of. But in late April 2014, Lal was directed to return to the United States and Atul Agrawal was directed to stop running the day-to-day operations of the FDA's India office. By June, Lal had been terminated.

Their removal came as a shock to some of Lal's investigators, who had been summoned to New Delhi to be interviewed by the U.S. embassy's human resources staff about office tensions. They viewed what occurred as a "coup," orchestrated both by those in the office who hadn't wanted to work and by the powerful generic drug lobby. They felt that they had been on the brink of changing India's drug industry for good, progress that would now be halted without Lal.

As rumors swirled, it turned out that Lal himself had fallen under the suspicion of the FDA's internal affairs office, which, among other things, had begun examining Lal's transfers of money to the United States related to his ownership of property in India, which he contends that he disclosed to both the FDA and the U.S. embassy on taking the job. Lal vehemently denied any impropriety and alleged that he'd been terminated because he'd exposed incompetence and misconduct at the FDA's India office. His dispute with the agency was resolved, he was reinstated, and he ended up retiring from the FDA in 2015.

In the murky battle over his ouster, one thing seemed clear: his departure would set back the cause of public health and drug safety for American consumers. In turn, Lal was left deeply troubled by his experience: "I see faces in the U.S. that consume these drugs. They're not just numbers to me."

Peter Baker remained in India, but a dark cloud hung over him, and not just fear that his life was in danger, after some of the threats he'd faced. His greater fear was for every American consumer he was there to protect.

He felt scarred from dealing with combative and nonresponsive executives who had purposely made low-quality drugs and tried to conceal the evidence with faked records. And from the plant managers he caught red-handed, who disputed that the evidence he placed in front of them even existed. The fact that they knowingly imperiled patient lives and showed no remorse seemed to Baker nothing short of *evil.*

Most of the plants he inspected in India made finished doses—completed capsules, pills, and tablets that were ready for patients to take. Many were aseptic plants, which meant that they had to operate with perfect sterility. Each vial represented a patient's life. In inspection after inspection, he—and he alone—stood between the American public and potentially dangerous medicine bound for the U.S. market. The stress was unrelenting.

While he was still in New Delhi, symptoms began to overtake him. He was dizzy, suffering from anxiety, and experiencing vertigo. He went to the embassy psychiatrist, who diagnosed him with post-traumatic stress disorder. The twenty months of psychic combat had taken their toll.

CHAPTER 28

STANDING

SEPTEMBER 17, 2014

New Delhi, India

One could hardly have blamed Dinesh Thakur if he'd taken the money he'd earned as a whistleblower and retired to enjoy time with his family. But more than a year after the U.S. government's settlement with Ranbaxy, Thakur was waiting impatiently in the shabby outer office of Room 348-A of India's Ministry of Health and Family Welfare. For three months, he'd called, emailed, and even sent letters by registered mail to schedule an appointment with the country's health minister, Dr. Harsh Vardhan. When that failed, he'd turned to Sonal's uncle, the chief minister of Chattisgarh, to arrange the meeting. Finally, the day had arrived. But thirty minutes, an hour, then two hours crept by as Vardhan tended to other matters.

In his clamor to see Vardhan, Thakur didn't have a clear plan. He had a belief: that the campaign he'd begun when he'd first contacted the FDA nine years earlier remained unfinished. No one who worked at Ranbaxy had been prosecuted. The former executives who had overseen the fraud had fanned out to companies across the industry. Thakur's allegations against Ranbaxy had forced the FDA to better scrutinize India's drug companies. But what the agency uncovered—the widespread, willful manufacturing of poor-quality drugs—still persisted, largely unchecked.

To Thakur, it seemed logical that he should play some role in helping India to help itself. He knew more than almost anyone about the perilous shortcuts India's drug companies were taking. He'd thought long and hard about possible solutions. He longed to be of use. Though not a celebrity, he had gained notoriety—as a whistleblower who'd not only made a fortune by doing the right thing but was still alive, in a country where whistleblowers often wound up dead. But not everyone saw him as a positive change agent. In the wake of the Ranbaxy settlement, pharmaceutical lobbyists had maligned him to journalists, accusing him of being "anti-national," and suggested that his efforts reflected a "foreign hand" trying to bring down Indian companies. Thakur suspected that their accusations had made it even harder for him to get on Vardhan's calendar.

When Thakur was finally ushered into Vardhan's office, the minister made his lack of interest clear. Vardhan kept one eye on the news, watching reports of flooding in Kashmir, while his secretary helped him arrange travel. After more waiting, the brief meeting ended with Vardhan asking Thakur to send whatever it was he wanted to say in writing. Within a month, Thakur sent Vardhan a biting three-page letter. "I truly appreciated your attention to what I had to say while you discussed your schedule with your secretary in those five minutes," he began. He went on to explain that, despite sanctions by international regulators, Indian drug companies had done little to change their practices or attitudes and were being egged on in their defiant stance by a regulator, the Central Drugs Standard Control Organization, which Vardhan himself had called out publicly as a "snake pit of vested interests."

Without a new mind-set, Thakur warned, "this once prosperous industry will wither away taking with it the prospects of thousands of well paying jobs for the people of India." The first step, he advised, would be for India to acknowledge the quality problems with some of its drugs—the same suggestion that his mentor Raj Kumar had given Ranbaxy's executives years earlier: *come clean*. To that end, Thakur offered his services. "As a person of Indian origin, an avid public health worker and someone who wishes to see the pharmaceutical industry grow and prosper in India, I came to your office to offer my service, my knowledge, my experience and my commitment to help you solve this problem."

He never received a response.

Thakur moved through the Indian bureaucracy, trying to find someone interested in reforming the drug industry. Instead, he was greeted with silence, indifference, or downright hostility. He was repeatedly told that because he was a U.S. citizen, his efforts would be viewed as anti-Indian and as a conspiracy by the West to defame the Indian pharmaceutical industry.

Even in the best circumstances, fixing anything in India was a formidable task, and Thakur was operating from a no-man's-land familiar to many whistleblowers. No company would hire him. Drug makers viewed him as an enemy. The government wanted him to go away. He had become a professional exile. He was also essentially stateless. A naturalized American citizen and an overseas Indian citizen with a lifelong visa, he felt that he belonged to both countries and to neither. He shuttled back and forth between Sonal and the children in New Delhi and the family condo in Tampa, more out of habit than necessity. Each time he took the journey, it seemed less clear to him where he belonged—except among other whistleblowers.

Five months after the Ranbaxy settlement, Thakur had donned a dark suit and gray silk tie and stood before a crowd of hundreds at the Grand Hyatt Hotel in Washington, D.C., to receive a Whistleblower of the Year Award from the Taxpayers Against Fraud Education Fund (TAFEF). It was the same group that had first helped him find a lawyer back in 2007. He used the event to speak about the importance of role models in public life, pointing to the example that Raj Kumar had set at Ranbaxy and the fight that Debbie Robertson had waged at the FDA.

An event several weeks earlier, also organized by TAFEF, had helped Thakur feel more at home in a community he could rightfully call his own. For the first time in its history, TAFEF had invited all the whistleblowers it had helped over the years to spend a weekend together in the Florida Keys. About eighteen of them came. Some were unknown, while others were marquee names, like the financial sleuth Harry Markopolos, who had first alerted the Securities and Exchange Commission to Bernie Madoff's Ponzi scheme. Cheryl Eckard, a former quality assurance manager at GlaxoSmithKline, helped host the event. She'd been granted $96 million, the largest whistleblower recovery ever, for exposing nonsterile manufacturing at a GSK plant in Puerto Rico.

Patrick Burns, the acting executive director of Taxpayers Against Fraud, had been worried about getting whistleblowers together: "They don't herd well," he later said. But they all shared something in common. As he put it, "These are people who have chosen integrity, and they've paid a price for it." The whistleblowers spent the weekend fishing and sharing meals, and they visited the house where the writer Ernest Hemingway had lived and worked. Famously, dozens of six-toed cats lived at the house, and it seemed about right that the whistleblowers, often maladapted themselves, should be visiting the polydactyl creatures.

Thakur found the weekend cathartic. He became fast friends with Cheryl Eckard, who announced delightedly to Burns, "These are my people." As Burns later reflected: "Whistleblowers are like polar bears at a roadside zoo in Kansas. They're pretty sure there's another polar bear in the world, but they've never seen one."

In October 2014, the Thakur family moved to a larger and grander home within the gated community of World Spa West in Gurgaon. Sonal hoped it would be a joyful place to restore her fractured family. She had taken exquisite care with the interior: light, patterned window trimmings, dark wood furnishings, and whimsy in the children's rooms, with a princess theme for Mohavi and stars and planets for Ishan. Downstairs, a sunken living room with glass doors led into a fenced-in backyard. The Thakurs took their morning coffee together in an open family area. Thakur had set up a home theater in the basement where he could watch American movies. Next to that was his home office, where he'd hung framed articles and photos, celebrating his work as a whistleblower.

Even as they moved in, Sonal and Thakur bickered. She felt alone and unappreciated in the effort she'd made. He felt restless and ill at ease in their new world of comfort. The question of why he'd chosen to intervene at Ranbaxy hovered over their marriage. "Ranbaxy employed twenty thousand people," Sonal argued. "Why did you have to put me and the family through all of this?" She still felt keenly her lack of agency, even in the decision to get married at all, since it had been arranged by their parents.

Thakur's answer never changed: "I couldn't sleep at night if I didn't do it."

The children softened his rough edges, but too often he broke off moments with them to retreat to his office, where he wrote blog posts about drug quality and took calls from reporters with whom he'd cultivated relationships.

He fought with the workmen who came and went at his new home and complained about the low standards that permeated the country, at one point fearing that his office equipment was in peril due to their shoddy electrical work. As he tried to explain to the workmen the right way to install the equipment, Sonal observed to a visitor, "He's trying to do it right, the American way in India." For years the idea of America—with its exacting standards and promise of equal justice—had sustained Thakur. He'd spent months at a time away, using FaceTime to watch his kids grow up, because he believed that the U.S. system would expose the truth and protect patients. It had worked, to a point. But now, as he looked around his reclaimed country, the low standards gnawed at him. It was not just the wiring in his own home, but the disparate impact of India's low standards on the nation's poorest. "You can't even go a mile outside your house," he said, without seeing "how hard it is to make ends meet." The workmen laboring over his new home biked miles to work each day. If they needed medicine, they would spend a day's wages to buy it. The fact that their medicine was of the lowest quality and was barely regulated filled him with rage.

When he looked around his privileged redoubt, he felt little satisfaction. Instead, he felt troubled and drawn to a battle that lay before him. As he tried to explain to an acquaintance, "There is some level of onus on me to try and do something with public health. I am in this position that I know things. It's a cop-out to say it's not my problem."

As Thakur pursued his quest to change the generic drug industry, a band of unlikely colleagues coalesced around him. One was Dr. Harry Lever, the Cleveland Clinic cardiologist. Joe Graedon, the host of NPR's *The People's Pharmacy*, became an ally, as did Roger Bate, an economist specializing in health policy at the American Enterprise Institute, a conservative D.C. think tank. The men were joined by the Canadian lawyer and biologist Amir Attaran, who had been studying the failure of international law to address substandard medicine.

Before long, a paper published in the *Journal of Clinical Lipidology* caught their attention. From 2011 to 2013, one of its coauthors, Preston Mason, a Harvard-affiliated scientist, collected thirty-six samples of generic Lipitor from fifteen countries, manufactured by more than two dozen generic drug companies. When he tested the chemical composition of each, Mason was stunned by his findings: thirty-three of the samples had impurity levels high enough to render them ineffective. Even samples manufactured by the same company but sold in different countries contained widely different impurity levels—proof that some generic companies were making different versions of the same product, a high-quality drug for the West and a poor-quality one for low-income countries.

Before long, Mason had joined ranks with Thakur and his new colleagues. What started as a group of like-minded professionals connecting over email briefly became a formal advocacy group, the Safe Medicines Coalition. They tried to alert the public that America's best public health bargain—low-cost drugs made overseas—had been dangerously compromised through negligent manufacturing and haphazard regulation.

The coalition held panel discussions, wrote editorials, aided journalists, and even spent whole days meeting with U.S. congressional staff. They organized Capitol Hill briefings that ranged from well attended to desolate. Separately, each member of the group had spent years teasing out pieces of the elaborate puzzle. At every opportunity, Thakur made the point that the incompetence and corruption of India's regulators was not a distant problem, but one with a direct impact on the quality of U.S. drugs. Without a functioning local regulator to partner with, the FDA could get little purchase on the willful manufacturing of low-quality drugs by Indian companies.

The group's efforts got a smattering of press coverage and even drew defensive attacks from the FDA. But with public fury fixed on ever-rising drug prices, their message—that the nation's most affordable drugs were compromised—was unwelcome.

Thakur persisted. He contacted the nongovernmental organizations that purchased medicine in bulk for the world's poorest patients: the Clinton Foundation, the Global Fund, the Gates Foundation,

Doctors Without Borders. These organizations focused on the cost of medicine and the world's access to it, but in Thakur's view did not prioritize issues of quality in their purchasing. Thakur asked for meetings. Most did not respond. One operations officer at the Global Fund replied, and Thakur flew from New Delhi to Geneva at his own expense to meet with him. There, he urged the Global Fund to add language to its purchasing contracts stating that medicine had to be of a certain quality. Nothing came of the meeting.

On January 26, 2015, in New Delhi, India held its annual Republic Day Parade, a lavish event that showcased to the world the nation's sophistication and military might. The gleaming display of missiles, tanks, Indian dancers, and military officers riding garlanded camels stretched for miles. The main message of the parade was not military, but commercial. Its centerpiece was a float bearing a spectacular metal lion made of thousands of turning cogs and gears. It bore the slogan "Make in India," reflecting Prime Minister Narendra Modi's signature effort to sell India as the world's next engineering and manufacturing center. The message was aimed, in part, at the parade's guest of honor, U.S. president Barack Obama.

The lion was the apex of a campaign that Modi had announced from the ramparts of Red Fort in New Delhi six months earlier called "Zero Defect, Zero Effect," an effort to make the quality of India's goods a point of patriotic pride. As Modi said at the event, "We should manufacture goods in such a way that they carry zero defect, that our exported goods are never returned to us." By zero effect, he meant that the manufacturing should not have a negative impact on the environment. But just three days before the Republic Day Parade and the debut of the "Make in India" lion, Modi's zero defect campaign suffered a serious blow. The European Medicines Agency, Europe's top drug regulator, recommended suspending from the European market seven hundred drugs made by a wide array of manufacturers, which all had one thing in common. The data proving that they were bioequivalent had come from a single Indian company, GVK Biosciences, a contract research organization hired by drug makers to test their medicine on patients.

In May 2012, a former employee of GVK Biosciences had written to five inspection agencies around the world, including the U.S. FDA, alleging that the company had routinely manipulated data from patient blood tests to make the drugs appear bioequivalent. The allegations were so detailed and alarming, and implicated so many drugs on world markets, that six weeks later investigators from four regulatory agencies, including the U.S. FDA, visited the company's clinical pharmacology unit in Hyderabad. Among them was a pioneering French investigator, Olivier LeBlaye, who, eight years earlier, had been the first to detect the fraud at Vimta Labs, the contract research organization used by Ranbaxy. His findings had sparked Dr. Kumar's suspicions and set the Ranbaxy case in motion.

As LeBlaye inspected GVK, he suspected fraud but couldn't prove it. Over the next two years, as he and French regulators scrutinized applications filed with GVK data, they found that electrocardiograms, which test patients' heart rhythms, appeared identical and were likely falsified in nine different studies. LeBlaye laid out his findings in an explosive 2014 report. Though GVK officials tried to refute the allegations, European regulators sided with LeBlaye and drew an obvious conclusion: GVK's willful falsification of *some* data made it impossible to trust *any* of its data.

As the scandal spiraled out of control, Indian government officials directed their ire not at GVK but at the whistleblower, Konduru Narayana Reddy, and ultimately at the European Union. By the time European regulators announced their decision to withdraw the seven hundred drugs, the whistleblower was sitting in a jail cell, having been accused by GVK of data theft, tampering and fabrication, breach of company trust, and threatening company staff.

The whistleblower Reddy had neither the aplomb nor the mental discipline of Dinesh Thakur. On his release, he sent meandering emails to dozens of investigators, politicians, and journalists around the world, alleging that his imprisonment had devastated his career, family, and livelihood. He had not found a protected pathway for making his allegations, in part because there wasn't one in India. But he had not necessarily been wrong.

At the highest levels, the Indian government accused European regulators of ulterior motives. India's top drug regulator, G. N. Singh, told an Indian newspaper there was a "bigger game being played out here": Big Pharma companies, he claimed, had contrived the GVK incident to malign Indian generic drug makers. In this instance, the well-worn allegation made no sense. The companies hurt by the European Union's decision were not just Indian companies but GVK's clients around the world, the drug companies whose products had been pulled from the market. Nonetheless, the Indian government canceled upcoming negotiations with the European Union over a new free trade agreement, vowing not to reinstate talks until the ban on the drugs linked to GVK was lifted. Prime Minister Modi even personally lobbied German chancellor Angela Merkel to lift the ban.

As the conflict exploded, Thakur and his band of colleagues followed it closely. "Hold onto your hats! This is incredible," Graedon emailed the group. Thakur was not surprised in the least. Instead of serving as a check on the pharmaceutical sector, Indian regulators were serving as its Praetorian Guard. Even less surprising was that the GVK company's chairman, D. S. Brar, had been the managing director and CEO of Ranbaxy during its most feverish development, from 1999 through 2003. He had presided over the meeting in Boca Raton, when top company executives had decided to launch the company's dangerous Sotret drug onto the U.S. market, despite knowing it was defective. Nonetheless, Brar, a titan of Indian industry, had emerged unscathed. He sat on corporate boards across the world, from the Wall Street investment firm Kohlberg Kravis Roberts to the Indian subsidiary of Suzuki, the Japanese car company.

Later on, GVK's CEO, Manni Kantipudi, lamented what he viewed as the unfairness of LeBlaye's conclusions, but did not directly deny them, pointing instead to "a difference of opinion between auditors." By mid-2016, GVK had shut down both of its laboratory units where LeBlaye had found his evidence and quietly left the business of bioequivalence testing.

Meanwhile, as the FDA barred over three dozen Indian drug plants from shipping their medicine to the United States, India's drug regulators defended the industry.

With few allies in India's government, Thakur studied the country's broken regulatory system on his own. It was governed by a seventy-year-old law that divided oversight of drug approval and manufacturing between a central authority and thirty-six state and territory regulators, each of whom pursued enforcement differently. It seemed obvious to Thakur that only a new law, or a total overhaul of the existing one, could begin to fix the problems. Decades of parliamentary standing committee and expert reports had urged the same changes, but none had been heeded.

As Thakur consulted his newfound cohorts and evaluated his resources, he came to believe that reforming India's regulatory system might be the best way not just to save India's industry but to improve the entire world's drug supply. If India overhauled its own standards and enforced them, then everyone in the world who bought India's drugs would benefit. After failing to find like-minded partners inside of India's government, he settled on a new path to his goal: he would get there by suing India.

Three days after the Republic Day Parade, the Thakur household in Gurgaon was abuzz with preparations for a long-planned housewarming party. Electricians and gardeners came and went.

Caterers and florists made deliveries. Sonal's closest friends cycled through the house to consult on outfits and food and practice a dance they had choreographed for the event.

Sonal viewed the party as a signal to her community that, after a long absence, the Thakur family was back. She had invited a wide swath of the World Spa residents, many of them top executives at the global companies that crowded Gurgaon. But as Sonal's hopes for the event rose, Thakur's spirits sank. It was not simply that he despised parties and always had. He also questioned the very notion of celebrating in their bubble of good fortune when all around them lay misery. Nonetheless, Thakur donned a crisp white tunic and matching pants, known as a *kameez shalwar.* Sonal wore a blue-green silk sari and sheer red scarf banded in gold. The caterers set out a lavish spread. In the garden, silk pillows adorned the risers and a small band set up at one end.

Guests began to arrive. The women wore sequins, jeweled dresses, and tunics of crushed velvet, their hair done perfectly. A photographer darted among them. As the party grew and Sonal and her friends danced in a circle, a striking woman made her way through the crowd. She was poised, with dark hair and red lips, and wore an impeccable white sari, edged in gold. It was Abha Pant, Ranbaxy's former vice president for regulatory affairs. Thakur, as was typical, had not complained when he learned that Sonal had invited her. She was their neighbor and seemed to belong in their circle as much as anyone. Yet Thakur's information had almost led to her indictment. Now she was a guest in Thakur's elegant home, paid for with the proceeds from his blowing the whistle on the company she'd helped to oversee. Nonetheless, they chatted cordially. She readily joined in as he gave a tour of his home to a handful of guests.

In his basement office, he stood silently as Pant studied the walls: the 2014 medal from the Association of Certified Fraud Examiners for "Choosing Truth over Self"; the picture of Thakur on the cover of *Fraud* magazine under the headline "Fighting a Culture of Fraud";

the framed text of his Joe A. Callaway Award for Civic Courage; a photograph of him surrounded after the settlement by jubilant lawyers from Beato's firm, who had all autographed the edges. She took it in silently. Upstairs, as the party extended late into the night, Thakur left his guests and retreated back into the basement.

Pant remained in the living room, sipping a glass of wine and chatting with a guest as she reflected on changes in the generic drug industry. She spoke of an important shift in FDA regulations. No longer did it matter who was first, second, or third in line in the FDA parking lot. Now any company that filed on a given day, regardless of the time stamp on their application, could be considered a "first filer" and share in the profits of a drug launch. This had taken a good deal of heat out of the market, with no more pitched tents or limousine campouts. Pant observed, "Too bad all the fun is gone."

Throughout 2015, Thakur poured his frustrated energy into building a legal case against India and its dysfunctional regulatory system. He hired a team of lawyers, who submitted well over one hundred requests for information to government agencies, using India's "right to information" law. It was a hunt-and-peck method, but Thakur believed that it was the best way to gather irrefutable proof that India's regulators had failed to protect citizens. The queries turned up evidence of a corrupt and threadbare regulatory system, with entirely different standards and rationales across India's thirty-six states and territories. Dangerous or barely effective drugs were approved for no rhyme or reason. Paperwork related to controversial decisions went missing. Even when foreign regulators found dangerous conditions in India's plants, the Indian government either ignored those findings or assailed the regulators, rather than investigate themselves.

By January 2016, his lawyers were ready. They had crafted two lengthy petitions stating that the fragmented structure of India's drug regulation system was not just broken but unconstitutional. They filed these with the Indian Supreme Court as public interest lawsuits, a legal mechanism that allows citizens to petition the nation's highest court on matters of social justice. Now the court would need to decide whether to grant a hearing.

As local news programs and newspapers reported on the lawsuit, Sonal begged Thakur not to proceed and accused him of doing "whatever you want, without considering the impact it has on us."

Thakur tried to defend himself, "Somebody's got to do it."

Even Thakur's son, Ishan, now a teenager, demanded, "Why are you trying to do this? You're going out there and drawing all the wrong kinds of attention."

Thakur not only remained undeterred but, as he worked on the case, his energy and confidence seemed to come back. His frustrations ebbed. Friends noticed that he appeared more at ease and better rested. In that time, he often thought back to his grandmother, who had drilled into him the teachings of the Bhagavad Gita, the ancient Hindu text that called for selfless action. While reading to him and his siblings at night, she had emphasized that emotions—fear, excitement, anxiety, joy—while all part of life, were transitory. One's responsibility or duty was a better guide for deciding what actions to take.

On the morning of March 10, 2016, a Thursday, Thakur awoke in his condo in Tampa, Florida, having traveled to the United States for meetings that he could not reschedule. He made coffee and opened the patio doors to a sweeping view of the Gulf of Mexico. He loved watching the birds swoop across the water. Sometimes, early in the mornings, he could even catch a glimpse of dolphins. That evening, he dozed off watching the primary debate among Republican presidential candidates, which was broadcast from nearby Miami.

As Thursday night fell in the United States, Friday morning

was beginning in New Delhi. It was a day that would shape up to be a referendum on Thakur's ability to create even modest change in India. His lawyers would appear before a small bench of Supreme Court justices and petition them to grant a hearing in the case of *Dinesh S. Thakur v. Union of India*. Already, India's chief drug regulator, G. N. Singh, whose agency was a defendant in the lawsuit, had publicly attacked Thakur, telling *Reuters*, "We welcome whistleblowers, we have got great respect, but their intentions should be genuine, should be nationalistic. . . . I don't have any comment on this guy." Still, Thakur had to admit that he felt an inkling of hope.

In New Delhi on Friday morning, Prashant Reddy, a vigorous intellectual property lawyer, ascended the steep steps of India's Supreme Court, flanked by several other attorneys. Thakur had spared no expense. In addition to Reddy, his team included one of India's most accomplished constitutional lawyers, senior advocate Raju Ramachandran. His lawyers would have only a brief opportunity to convince two Supreme Court justices that they should grant Thakur's lawsuit a hearing.

Courtroom 1 was crowded. Reporters from every major Indian newspaper were there to see whether Thakur would prevail. The two black-robed justices stared down sharply at the team of lawyers as they began.

"An overseas citizen has come all the way to challenge a rule. What is your locus?" the chief justice asked.

"Locus" referred to Thakur's "standing," or right to bring a case—in essence, the question of where he belonged. It was an issue that the lawyers had anticipated. Ramachandran explained that the Indian constitution did not restrict the nationality of who could bring a public interest lawsuit. Nonetheless, their client paid taxes in India, and so was entitled to judicial relief.

A justice then questioned whether the suit was an effort to seek publicity: "You are coming with academic issues when people are languishing in jails. Our hands are full."

"That is very uncharitable," Ramachandran objected, explaining that the matters in the petition were of critical concern, life and death. Within fifteen minutes, however, it was over. The justices refused to grant a hearing.

Prashant Reddy called Thakur. It was 2:00 a.m. in Tampa when Thakur picked up. Reddy broke the devastating news: the justices had blocked them from proceeding. Thakur got up in the darkness. He made coffee and sat down at his computer. He began to type out a passionate blog post to explain why he had undertaken the lawsuit. He described India's regulatory system as a "spectacular failure" that had done little to safeguard the public health of the nation's and the world's most vulnerable people. He savaged the excuses routinely offered by India's regulators and manufacturers:

> Unfortunately, no one yet has been able to point out where in Indian law does it say that it is ok to cheat, to destroy results from tests that fail, repeat tests until you get the result that you want, release substandard product knowingly to the market. . . . If the largest and most respected members of the Indian pharma industry function in this manner, what confidence do you have in the ability of the small and medium size companies to do things right? Is anyone bothered about this?

After four cups of coffee, he posted what he'd written, with the headline "A Sincere Attempt to Improve the Quality of Medicine for People around the World." Then he called Sonal, hoping for some comfort. She was simply glad it was over and reminded him, "I told you not to do this." He said little in response. Later, he learned that the drug industry's main lobbying group, the Indian Pharmaceutical Alliance, was jubilant.

In the months that followed, the darkness did not lift. Two years of work and about $250,000 in legal expenses had come to naught. He could no longer explain his crusade to his wife, and often he could barely explain it to himself. His efforts had been overpowered by the forces against reform and those in favor of inertia. His family, once an emotional anchor, seemed to be coming apart. It might have been logical for him to feel some sort of regret: for taking the job at Ranbaxy, for feeling forced to make a choice over the right path to take, for becoming a whistleblower, for starting a process that had torn his family to shreds while he continued the fight long after he'd had to. But regret was never part of the heartbreak he felt. As he told an acquaintance, "Why would I regret something that I knew was true and correct and right?" As days went by, he returned again to his grandmother's teachings—he must discharge what he viewed as his responsibility, knowing that the outcome was not entirely in his hands. This meant that he would have to accept the high court's decision and focus instead on what to do next.

Less than two weeks later, he sent an email to the joint secretary of the Ministry of Health and Family Welfare, K. L. Sharma, introducing himself and, in effect, starting all over again: "From what little I have read, you seem to have a good sense of right and wrong when it comes to public health and I am writing today to seek an appointment to come and see you in Delhi at your office."

EPILOGUE

In October 2017, an international scandal caught Peter Baker's attention.

Kobe Steel, a Japanese steel maker, had been caught falsifying data. The company had misrepresented some of its products' tensile strength, meaning that its aluminum, copper, and steel could not bear loads as heavy as the company had claimed. Instantly, warnings sounded around the world. Were the bridges, railways, cars, and planes that had been built with Kobe steel actually safe?

The public attention surprised Baker. Every day he was out in the world's drug manufacturing plants, exposing fake data. His inspection reports were publicly available. His findings had harrowing implications for the safety and effectiveness of America's medicine, not to mention the world's. Yet his discoveries seemed stuck just below the public radar. Perhaps it was easier to understand the implications of a 500,000-ton bridge collapsing. But what about the consequences of a drug that didn't work properly? Or that contained harmful impurities or ingredients that hadn't been tested or disclosed? Of a capsule that released its ingredients in a surge rather than over hours, or degraded too quickly in the heat? To Baker, Thakur, and others who had tried to sound the alarm, poor-quality drugs were the equivalent of falling bridges. The only difference: the collapse was taking place invisibly inside human bodies, with potentially life-or-death consequences.

Those who had spent years investigating the generic drug industry, its trade-offs, and the potential danger to patients resorted to imperfect strategies to safeguard their own health: they tried to avoid taking the drugs that they suspected were compromised. At an industry conference, Carmelo Rosa, head of the FDA's Division of International Drug Quality, told the audience that he was injured after his boiler exploded and wound up in the hospital. He refused to take generic drugs from several different manufacturers, all of whom were under scrutiny by the FDA for falsifying data. "I like to pray," he told his audience. "But we shouldn't have to be praying for this batch to be a good batch."

An FDA investigator who'd visited Indian plants acknowledged, "Every time I fill a prescription, I think about it." He believed that low-quality medicine posed the greatest risk for people who took the drugs "day in, day out" for chronic conditions. "The probability is that one of those pills is going to be tainted," he observed. "You don't want those impurities in your system."

Thakur's lawyer Andrew Beato said that, before he got involved in the Ranbaxy case, "I never once looked at the bag or read the damn label." But that changed after he began to represent Thakur. "From 2007 on, we had a rule in our house going forward. We'd pay I don't care what" to avoid generic drugs made overseas. Debbie Robertson had come to the same conclusion. "As soon as I started the [Ranbaxy] case, I wouldn't let anyone in my family take any Indian generics," she recalled. In Congress, investigator David Nelson asked seven different FDA investigators who'd been in Ranbaxy's facilities whether they'd be willing to take Ranbaxy drugs. "They all said no," he recalled.

Baker swore off all low-cost generics made overseas after his extraordinary inspections at Wockhardt. "If people actually understood," he told a colleague, "then no one would take [these drugs]." After he left India in 2015, he got a new tattoo, a word written in cursive on the inside of his arm that summed up the requirement he'd tried to enforce: "Integrity."

Even as the public remained largely in the dark, the invisible battle over drug quality, and the fallout from it, continued. In India, Dinesh and Sonal Thakur never recovered from the strain on their marriage during the Ranbaxy case. By the summer of 2016, they separated and began bitter divorce proceedings. But in another realm, Thakur seemed to find his voice. He wrote increasingly pointed blog posts and opinion pieces about the silence, self-interest, and corruption in India that had allowed bad medicine to flourish.

In February 2018, in a column for an online publication, *The Wire*, Thakur revisited the self-serving and false statements made about the Ranbaxy scandal by former CEO Malvinder Singh, the company's board of directors, and "arm-chair" public health experts, who all publicly claimed that Ranbaxy's failures lay in minor record-keeping violations. He singled out India's top regulators, who had done little to sanction the company in the wake of the U.S. FDA's findings. By giving the company a "clean chit," he wrote, they had acted as a "shield to the fraud at Ranbaxy."

That same month, India's top drug regulator, G. N. Singh, was removed from his job, along with his deputy and several other mid-level officers, in response to allegations that they'd been allowed to stay on beyond the five years permitted by statute.

Today the Ranbaxy company no longer exists. In April 2014,

Daiichi Sankyo, eager to unload the entity that had caused so much grief, sold the tarnished company cheaply to an Indian generic drug company, Sun Pharma. It did so just as another Ranbaxy whistle-blower alerted Daiichi Sankyo to ongoing and elaborate deceit at two of Ranbaxy's manufacturing plants, Dewas and Toansa. The whistleblower claimed that the plants were swapping out high-quality ingredients for lower-quality, lower-cost ones and maintaining records of the substitutions in a second set of books. He alleged that the company was concealing high impurity levels in some of its medicine by using charcoal to bleach yellow pills white. It hid the degraded ingredients behind ceiling tiles and dumped evidence in a river. The whistleblower also reported nesting sparrows and monkey infestations inside plants where Ranbaxy purchased its ingredients.

Today, under Sun Pharma's ownership, the Mohali plant has cleared FDA inspections and exports drugs to the U.S. market, but the Dewas, Paonta Sahib, and Toansa plants are still under sanction and the company is "evaluating whether to supply products to the U.S. market in future from these three plants," according to a Sun Pharma spokesperson.

Meanwhile, Daiichi Sankyo emerged victorious from its arbitration in Singapore against former Ranbaxy CEO Malvinder Singh. Remarkably, former company lawyer Jay Deshmukh, as well as several of Ranbaxy's outside lawyers, lined up to testify against Malvinder. They detailed the lengths to which he and his allies had gone to hide the Self-Assessment Report from Daiichi Sankyo executive chairman Tsutomu Une. In April 2016, Singapore's International Court of Arbitration directed the Singh brothers to pay Daiichi Sankyo $550 million in legal damages, a calculation based on how much the duped Japanese company had overpaid for Ranbaxy's stock, due to the concealment of the SAR.

The Singhs fought back. They challenged the size of the judgment in a Singapore court, in a case they ultimately lost. By 2018, their lawyers had turned to the Indian Supreme Court and argued that the Singapore judgment was not enforceable in India. They lost there also. The decision led the Singh brothers to relinquish their posts at the head of another family business, a hospital chain called Fortis Healthcare, where they faced new allegations that they'd siphoned $78 million from the publicly traded company into private family accounts. They also faced a similar allegation from a New York private equity firm, which alleged in a lawsuit that they were "systematically plundering" a division of their publicly traded financial firm, Religare Enterprises, Ltd., to meet personal debts of close to $1.6 billion. Malvinder denied any "mismanagement or misuse of funds and position at Fortis" and said that any intercorporate deposits between companies were collectively made decisions.

By September 2018, warfare between the Singh brothers broke out into the open. Shivinder filed a petition against his older brother with India's National Company Law Tribunal, alleging that Malvinder had committed fraud and mismanaged the family businesses, leading to an "unsustainable debt trap." Just days later, however, Shivinder withdrew the petition, claiming that he'd done so after his mother demanded that her sons enter mediation instead. But the truce did not last long. Within a year, Malvinder Singh would publicly accuse his brother, Shivinder, of physical assault, allegations his brother called "fake" and a "lie."

Some of those involved in the bitter fight over generic drug quality landed on their feet. With the Ranbaxy settlement completed, Andrew Beato became a partner at his law firm, now called Stein Mitchell Beato & Missner LLP. And Debbie Robertson, after her retirement from the FDA, worked briefly for Beato's firm as an investigator, but then retired for good.

Others found their way to new posts. In September 2018, Jay Deshmukh, the Ranbaxy lawyer who'd been driven out of the company by Malvinder Singh, became a partner at the law firm of Kasowitz, Benson, Torres LLP, practicing patent law. Altaf Lal, the crusading former head of the FDA's India office, became a senior adviser on global health and innovation at Sun Pharma, the drug company that bought Ranbaxy. From that perch, he oversees the company's tropical disease program, including the development of medicine to combat drug-resistant malaria.

By 2015, Jose Hernandez, Doug Campbell, and Mike Gavini had all retired from the FDA. Separately, they each set up shop as consultants, advising companies on compliance with FDA regulations. Hernandez liked to say that Peter Baker—and companies' fear of him—was making him rich.

Tom Cosgrove, the FDA bureaucrat who in 2017 had downgraded findings against Mylan's Morgantown plant, allowing the company to avoid a warning letter, left the agency shortly afterward to represent drug companies at a D.C. law firm.

In October 2017, Mylan's president, Rajiv Malik, faced a new and serious allegation. In an expanded civil complaint that was announced and later filed in federal court, forty-seven state attorneys general accused eighteen generic drug companies of colluding with one another to keep the prices of their drugs artificially high. The complaint, the result of a multi-year investigation, singled out two executives from rival companies as having colluded with one another to fix prices: the CEO of India's Emcure Pharmaceuticals and Mylan's Rajiv Malik. In a statement, the company rushed to Malik's defense: "Mylan has deep faith in the integrity of its president, Rajiv Malik, and stands behind him fully." It vowed to fight the allegations.

Meanwhile, the FDA continued to unearth problems at Mylan's plants. In March 2018, the FDA sent eight investigators back to the Morgantown, West Virginia, plant for twenty-five days of inspection. The scrutiny, prompted by whistleblower allegations, uncovered cleaning lapses so serious that the FDA grew concerned about product contamination and possible cross-contamination between drugs.

As the agency weighed the prospect of a warning letter that

might halt new applications from the Morgantown plant, Mylan worked urgently behind the scenes to try and establish a back channel to top FDA officials. In June 2018, Mylan's head of regional quality compliance called an FDA division director on his personal cell phone and asked to meet informally over coffee, explaining that he was doing so at the behest of Mylan's president, Rajiv Malik. The FDA official sternly rejected the request and documented the encounter in an email to colleagues: "I explained it was inappropriate for him to contact me with this request," the official wrote, and "I don't have private meetings with industry, especially when we are in the middle of a review."

Roughly six weeks later, in August, Rajiv Malik tried directly, reaching out to the same wary official, this time to request a formal meeting. But none of that worked. In November 2018, the agency issued a warning letter for Mylan's Morgantown plant, citing cleaning failures, the unaddressed risk of cross-contamination between drug products, and a failure to adequately investigate anomalous test results. It was a major rebuke to the company and its flagship plant, both of which had once been industry role models. Mylan issued a response to the warning letter, explaining that it had implemented a "comprehensive restructuring and remediation plan at our Morgantown facility," and pledged to satisfy the FDA's concerns.

Included in its plan—which Malik spelled out to the FDA official in the August email where he'd requested a meeting—was to slash manufacturing at Morgantown to "less than half the volume in doses and products generated prior to 2018." He didn't mention whether the company would move that manufacturing out of the United States, which would effectively put more distance between its operations and FDA scrutiny.

At Cleveland Clinic, Harry Lever's instincts continued to prove accurate as he shifted his patients off generics that the FDA had approved—and that it continued to defend. In March 2014, fifteen months after Lever had reported his concerns about generic versions of the beta blocker metoprolol succinate to the FDA, a senior FDA official contacted him to explain that an extensive "multidisciplinary investigation" had determined that the generic drugs were bioequivalent to the brand. The FDA had reached this conclusion in part by reviewing the initial data the companies had submitted in order to get approval (apparently without considering whether the company data might be false).

Less than a month later, however, Wockhardt, followed by Dr. Reddy's, recalled their metoprolol succinate from the market, with the admission that the drugs were not bioequivalent. Lever had been right after all.

In June 2018, a woman arrived at Cleveland Clinic's emergency room, suffering from chest pain and shortness of breath. Kristy Jordan, thirty-five, had had a successful heart transplant three years earlier and since then had taken the daily immunosuppressant Prograf to prevent organ rejection. But six months earlier, a CVS pharmacy refilled her prescription with generic tacrolimus, made by Dr. Reddy's. In the time she took it, she felt progressively worse. At the Cleveland Clinic emergency room, tests showed that she was suffering from organ rejection and had lower-than-expected levels of tacrolimus in her blood, which meant that the Dr. Reddy's drug wasn't working sufficiently. Doctors stabilized her.

This time Lever and his colleague Randall Starling were prepared to connect the dots. Armed with the patient's blood test results, they retrieved her tacrolimus capsules and sent them for testing to a Massachusetts laboratory. Meanwhile, Jordan continued to feel sluggish and never fully recovered her health. In September 2018, she died of a heart attack. Starling said that there was no way to know whether the tacrolimus made by Dr. Reddy's, and the medical setback that Jordan suffered, contributed to her death. But he pointed out that hospitals are now required to pay penalties to Medicare when their patients are readmitted. "If we learn that part of our efforts to prevent readmissions are thwarted by bad drugs, it's going to be a very important discovery," he noted. By February 2019, the Massachusetts laboratory had one preliminary finding: the Dr. Reddy's tacrolimus released its active ingredient very rapidly, compared with the brand. Testing continued.

In February 2015, Peter Baker left India and relocated to Beijing, where he became the FDA's sole drug investigator stationed in China, responsible for inspecting over four hundred factories approved to export drugs or drug ingredients to the United States. The Chinese government, operating under the suspicion that any investigator could be a spy, had approved only a handful of visas.

Though Baker knew his every move, email, and phone call would be monitored by the Chinese government, he felt relieved to have left New Delhi behind. He would no longer be the last man standing between American consumers and unsafe drugs. The manufacturing plants he'd be inspecting in China made mostly active ingredients, not finished doses. Most were not sterile facilities. If something slipped past him, an investigator at the next plant could still catch the problem—at least in theory.

Within a month, he arrived at the massive Zhejiang Hisun plant in Taizhou, two hundred miles south of Shanghai, the site of a joint venture with Pfizer, started in 2012, to create high-quality, low-cost medicine under the umbrella of Hisun-Pfizer Pharmaceuticals. The company had seemed like a safe bet: it was China's largest exporter of drug ingredients to the United States.

Pfizer had a formidable apparatus dedicated to maintaining quality at the two hundred plants that the company operates, or contracts with, around the world. Hundreds of people, from trained auditors to laboratory analysts, worked to safeguard the company's drugs. Pfizer usually stations an employee in any plant where it has a stake, to ensure that its standards are met, and the Zhejiang Hisun plant was no different. Brian Johnson, formerly Pfizer's senior director of supply chain security, acknowledged that detecting outright fraud was a challenge. But because of his confidence in Pfizer's layered system, he didn't view outsourcing as additionally hazardous. "If you have the appropriate controls," he said, "I don't think it's adding risk." The FDA's investigators had been at the Zhejiang Hisun plant over a dozen times and had found little to concern them.

Baker, on arriving there, went first to the quality control laboratory. Using the rudimentary Mandarin that he learned in college, he hunted through the forest of Chinese symbols in the computer audit trails for the words "trial injection" and "experimental sample." Despite Pfizer's three-year head start, it took him about a day to figure out that the plant was running an alternate and hidden laboratory operation.

The plant was secretly pretesting its drug samples and then masking the results, in part by turning off audit trails to leave no evidence of the tests. In one instance, Baker found that technicians had turned off the audit trail on February 6, 2014, at 9:09 a.m., then proceeded to run eighty secret tests. The audit trail was turned back on two days later at 8:54 a.m., and the tests—now rigged and with the outcomes assured—were repeated. Baker found the telltale evidence in the software's metadata.

By the third day of inspection, the plant managers and analysts were well aware of how devastating his inspection might be. When Baker returned from a lunch break to the quality control laboratory, he saw an analyst quickly remove a thumb drive from one of the HPLC machines and slip it into his lab coat.

Baker demanded that he hand over the thumb drive, but the man "began running and fled the laboratory premises," he documented in his inspection report. Fifteen minutes later, a manager returned to offer him the thumb drive, but Baker had no idea whether it was the same one. He noted the incident as a refusal to share records—which was serious enough to get the plant's drug ingredients blocked from the United States.

Baker laid out his findings in a forty-seven-page report. It was a document unlike anything the Chinese drug industry had ever seen. Word of it whipped through China's manufacturing plants, where the prevailing attitude had long been "we can always fool a foreigner," as one Western drug executive put it. But Baker was no ordinary foreigner. Two and a half years later, Pfizer ended its partnership with Zhejiang Hisun.

Six weeks after the Zhejian Hisun inspection, Baker went to Dalian in the Liaodong Peninsula and inspected another plant; this one, owned and operated by Pfizer, was making finished doses for the U.S. market. There, too, he found manipulated tests, unreported results, and loose batch records that showed the plant using expired materials. One stack of documents disappeared entirely during his inspection; he found them later on an upper floor, tucked inside a wooden crate.

Before long, Baker had carved a wide swath through China. At thirty-four more plants he inspected across the country, he found violations—many of them similarly devastating instances of data manipulation. Baker's inspections cast a harsh light not just on Chinese drug manufacturing, where fraud was endemic, but also on the FDA's foreign inspection program. "Every time he puts a foot in a company he's finding more problems," as one senior FDA official said of Baker. "What does that say about an inspectional force that's not finding this?"

The FDA's investigators had been trained for a different era, when the data printed out on paper was the only data that existed. The agency had not significantly rethought or overhauled its training program in decades. As one FDA consultant put it, "People are using brains from 1990 to do their thinking" today.

Most of the FDA's investigators who were sent to China did

not speak the language. They couldn't read the manufacturing records. The FDA did not provide independent translators. Instead, the companies provided the translators who, more often than not, were company salesmen. Too frequently, FDA investigators simply gave plants a pass, deeming them to be No Action Indicated because they had no way to tell otherwise.

The investigators also couldn't read street signs, which made them vulnerable to wild manipulations. Companies steered them to phony "show" plants, where everything looked compliant, but the companies weren't manufacturing there. Sometimes a group of companies pooled their resources and invested in the same "show" factory, so that different FDA inspectors returned to the same plant at different times, each one thinking they were inspecting a different facility.

Baker often inspected alone. But when American or other global investigators worked alongside him, they also became detectives. They followed tire tracks off manufacturing sites to see where companies were taking their products. They took photographs through the windows of dilapidated plants, documenting the labels on the boxes inside, to prove that these were the actual factories where companies were doing their manufacturing. As Baker collaborated with investigators from other countries, his reputation spread around the world. At conferences, regulatory agencies from ANVISA in Brazil to the European Medicines Agency invited Baker to train their investigators.

In Baker's view, only a critical mass of investigators who knew exactly where to look and what to look for could truly protect consumers and change the industry for good. In December 2015, at an hour-long meeting with the acting FDA commissioner, Baker proposed a program for training the FDA's investigators to detect data fraud.

At the FDA's headquarters in Silver Spring, Maryland, Baker's proposal landed in a teetering system. If investigators looked too hard, if reviewers scrutinized applications too much, if the agency didn't approve enough drugs, the entire system could topple.

Already, Peter Baker's razor-sharp inspections had halted imports and contributed to drug shortages rippling across the country. His findings had placed a question mark over generic drug companies operating overseas. They had also slowed drug approvals. That, in turn, jeopardized FDA funding, which was based in part on how many drug applications the agency approved. Baker's proposal—to train the FDA's investigators to look harder and find more violations—threatened even more disruptions. Consequently, some in the agency began to target a different problem: Peter Baker himself.

Publicly, FDA officials gave every impression that they were taking the fight to companies that falsified data. As Thomas Cosgrove, the FDA's director of compliance at the Center for Drug Evaluation and Research, told a reporter for an industry newsletter, the FDA planned to make drug companies that "shade the facts . . . increasingly uncomfortable." They could expect additional penalties, he said. To the world, it looked like the FDA had launched a global crackdown on data fraud.

Yet Baker watched as the FDA, despite clear authority and well-written laws, chose to undercut inspections and soften investigators' findings. From 2012 to 2018, the agency downgraded 112 inspections in India to make the final classifications less severe. For company after company—Mylan, Cipla, Aurobindo, Dr. Reddy's, Sun Pharma, Glenmark—findings of Official Action Indicated (OAI) became Voluntary Action Indicated (VAI). These downgrades essentially nullified the judgments made by its investigators in the field and replaced them with judgments made by bureaucrats in Maryland. Cosgrove and other officials waived import restrictions. They chose to communicate confidentially with some firms through so-called untitled letters instead of issuing public reprimands. Politics seemed to guide the agency's enforcement actions. As the former compliance officer Doug Campbell observed, "They want to weaken compliance at CDER because they want to approve applications. Compliance only messes it up."

After Baker's inspection at the Pfizer-affiliated Zhejiang Hisun plant, the FDA restricted the import of thirty of the plant's drug products. But fifteen of the drug ingredients were in short supply in the United States, so the agency lifted the restriction on about half of the drugs, including a crucial chemotherapy drug for treating leukemia and breast and ovarian cancers.

To Baker, the decision made no sense. According to regulations, the drugs had no place in the U.S. supply. They weren't good or safe enough. Shortages didn't change that fact. Drug shortages had become a game, and the FDA was getting played. Companies committing fraud could still protect their bottom line by making drugs in short supply. Those would not be restricted, whether made with dubious methods or not, and could serve as a steady source of business, even if companies were caught making unsafe drugs. "There are no consequences for companies that are shipping substandard product," Baker observed to a colleague. "It's a win-lose situation—and [patients] are the losers."

Baker left China in March 2018 and relocated to Santiago, Chile, to become the FDA's country director there. But the FDA stopped sending him to do inspections, a factor in his resignation from the agency a year later.

In July 2018, a safety crisis rocked the global drug supply—and seemed to prove Baker's point. Regulators in Europe announced a harrowing discovery: the widely used active ingredient for valsartan, a generic version of the blood pressure drug Diovan, contained a cancer-causing toxin known as NDMA (once used in liquid rocket fuel). The drug had been made by the Chinese company Zhejiang Huahai Pharmaceuticals, the world's largest manufacturer of valsartan active ingredients. In the United States, over a dozen drug manufacturers, all of which used the Chinese ingredient, recalled their products, as did dozens more manufacturers around the world.

The Chinese company tried to defend itself by explaining that it had altered its production process in 2012 to increase yields of the drug, a change that had been approved by regulators. In short, the change had been made to maximize profit. Some patients had been consuming the toxin for six years.

As the FDA tried to reassure consumers that the risk of developing cancer, even from daily exposure to the toxin, was extremely low, a second cancer-causing impurity was detected in the ingredients. Though the valsartan catastrophe seemed to take the FDA by surprise, it shouldn't have. In May 2017, an FDA investigator had found evidence at the plant in Linhai, China, that the company was failing to investigate potential impurities in its own drugs, which showed up as aberrant peaks in its test results. The investigator designated the plant as Official Action Indicated, but the agency downgraded that to VAI. In short, the company was let off the hook—only to wind up in the middle of a worldwide quality scandal less than a year later.

By 2017, Baker had gotten a limited green light to hold occasional workshops for his colleagues. But even as he educated them to detect data fraud, the FDA had already taken its most significant step to undermine the inspections that might reveal it.

Under the India pilot program of short- and no-notice inspections that Altaf Lal had launched, the rate of inspections resulting in the FDA's most serious finding, OAI, had increased by at least 50 percent. It seemed logical for the FDA to expand this model and make unannounced inspections the norm in every country in the world. But agency bureaucrats made a different decision.

On the morning of November 3, 2016, the FDA's top officials in India gathered with senior Indian drug regulators, the Indian drug industry's chief lobbyist, and three Indian generic drug executives including from the companies Cadila and Dr. Reddy's, both of which had endured scorching FDA findings at their facilities. The meeting lasted an hour and was led by Dr. Mathew Thomas, who'd taken over the job of running the FDA's India office from Altaf Lal.

The men discussed collaboration and capacity-building and the schedule of future workshops that the FDA was holding on good manufacturing practices. Dilip Shah, secretary-general of the Indian Pharmaceutical Alliance, made clear that Indian drug makers were doing their part to address quality issues, and that his trade group would release guidelines to the industry, emphasizing the importance of data reliability. Mathew Thomas then told those gathered the words they'd been waiting to hear. The experiment was over. From now on, for all routine inspections, the FDA would notify India's companies in advance.

ACKNOWLEDGMENTS

Writing this book has been a journey, and many people helped along the way.

Self magazine published my first article about generic drugs in 2009, where I was lucky enough to be edited by Sara Austin, now executive editor at *Real Simple*. At *Fortune* magazine, my May 2013 article on Ranbaxy, "Dirty Medicine," which served as the jumping-off point for this book, benefited greatly from the exceptional skill, judgment, and dedication of my editor, Nick Varchaver, now a senior editor at ProPublica.

In 2014, as I began work on this book, I needed reporting help all over the world. I am grateful to David Kaplan, the executive director of the Global Investigative Journalism Network (GIJN), who connected me with talented journalists in India, Ghana, China, and elsewhere. The GIJN conferences opened the door to an international community of brave and talented journalists, who inspired and helped me throughout this project. I am indebted to Mark Lee Hunter of Story-Based Inquiry Associates for his advice on how to turn years of reporting and mountains of information into an actual story, at a moment when I felt stuck.

At the International Consortium of Investigative Journalists (ICIJ), deputy director Marina Walker Guevara generously allowed access to the offshore banking records of the Panama and Paradise papers, while Emilia Diaz Struck patiently provided instructions on how to navigate within the records. At the Freedom of the Press Foundation, director of newsroom digital security Harlo Holmes and digital security trainer Olivia Martin offered valuable guidance on digital file encryption, risk assessment, and secure communication with sources.

Throughout my reporting, I relied on a database called FDAzilla, which captures every inspection the FDA has performed around the world, and also provides important ancillary data. As my expenses mounted and my budget tightened, the site's cofounder Tony Chen and CEO Michael de la Torre allowed me continued use of the site, and even provided customized data. Other outside organizations also provided invaluable help. Dr. Rebecca Goldin, the director of STATS.org, helped me to translate statistical concepts into plain English. Peter Sorenson of the Sorenson Law Office and Daniel J. Stotter of Stotter and Associates LLC excel at litigating stalled Freedom of Information requests and helped me get records from the FDA.

I am also grateful to so many who hosted me along the way or shared their local knowledge during my reporting trips: Sophy Burnham, Kathy Sreedhar, Vivienne Walt, Anton Harber, and Rimjhim Dey.

I am indebted to a number of organizations whose generous support allowed me to complete my reporting. The Carnegie Corporation selected me as a 2015 Andrew Carnegie Fellow and provided significant funding. The Alfred P. Sloan Foundation provided a grant through its Public Understanding of Science, Technology, and Economics program. I am particularly grateful to Sloan's vice president and program director Doron Weber for his faith in this project. The McGraw Center for Business Journalism at CUNY's Craig Newmark Graduate School of Journalism provided me with a McGraw Fellowship for Business Journalism. The McGraw Center's executive director Jane Sasseen was generous with her time and advice. A George Polk Award for Investigative Reporting also provided essential support.

Some very fine journalists worked on this book. Ariel Bleicher brought her elegant writing and sharp reporting skills to work as a research associate on the book for a year. In India, Syed Nazakat helped me to navigate the complex world of India's business networks and government bureaucracy. Kent Mensah helped me on the ground in Ghana. Sunny Yang assisted me in China. Doris Burke and Andrew Goldberg delved into legal and financial records. Without the talent and hard work of Sony Salzman, a science journalist adept at analyzing data who worked by my side for three years, this book would not have gotten across the finish line. Kelsey Kudak fact-checked this book with remarkable skill. Any remaining errors are my own.

I am also grateful to a number of top-flight editors. Hilary Redmon, now at Penguin Random House, acquired this book for Ecco/HarperCollins, and gave vital early guidance. Domenica Alioto, a master of narrative nonfiction editing, helped to forge the book's structure. At Ecco, I am indebted to Emma Janaskie, who skillfully edited the book and moved it into production. Additional thanks to the great team there: Daniel Halpern, Miriam Parker, Gabriella Doob, Meghan Deans, Caitlin Mulrooney-Lyski, and Rachel Meyers. I am particularly grateful to William S. Adams at HarperCollins for his sterling legal review and unflagging patience.

This book could not have happened without the unstinting wisdom, encouragement, and vision of my agent and friend, Tina Bennett. Even at the most difficult of times, her support never wavered.

I am grateful to the friends, colleagues, and family members whose close reading, consultation, and first-rate editorial comments made this book better in every way: Nick Varchaver; Jennifer Gonnerman; my brother, Matthew Dalton; Philip Friedman (who also came up with the book's title); Sony Salzman; my mother, Elinor Fuchs; and my father, Michael O. Finkelstein, all served as early and insightful readers.

Maryam Mohit and Erik Blachford helped decipher financial records. Bryan Christy gave prescient editorial advice. Maureen N. McLane let me tap her unsurpassed wordsmithing skills. Vivian Berger, my stepmother, gave me expert advice on dispute resolution.

Thank you to my friends and family—including Lindy Friedman, Tracy Straus, and my sister, Claire Finkelstein—for their continued support during this process. My daily talks with Julia Freedson sustained me. My cherished friend Karen Avenoso (1967–1998) lives on every page.

My beloved children, Amelia and Isobel, made sure that I never lost touch with the world beyond these pages. They were patient, supportive, and funny during a project that seemed to never end. (They recommend that I write children's books next.) My husband, Ken Levenson, helped in every way. He encouraged me at each step, took care of the family during my long reporting trips, offered wise counsel through thorny reporting and writing problems, and read many drafts.

Finally, a special thank-you to my many sources, who cared enough about the integrity of our medicine and the well-being of patients to entrust me with their information. Many of them spent hours—even years—patiently fielding my questions and helping me to understand complex processes. Without them, there would be no book.

GLOSSARY

Accutane/Sotret: An anti-acne medication. The FDA approved Roche's brand-name Accutane in 1982. In 2002, the FDA approved a version called Sotret, made by Ranbaxy Laboratories. The active ingredient is isotretinoin.

AIP: Application Integrity Policy. A sanction imposed by the FDA when it suspects fraud in a company's applications. Under an AIP, the FDA will stop reviewing a company's pending applications until a company can prove the accuracy of its data.

ANDA: Abbreviated New Drug Application. An application compiled by a generic drug company and submitted to the FDA to request approval of a generic drug. An ANDA is known in industry parlance as a "jacket."

API: The active pharmaceutical ingredient is the biologically active component of a drug. This is the most important, and often most expensive, ingredient in a finished drug.

AUC: The area under the curve, as charted in a graph. The AUC reflects the total concentration of a drug in the blood of a patient over time.

bioequivalence: A standard the FDA uses to determine whether a generic drug works similarly in the body to the brand-name version. According to the FDA's statistical formula, a generic drug's concentration in the blood should not fall below 80 percent or rise above

125 percent of the brand-name concentration, using a 90 percent confidence interval.

brand-name drug: A drug that is discovered and developed by a pharmaceutical company, and typically protected by a patent. Manufacturers of brand-name drugs are sometimes referred to as innovator companies.

bulk drugs: The primary ingredients used in making active pharmaceutical ingredients or finished drugs.

CDER: The Center for Drug Evaluation and Research. A branch of the FDA that regulates brand-name and generic drugs by reviewing new drug applications and monitoring the safety of drugs after they are approved.

cGMP: Current good manufacturing practices, as outlined in Title 21 of the Code of Federal Regulations, lay out the FDA's expectation of how manufacturing facilities should operate.

chromatogram: A graph that is often produced using an HPLC machine, displaying the results from separating the components of a drug sample.

Coreg/carvedilol: A medication used to treat high blood pressure and heart failure. The FDA approved brand-name Coreg, made by GlaxoSmithKline, in 1997. Subsequent generic versions are known by the drug's active ingredient, carvedilol.

Demadex/torsemide: A medication used to treat fluid retention for patients with congestive heart failure. The FDA approved Roche's brand-name Demadex in 1993. Generic versions are named for the drug's active ingredient, torsemide.

excipients: The inactive ingredients in a drug that may include coloring agents, preservatives, and fillers.

FDA: The U.S. Food and Drug Administration. The FDA is the federal agency that regulates the safety and quality of the nation's food, medicine, and medical devices.

FDA investigator: An FDA employee trained to inspect manufacturing facilities. Investigators are also sometimes referred to as "inspectors" or "consumer safety officers."

Form 483: When an FDA investigator uncovers violations of current good manufacturing practices during an inspection at a manufacturing facility, those findings are recorded in a form called a 483.

furosemide: A generic drug used to treat fluid retention in patients with congestive heart failure. The brand-name version, Lasix, was first approved for use in the 1960s.

gabapentin: An antiseizure medication named for the drug's active ingredient, gabapentin. The brand-name drug Neurontin, made by Pfizer, was approved in 1993.

generic drug: Medicine manufactured to work similarly in the body as a brand-name drug, typically sold after the original drug's patent expires. To be approved as bioequivalent, a generic drug must have the same "dosage form, safety, strength, route of administration, quality, performance characteristics, and intended use" as a brand-name drug, according to the FDA.

heparin: An anticoagulant drug used to prevent blood clotting.

HPLC: High performance liquid chromatography. A common technique used to separate out and quantify components within a drug sample. Drug manufacturers use HPLC machines to identify and measure impurities in a drug.

import alert: A public announcement issued by the FDA to its field staff to detain shipments of products deemed unsafe at ports of entry. Import alerts can be issued for specific types of products or for products from specific manufacturing plants.

Lipitor/atorvastatin: A medication that helps lower cholesterol. The FDA approved Pfizer's brand-name Lipitor in 1996, and later approved Ranbaxy's generic version, atorvastatin, named for the active ingredient atorvastatin calcium.

NAI: No Action Indicated. One of three designations FDA investigators use to rank a facility after inspection. NAI means the investigator did not observe deviations from current good manufacturing practices and the facility does not need to take corrective action.

OAI: Official Action Indicated. One of three designations FDA investigators use to rank a facility after inspection. OAI means the investigator observed significant deviations from current good manufacturing practices and has recommended the facility take prompt corrective action or face further regulatory consequences.

PEPFAR: The President's Emergency Plan for AIDS Relief, launched by President George W. Bush in 2003. The program, ongoing today, funds the purchase of low-cost generic drugs and provides them to AIDS patients in Africa and elsewhere.

Pravachol/pravastatin: A medication used to lower cholesterol. The FDA approved Bristol-Myers Squibb's Pravachol in 1991. Generic versions are named for the drug's active ingredient, pravastatin sodium.

Prograf/tacrolimus: A medication that suppresses the immune system to prevent organ rejection in transplant patients. The FDA approved Astellas' brand-name version, Prograf, in 1994. Generic versions are named after the drug's active ingredient, tacrolimus.

SAR: The Self-Assessment Report. A confidential document compiled inside Ranbaxy Laboratories in 2004 that details extensive data fraud at the company.

substandard drug: A drug that does not meet quality standards set by the FDA or other established regulatory authorities.

Toprol XL/metoprolol succinate extended-release: A long-acting beta-blocker used to treat chest pain and high blood pressure. The FDA approved AstraZeneca's brand-name Toprol XL in 1992. Subsequent generic versions are named after its active ingredient, metoprolol succinate.

USP: The United States Pharmacopeia. An independent nonprofit organization that establishes and reconciles global standards for how prescription drugs are made.

VAI: Voluntary Action Indicated. One of three designations FDA investigators use to rank a facility after inspection. VAI means the investigator observed some deviations from current good manufacturing practices and recommends the facility make voluntary corrections.

warning letter: This official message from the FDA warns a company that its facility has violated FDA regulations and must promptly address the issues detailed in the letter or face additional enforcement actions.

Wellbutrin XL/Budeprion XL: A long-acting medication used to treat depression. The FDA first approved GlaxoSmithKline's brand-name Wellbutrin XL in 1985, and later approved Teva's generic version, Budeprion XL. The drug's active ingredient is bupropion hydrochloride.

NOTES

PROLOGUE

2 *after completing eighty-one inspections:* FDA, Establishment Inspection Report, Wockhardt Ltd., Aurangabad, India, March 18–22, 2013. A full list of Peter Baker's inspections, and those of all FDA investigators, can be found on the website https://fdazilla.com/.

2 *than it did within U.S. borders:* The figures related to the growth of imported drugs and overseas drug plants come from Pew Charitable Trust, Pew Health Group, "After Heparin: Protecting Consumers from the Risks of Substandard and Counterfeit Drugs," July 12, 2011, 22.

3 *about a quarter of U.S. patients:* Ketaki Gokhale, "Urine Spills Staining Image of Wockhardt's Generic Drugs," *Bloomberg,* September 27, 2013.

4 *was walking toward him just a little too quickly:* Documents that helped to re-create this scene include: FDA, Form 483, Inspectional Observations, Wockhardt Ltd., Aurangabad, India, March 18–22, 2013; FDA, Establishment Inspection Report, Wockhardt Ltd., Aurangabad, India, March 18–22, 2013; FDA, Warning Letter (WL: 320-13-21), July 18, 2013; news stories, including Pallavi Ail, "USFDA Says Team Threatened during Wockhardt Inspection," *Financial Express,* May 28, 2014.

5 *contained black particles:* FDA, Establishment Inspection Report, 13–18.

6 *restricted the import of drugs:* Reuters, May 23, 2013. Accessed December 14, 2018. https://in.reuters.com/article/wockhardt-fda-revenue-loss/wockhardt-hit-by-fda-import-alert-on-drug-plant-idINDEE94M09320130523.

6 *"in a month or two months maximum":* Wockhardt Ltd. did not respond to repeated emails and phone calls seeking comment. However, after the FDA issued its import alert, Wockhardt CEO Habil Khorakiwala held an emergency conference call on May 24, 2013. During the call, he sought to reassure investors that the company was working to satisfy

the FDA's concerns and planned to hire a U.S. consultant to help bring "this facility into compliance in a month or two months maximum." When pressed by an investor, Khorakiwala said that the reason the FDA found so many lapses in Waluj was because the bulk of products manufactured at one facility there were destined for non-U.S. markets. The FDA wound up inspecting this facility, which wasn't explicitly set up to meet rigorous U.S. standards. Habil Khorakiwala, "Wockhardt Conference Call to Discuss U.S. FDA Report on Waluj Facility," May 24, 2013, http://www.wockhardt.com/pdfs/Wockhardt-Investor-Call-USFDA-Import-Alert-version-final.pdf (accessed December 3, 2018).

CHAPTER 1: THE MAN WHO SAW FURTHER

9 *Bristol-Myers Squibb's research and development center:* This description of the campus comes from a reporting trip there in November 2015.

12 *In 2001, it was on track to clock:* Detailed information regarding Ranbaxy and the development of its business in the United States can be found in: Bhupesh Bhandari, *The Ranbaxy Story: The Rise of an Indian Multinational* (Delhi: Penguin Books India, 2005); *Legends Are Forever: The Story of Ranbaxy* (Ranbaxy Global Corporate Communications, 2015); P. Indu, *Ranbaxy's Globalization Strategies and Its Foray into the U.S.* (ICMR Center for Management Research, 2005).

12 *The FDA had already approved over a dozen:* By 2001, the FDA had approved seventeen of Ranbaxy's drug applications, as listed in *Orange Book: Approved Drug Products with Therapeutic Equivalence Evaluations* (Rockville, MD: U.S. Department of Health and Human Services, Food and Drug Administration, Center for Drug Evaluation and Research, Office of Pharmaceutical Science, Office of Generic Drugs).

12 *comprised half of the U.S. drug supply:* Ann M. Thayer, "30 Years of Generics," *Chemical and Engineering News,* September 29, 2014.

13 *a doorstop of a master's thesis:* P. T. Vasudevan and D. S. Thakur, "Soluble and Immobilized Catalase," *Applied Biochemistry and Biotechnology,* 49, no. 3 (1994): 173–89, doi:10.1007/bf02783056.

CHAPTER 2: THE GOLD RUSH

17 *what would become an Abbreviated New Drug Application or ANDA:* U.S. Department of Health and Human Services, Food and Drug Administration, "Abbreviated New Drug Application (ANDA)," updated May 17, 2018, https://www.fda.gov/Drugs/Development

ApprovalProcess/HowDrugsareDevelopedandApproved/ApprovalApplications/AbbreviatedNewDrugApplicationANDAGenerics/default.htm (accessed January 10, 2018).

18 *first $10-billion-a-year drug:* John Simons, "The $10 Billion Pill," *Fortune,* January 20, 2003; Katherine Eban, "The War over Lipitor," *Fortune,* May 6, 2011.

18 *the final submission was ready:* Keith Webber, FDA CDER, letter to Scott D. Tomsky, Ranbaxy, November 30, 2011.

18 *"RECEIVED: August 19, 2002":* Abha Pant, Ranbaxy Laboratories Ltd., letter to Office of Generic Drugs, August 19, 2002, vi.

19 *Abbreviated New Drug Application 76-477:* U.S. Food and Drug Administration, Center for Drug Evaluation and Research, *Approval Package for Application Number: ANDA 076477Orig1s000,* November 30, 2011, https://www.accessdata.fda.gov/drugsatfda_docs/anda/2011/076477Orig1s000.pdf (accessed May 24, 2018).

20 *Lipitor is as moody as the slate-gray landscape:* This description of Pfizer's Ringaskiddy plant comes from a reporting trip there in August 2014.

20 *the battle that lay ahead:* A record of the litigation between Pfizer and Ranbaxy can be found at Pfizer Inc. et al. v. Ranbaxy Laboratories Ltd., et al. (U.S. District Court for the District of Delaware, August 2, 2006), Pacer Case Locator Case 06-1179, https://ecf.ded.uscourts.gov/cgi-bin/HistDocQry.pl?363128528119674-L_1_0-1 (accessed May 23, 2018).

21 *a CNN business reporter assessed it as:* Aaron Smith, "Investors Biting Nails over Lipitor," *CNN Money,* August 2, 2005.

21 *"extra gland that produces publicity":* Nora Ephron, "Oh Haddad, Poor Haddad," *New York,* November 25, 1968.

23 *Touting the benefits of lower-cost drugs:* Ronald Reagan, "Remarks on Signing the Drug Price Competition and Patent Term Restoration Act of 1984," September 24, 1984, Reagan Library, https://www.reaganlibrary.gov/research/speeches/92484b.

23 *"a place where you put raw materials into a mixing vat":* This quote attributed to Seife comes from Herbert Burkholz, *The FDA Follies* (New York: Basic Books, 1994), 26. In an endnote, Burkholz attributes it to the personal communication of David W. Nelson, former chief investigator for the House Subcommittee on Oversight and Investigations for the Committee on Energy and Commerce (personal communication).

24 *On the cold, clear night:* An account of the competition for Provigil and the waiting in the FDA parking lot appears in the lawsuit Federal Trade Commission v. Cephalon, Inc., Civil Action No. 2:08-cv-2141-MSG (U.S. District Court for the Eastern District of Pennsylvania, August 12, 2009), www.ftc.gov, https://www.ftc.gov/sites/default

/files/documents/cases/2009/08/090812cephaloncmpt.pdf (accessed June 15, 2018).

25 *In written guidance to the industry:* U.S. Department of Health and Human Services, Food and Drug Administration, Center for Drug Evaluation and Research, Office of Generic Drugs, "Guidance for Industry: 180-Day Exclusivity When Multiple ANDAs Are Submitted on the Same Day," July 2003, 4.

25 *As one of Ranbaxy's CEOs, Davinder Singh Brar, explained: Legends Are Forever,* 54.

27 *In May 2003, Ranbaxy's top executives:* Several documents helped to recreate this scene. Rajiv Malik described the events in Boca Raton to FDA criminal investigators during the Ranbaxy investigation. They detailed his statements, which he made on February 26, 2010, in Rajiv Malik, "Memorandum of Interview," Food and Drug Administration, Office of Criminal Investigations, February 26, 2010. Ranbaxy officials summarized the results of their Sotret tests in a four-page document entitled "Sotret-Investigation Report."

29 *had killed himself while taking it:* Jennifer Frey, "A Double Dose of Heartache," *Washington Post,* January 10, 2001. Frey chronicles the suicide of BJ, the son of Representative Bart Stupak, a Democrat from Michigan.

CHAPTER 3: A SLUM FOR THE RICH

31 *shopping malls, followed:* This description is supported by 2011 census data released by the Directorate of Census Operations in Haryana (accessed December 20, 2017) and multiple news articles, including Vidhi Doshi, "Gurgaon: What Life Is Like in the Indian City Built by Private Companies," *Guardian,* July 4, 2016.

32 *Donkeys and pigs wandered:* Some observations of Gurgaon, the Mehrauli-Gurgaon road, Ranbaxy Laboratories, and Dinesh Thakur's first home in Phase 1 of Gurgaon come from a reporting trip made in January 2015.

32 *Instead, the BBC suggested:* Shalu Yadav, "India's Millennium City Gurgaon a 'Slum for the Rich'?" *BBC Hindi,* August 17, 2012.

34 *The Indian press had dubbed Parvinder:* "Cover Story: India's Best Managed Company," *Business Today,* March 13, 2005.

37 *He declared to adoring throngs:* Celia W. Dugger, "Whatever Happened to Bill Clinton? He's Playing India," *New York Times,* April 5, 2001.

37 *Even in Africa, their drugs: Legends Are Forever,* 46.

38 *at almost two dollars a pill:* Shankar Vedantam and Terence Chea, "Drug

Firm Plays Defense in Anthrax Scare," *Washington Post,* October 20, 2001.

38 *was one-fifth that:* Manu Joseph, "Indian Cipro Copies Don't Pay Off," *Wired,* November 8, 2001.

38 *"It's very important to give these companies":* Biman Mukherji, "No AIDS Progress without Affordable Medicine, Clinton Says in India," *Agence France-Presse,* November 21, 2003.

39 *As Ranbaxy's next managing director, Dr. Brian Tempest:* Randeep Ramesh, "Benign Buccaneer: Interview Brian Tempest, Chief Executive Designate of Ranbaxy," *Guardian,* March 27, 2004.

39 *"Our humanitarian effort has been":* Rohit Deshpande, Sandra J. Sucher, and Laura Winig, "Cipla 2011," Case Study N9-511-050, Harvard Business School, May 3, 2011.

40 *the electric bus stalled:* "Indian Officials Red-Faced after Clinton's Taj Mahal Bus Breaks Down," *Agence France Presse,* November 23, 2003.

CHAPTER 4: THE LANGUAGE OF QUALITY

41 *"Hazard Analysis Critical Control Point":* The CFR Title 21 covering food and drugs can be found on the FDA website, https://www.accessdata.fda.gov/scripts/cdrh/cfdocs/cfcfr/cfrsearch.cfm (accessed June 15, 2018).

43 *a pan on the stove with pieces of meat in it—dog meat:* This scene was re-created based on interviews and secondary sources. I requested primary inspection documents from the FDA under the Freedom of Information Act, but all documents related to Hernandez's inspection, the FDA explained, including the FDA Form 483, had been destroyed in Hurricane Katrina in 2005.

44 *his inspection at the Sherman Pharmaceuticals plant in Abita Springs:* The FDA purged records of the 1994 Sherman Pharmaceuticals inspection in accordance with its documentation retention schedule, according to a Freedom of Information Act request for the Establishment Inspection Report and subsequent Form 483 related to this inspection. Thus, I've re-created scenes based on interviews and secondary sources. Sherman Pharmaceuticals also appears on a list of companies found to be in violation of the FDA's Application Integrity Policy; see U.S. Food and Drug Administration, "Application Integrity Policy—Application Integrity Policy List," updated October 7, 2011, https://www.fda.gov/ICECI/EnforcementActions/ApplicationIntegrityPolicy/ucm134453.htm (accessed June 19, 2018).

46 *debut in a 1962 amendment:* Garnet E. Peck, "Historical Perspective," *Food Drug Cosmetic Journal,* August 1979.

46 *laid out seven rules:* Mona Nasser et al., "Ibn Sina's *Canon of Medicine:* 11th Century Rules for Assessing the Effects of Drugs," *Journal of the Royal Society of Medicine* 102, no. 2 (2009): 78–80, https://www.ncbi.nlm.nih.gov/pmc/articles/PMC2642865/ (accessed December 28, 2017).

47 *the Assize of Bread:* Peter Cartwright, *Consumer Protection and the Criminal Law: Law, Theory, and Policy in the U.K.* (Cambridge: Cambridge University Press, 2001), 152, http://assets.cambridge.org/97805215/90808/frontmatter/9780521590808_frontmatter.pdf (accessed December 28, 2017).

47 *known as pharmacopoeias:* Lembit Rägo and Budiono Santoso, "Drug Regulation: History, Present and Future," in *Drug Benefits and Risks: International Textbook of Clinical Pharmacology,* 2nd ed., rev., edited by C. J. von Boxtel, B. Santoso, and I. R. Edwards (Amsterdam: IOS Press and Uppsala Monitoring Centre, 2008), 65–77, http://www.who.int/medicines/technical_briefing/tbs/Drug_Regulation_History_Present_Future.pdf (accessed December 28, 2017).

47 *was meant to rid the country:* The Authority of the Medical Societies and Colleges, *Pharmacopoeia of the United States of America* (Boston: Wells and Lilly, for Charles Ewer, 1820). See also Jeremy A. Greene, *Generic: The Unbranding of Modern Medicine* (Baltimore: Johns Hopkins University Press, 2014), 27.

47 *published a controversial book:* Friedrich Christian Accum, *A Treatise on Adulterations of Food, and Culinary Poisons. Exhibiting the Fraudulent Sophistications of Bread, Beer, Wine, Spiritous Liquors, Tea, Coffee, Cream, Confectionery, Vinegar, Mustard, Pepper, Cheese, Olive Oil, Pickles, and Other Articles Employed in Domestic Economy. And Methods of Detecting Them* (London: printed by J. Mallett, sold by Longman, Hurst, Rees, Orme, and Brown, 1820), https://trove.nla.gov.au/work/19480247?selectedversion=NBD4018878 (accessed December 28, 2017).

47 *square-jawed, meticulous doctor:* Dale A. Stirling, "Harvey W. Wiley," *Toxicological Sciences* 67, no. 2 (June 1, 2002): 157–58, https://academic.oup.com/toxsci/article/67/2/157/1635211 (accessed December 28, 2017).

48 *"hygienic table trials":* National Endowment for the Humanities, Chronicling America, "The Washington Times, December 14, 1902, Page 14, Image 14," https://chroniclingamerica.loc.gov/lccn/sn84026749/1902-12-14/ed-1/seq-14/ (accessed December 28, 2017).

48 *an epidemic of diphtheria:* U.S. Food and Drug Administration, Center for Biologics Evaluation and Research, Office of Communication, Training, and Manufacturers Assistance, "The St. Louis Tragedy and Enactment of the 1902 Biologics Control Act," Commemorating 100 Years of Biologics Regulation.

48 *throat contorted in painful spasms:* Paul A. Offit, *The Cutter Incident: How America's First Polio Vaccine Led to the Growing Vaccine Crisis* (New Haven, CT: Yale University Press, 2007), 58.

49 *as being worthless and deadly:* Samuel Hopkins Adams, "The Great American Fraud," *Collier's Weekly*, October 7, 1905, https://books.google.com/books?id=fd_S2Van52EC&printsec=frontcover&source=gbs_ge_summary_r&cad=0#v=onepage&q&f=false (accessed December 28, 2017).

49 *exhibition of hazardous food:* U.S. Food and Drug Administration, Center for Biologics Evaluation and Research, "The American Chamber of Horrors," Histories of Product Regulation.

49 *107 people, many of them children:* Carol Ballentine, "Sulfanilamide Disaster," *FDA Consumer*, June 1981.

50 *nearly three hundred people fell into comas:* John P. Swann, "The 1941 Sulfathiazole Disaster and the Birth of Good Manufacturing Practices," *PDA Journal of Pharmaceutical Science and Technology* 53, no. 3 (May/June 1999): 148–53, https://www.ncbi.nlm.nih.gov/pubmed/10754705 (accessed December 28, 2017).

51 *what a good control system should be:* Dale E. Cooper, "Adequate Controls for New Drugs," *Pharmacy in History* 44, no. 1 (2002); John P. Swann, "The 1941 Sulfathiazole Disaster and the Birth of Good Manufacturing Practices," *Pharmacy in History* 40, no. 1 (1999).

51 *to sell a drug called Kevadon:* Linda Bren, "Frances Oldham Kelsey: FDA Medical Reviewer Leaves Her Mark on History," U.S. Food and Drug Administration, *FDA Consumer* (March/April 2001), http://web.archive.org/web/20061020043712/http:/www.fda.gov/fdac/features/2001/201_kelsey.html (accessed December 28, 2017).

51 *known as the Kefauver-Harris Amendment:* Cornelius D. Crowley, "Current Good Manufacturing Practices," *Food and Drug Law Journal* (March 1996).

51 *the amendment redefined what it meant:* Cooper, "Adequate Controls for New Drugs."

52 *master formula and documentation: Federal Register* (June 20, 1963): 6385–87.

52 *the agency undertook a major survey:* Seymore B. Jeffries, "Current Good Manufacturing Practices Compliance—A Review of the Problems and an Approach to Their Management," *Food and Drug Law Journal* (December 1968).

53 *sites regulated by the FDA exceeded those in the United States:* Pew Charitable Trust, Pew Health Group, "After Heparin: Protecting Consumers from the Risks of Substandard and Counterfeit Drugs," July 12, 2011.

54 *advance notice became the jury-rigged solution:* In response to written

questions, the FDA stated, "There are many reasons that FDA inspections are preannounced such as ensuring that the appropriate personnel from the inspected firm are available during the inspection."

CHAPTER 5: RED FLAGS

59 *echoed in a management review:* Christopher King, "Management Development Report: Dinesh Thakur," Kelly & King, August 3, 2004.
63 *report from the World Health Organization:* World Health Organization, "Inspection Report," Vimta Labs, Hyderabad, India, July 26–27, 2004.

CHAPTER 6: FREEDOM FIGHTERS

70 *He left school and went to Sabarmati:* Many of the details from Dr. K. A. Hamied's life are drawn from his autobiography, K. A. Hamied, *A Life to Remember* (Bombay: Lalvani Publishing House, 1972).
70 *"His words were law unto us":* K. A. Hamied, "Oral History Reminisces of India's History Freedom Struggle from 1913 Onwards," interview by Uma Shanker, Centre of South Asian Studies, January 13, 1970.
71 *Gandhi urged Hamied to step into this void:* Peter Church, *Added Value: 30 of India's Top Business Leaders Share Their Inspirational Life Stories* (New Delhi: Roli Books Pvt., 2010), 85.
71 *enabled him to lease a palatial apartment:* Hamied, *A Life to Remember,* 111.
71 *In 1953, he was appointed sheriff of Bombay:* Ibid., 240.
72 *India followed the outdated British patent laws of 1911:* Y. K. Hamied, "Indian Pharma Industry: Decades of Struggle and Achievement," address on the occasion of Dr. A. V. Rama Rao's seventieth birthday, Hyderabad, April 2, 2005.
72 *Prime Minister Indira Gandhi was highly sympathetic:* Deshpande, Sucher, and Winig, "Cipla 2011," 2.
73 *The 1970 Indian Patents Act:* Ibid.
74 *Few got better results than Bhai Mohan Singh:* The history of Bhai Mohan Singh and Ranbaxy's early days is drawn from Bhandari, *The Ranbaxy Story*; and *Legends Are Forever.*
74 *Bhai Traders and Financiers Pvt. Ltd.:* Bhandari, *The Ranbaxy Story,* 29.
74 *His cousins Ranjit and Gurbax:* Ibid.
75 *When his alliance with an Italian drug company:* Ibid., 40.
75 *launched a generic version of Roche's Valium:* Ibid., 47.
76 *The dean there had written to Bhai Mohan:* Ibid., 51.

76 *the company held a prayer ceremony with sixteen local monks: Legends Are Forever,* 52.

76 *In 1987, two Ranbaxy executives traveled:* Ibid., 45.

77 *Varis was a firebrand:* Margalit Fox, "Agnes Varis, 81, Founder of Drug Company," *New York Times,* August 3, 2011.

77 *In 1989, at age seventy-one:* Bhandari, *The Ranbaxy Story,* 90.

78 *As the businesses inherited by the younger sons foundered:* For a fuller account of the Singh family rivalries, see ibid., 90–107. See also Shyamal Majudal, "The Ranbaxy Clash," in Majudal, *Business Battles: Family Feuds That Changed Indian Industry* (New Delhi: Business Standard Books, 2014).

78 *"chili up my back every day":* Bhandari, *The Ranbaxy Story,* 111.

78 *"liaison managers between the government and industrialists":* See Majudal, "The Ranbaxy Clash."

79 *Dozens of warring executives:* Bhandari, *The Ranbaxy Story,* 143–51.

CHAPTER 7: ONE DOLLAR A DAY

81 *annual subscription budget that topped $150,000:* Donald G. McNeil Jr., "Selling Cheap 'Generic' Drugs, India's Copycats Irk Industry," *New York Times,* December 1, 2000.

81 *"What is AIDS?":* There are numerous accounts of Dr. Yusuf Hamied's efforts to combat the AIDS epidemic. Some helpful accounts include Michael Specter, "India's Plague: Are Cheap Drugs Really the Answer to AIDS?" *The New Yorker,* December 17, 2001; *Fire in the Blood: Medicine, Monopoly, Malice,* documentary film directed by Dylan Mohan Gray, 2013.

81 *Just five years earlier, in 1981:* Deshpande, Sucher, and Winig, "Cipla 2011," exhibit 1, AIDS timelines.

82 *But it was brewing so forcefully:* Bob Drogin, "Bombay: Epicenter of Disaster," *Los Angeles Times,* November 26, 1992.

82 *AIDS was destroying Africa:* Mark Schoofs, "The Agony of Africa," *Village Voice,* November/December 1999, http://www.pulitzer.org/winners/mark-schoofs (accessed May 25, 2018). See also UNAIDS Joint United Nations Programme on HIV/AIDS, "AIDS Epidemic Update: December 1998," December 1998, http://data.unaids.org/publications/irc-pub06/epiupdate98_en.pdf (accessed December 8, 2018).

82 *biggest industry was making wooden coffins:* Neil Darbyshire, "Land Where Only Coffin Makers Thrive," *Telegraph,* June 24, 2002.

82 *90 million Africans by 2025:* Joint United Nations Programme on HIV/AIDS (UNAIDS), *AIDS in Africa: Three Scenarios to 2025,* January

2005, http://www.unaids.org/sites/default/files/media_asset/jc1058-aidsinafrica_en_1.pdf (accessed December 8, 2018).

82 *In 1991, Dr. Rama Rao:* Peter Church, *Added Value: 30 of India's Top Business Leaders Share Their Inspirational Life Stories* (New Delhi: Roli Books Pvt., 2010), 92.

83 *In 1997, under the leadership of Nelson Mandela:* Helene Cooper, Rachel Zimmerman, and Laurie McGinley, "AIDS Epidemic Puts Drug Firms in a Vise: Treatment vs. Profits," *Wall Street Journal,* March 2, 2001, https://www.wsj.com/articles/SB983487988418159849 (accessed May 25, 2018); see also Deshpande, Sucher, and Winig, "Cipla 2011," 5.

84 *On September 28, 2000, he took to the podium:* Y. K. Hamied, speech at the Round Table Conference, European Commission, Brussels, September 28, 2000. This scene is also recounted in Specter, "India's Plague."

85 *The resulting article, a detailed profile:* McNeil, "Selling Cheap 'Generic' Drugs, India's Copycats Irk Industry."

86 *one of the most devastating earthquakes ever:* R. Bendick et al., "The 26 January 2001 'Republic Day' Earthquake, India," *Seismological Research Letters* 72, no. 3 (May/June 2001): 328–35, doi:10.1785/gssrl.72.3.328 (accessed June 15, 2018).

86 *Bill Clinton, who'd just left office:* David Remnick, "The Wanderer: Bill Clinton's Quest to Save the World, Reclaim His Legacy—and Elect His Wife," *The New Yorker,* September 18, 2006.

86 *McNeil's story was published the next morning:* Donald G. McNeil Jr., "Indian Company Offers to Supply AIDS Drugs at Low Cost in Africa," *New York Times,* February 7, 2001.

87 *As the* Wall Street Journal *summed it up:* Cooper, Zimmerman, and McGinley, "AIDS Epidemic Puts Drug Firms in a Vise."

87 *"They are pirates":* Neelam Raj, "Cipla: Patients before Patents," in *The Politics of the Pharmaceutical Industry and Access to Medicines,* edited by Hans Löfgren (New York: Routledge/Social Science Press, 2018).

87 "Of course *I have an ulterior motive":* Deshpande, Sucher, and Winig, "Cipla 2011," 6.

87 *people rallied against the drug companies all over the world:* Adele Baleta, "Drug Firms Take South Africa's Government to Court," *The Lancet* 357, no. 9258 (March 10, 2001), doi:10.1016/S0140-6736(00)04158-1 (accessed June 15, 2018).

88 *The Clinton Foundation stepped in:* Celia W. Dugger, "Clinton Makes Up for Lost Time in Battling AIDS," *New York Times,* August 29, 2006, https://www.nytimes.com/2006/08/29/health/29clinton.html (accessed June 15, 2018).

88 *But on January 28, 2003, he stunned them:* Ethan B. Kapstein and

Joshua W. Busby, *AIDS Drugs for All: Social Movements and Market Transformations* (Cambridge: Cambridge University Press, 2013), 138.

89 *Randall Tobias, the former CEO:* John W. Dietrich, "The Politics of PEPFAR: The President's Emergency Plan for AIDS Relief," *Ethics and International Affairs* 21, no. 3 (Fall 2007): 277–93.

89 *In March 2004, six senators:* Senators John McCain, Russell D. Feingold, Ted Kennedy, Lincoln Chafee, Olympia Snowe, and Dick Durbin, letter to the Honorable George W. Bush, March 26, 2004.

90 *first Indian generics company to get approval from the PEPFAR program:* "Appendix VI: Generic HIV/AIDS Formulations Made Eligible for Purchase by PEPFAR Programs under the HHS/FDA Expedited Review Process, through December 10, 2006," United States President's Emergency Plan for AIDS Relief, 2006, https://www.pepfar.gov/press/82131.htm (accessed June 21, 2018); *The Power of Partnerships: The United States President's Emergency Plan for AIDS Relief: Third Annual Report to Congress on PEPFAR,* 2007, https://www.pepfar.gov/documents/organization/81019.pdf (accessed June 19, 2018).

CHAPTER 8: A CLEVER WAY OF DOING THINGS

92 *He submitted his resignation:* Rajiv Malik, "Memorandum of Interview," Food and Drug Administration, Office of Criminal Investigations, February 26, 2010. The MOI states, "He explained that he drafted his resignation letter on the plane to India after the Boca Raton meeting. Malik submitted his resignation on June 1, 2003 but stayed on for approximately another two months."

92 *in June 2003*: According to the MOI, Malik and his Ranbaxy team were hired by the generic drug company Sandoz, in Vienna, where they stayed for two years.

92 *advanced a theory:* Dr. Raghunath Anant Mashelkar served as director-general of India's Council of Scientific and Industrial Research. He outlined his concept of Gandhian innovation in an influential article, R. A. Mashelkar and C. K. Prahalad, "Innovation's Holy Grail," *Harvard Business Review* (July/August 2010), https://hbr.org/2010/07/innovations-holy-grail (accessed January 10, 2018).

93 *It could often mean "better":* Some of India's leading generic drug manufacturers have argued that they could make superior drugs but are restricted by regulations that require their products to be similar to the brand. At Cipla, Dr. Yusuf Hamied explained that though his chemists were the best in the world, "I have to be as bad as the original," he said.

93 *In 1961, two army vets:* The account of Mylan's early history comes from

John T. Seaman and John T. Landry, *Mylan: 50 Years of Unconventional Success: Making Quality Medicine Affordable and Accessible* (Canonsburg, PA: Mylan, 2011).

93 *"white as ever":* Ibid., 65.

94 *"Bresch and Coury saw":* Ibid., 114.

96 *resort to "shenanigans":* Carolyn Y. Johnson, "FDA Shames Drug Companies Suspected of Abusive Tactics to Slow Competition," *Washington Post,* May 18, 2018, http://www.highbeam.com/doc/1P4-2040528829.html?refid=easy_hf (accessed November 12, 2018).

97 *The results yield a graph that contains:* Sam H. Haidar, Barbara Davit, Mei-Ling Chen, Dale Conner, Laiming Lee, Qian H. Li, Robert Lionberger, Fairouz Makhlouf, Devvrat Patel, Donald J. Schuirmann, and Lawrence X. Yu, "Bioequivalence Approaches for Highly Variable Drugs and Drug Products," *Pharmaceutical Research* 25, no. 1 (2007): 237–41, doi:10.1007/s11095-007-9434-x. For more details of the FDA's bioequivalence standards, see also "Preface," in *Orange Book*; Lynda S. Welage, Duane M. Kirking, Frank J. Ascione, and Caroline A. Gaither, "Understanding the Scientific Issues Embedded in the Generic Drug Approval Process," *Journal of the American Pharmaceutical Association* 41, no. 6 (2001): 856–67, doi:10.1016/s1086-5802(16)31327-4.

97 *So, in 1992:* The FDA published a guidance in July 1992 that outlined the central concepts of bioequivalence: "Statistical Procedures for Bioequivalence Studies Using a Standard Two Treatment Cross-over Design" (Washington, DC: FDA, Center for Drug Evaluation and Research, 1992). That guidance recommended a statistical analysis for pharmacokinetic measures, including AUC and Cmax. It proposed a calculation of a 90 percent confidence interval for the ratio of averages, and a requirement that this confidence interval fall within a range of 80 to 125 percent bioequivalence. The FDA deemed this approach "average bioequivalence." However, the debate over bioequivalence wasn't put to rest with the 1992 guidance, as documented in Robert Schall's "Bioequivalence: Tried and Tested" (*Cardiovascular Journal of Africa* 21, no. 2 [April 2010]: 69–70), because, at the same time the FDA issued its guidance, the biostatisticians Sharon Anderson and Walter W. Hauk raised a new question about whether drugs deemed "bioequivalent" work the same way in different patients, with implications for patients switching from one drug to another. This concept, dubbed "individual bioequivalence," became a new area of research and debate. Yet there was doubt about whether the new concept of individual bioequivalence was a legitimate concern or an unnecessary precaution. According to Schall and other critics, the FDA essentially ignored the concept in its 2003 guidance, which instead reiterated the 1992 definition of av-

erage bioequivalence; see "Guidance for Industry: Bioavailability and Bioequivalence Studies for Orally Administered Drug Products: General Considerations" (Washington DC: FDA, Center for Drug Evaluation and Research, March 2003).

98 *Johnson & Johnson's epilepsy drug:* Natasha Singer, "J&J Unit Recalls Epilepsy Drug," *New York Times,* April 14, 2011, https://prescriptions.blogs.nytimes.com/2011/04/14/j-j-unit-recalls-epilepsy-drug/ (accessed July 16, 2018).

99 *the number of drug applications:* Seaman and Landry, *Mylan,* 121.

CHAPTER 9: THE ASSIGNMENT

102 *"Everyone knows," he said, by way of greeting:* The details of Dinesh Thakur's interaction with Arun Kumar come, in part, from an account that Dinesh Thakur drafted, starting one year after he left Ranbaxy. Reached in May 2013, Arun Kumar denied that he shared information with Thakur and claimed instead that he shared it with the company. He asserted that management launched an investigation because it didn't know about the misconduct. He did not respond to more recent attempts to contact him. Thakur's account of Arun Kumar sharing information for his review and working on his report is corroborated by an interview that Dr. Raj Kumar gave to FDA criminal investigators in which he told them: "Kumar asked the Director of Regulatory Affairs, Arun Kumar if there was exposure on other Ranbaxy products in addition to the ARVs and without hesitation Kumar responded, 'Yes.' Kumar instructed Arun Kumar and Thakur to provide a risk assessment and a summary of what the problematic issues were with Ranbaxy ARV products." Rajinder Kumar, "Food and Drug Administration Office of Criminal Investigations, Memorandum of Interview," April 10, 2007.

104 *by concealing unsold inventory:* "Ex-Bristol-Myers Execs Plead Not Guilty," Associated Press, June 22, 2005. Prosecutors dropped the criminal case against the two men after an April 2010 court decision restricted the testimony of an expert witness. The case ended in 2012 when the men reached a settlement with the U.S. Securities and Exchange Commission. Richard Vanderford, "Ex-Bristol Myers Exec Settles SEC Profit Inflation Suit," *Law 360,* April 2, 2012.

114 *Kumar distributed a spreadsheet:* The ten-page spreadsheet prepared by Dinesh Thakur came to be known as the Ranbaxy issues portfolio.

115 *Kumar showed the men a PowerPoint of twenty-four slides:* The PowerPoint presentation of October 14, 2004, prepared by Dinesh Thakur and shown to a board subcommittee by Raj Kumar, came to be known

by Ranbaxy executives as the Self-Assessment Report or "SAR." It also appeared in a PowerPoint presentation associated with *United States v. Ranbaxy,* shown by prosecutors to Ranbaxy lawyers on November 17, 2009. Dr. Kumar also described to FDA criminal investigators what occurred in the boardroom in Rajinder Kumar, "Food and Drug Administration Office of Criminal Investigations, Memorandum of Interview," April 10, 2007.

116 *broken down piece by piece:* Rajinder Kumar's memorandum of the interview states: "Tempest directed all who had a copy of the presentation to destroy it. He also directed that the computer used to generate the presentation be destroyed."

CHAPTER 10: THE GLOBAL COVER-UP

121 *They were determined to make the most of the soggy event:* Ramesh Adige, "Clinton Library Dedication," *Ranbaxy World: A Bi-Annual External Newsletter of Ranbaxy* (August 2005): 9. Donation amounts are taken from the Clinton Foundation's website, where they are listed as a range; see https://www.clintonfoundation.org/contributors.

122 *the company had big future plans: Ranbaxy World,* 2, 18.

123 *President of Pharmaceuticals Malvinder Singh:* Amberish K. Diwanji (deputy managing editor), "The Rediff Interview/Malvinder Singh, President, Ranbaxy," *Rediff,* November 25, 2004, http://www.rediff.com/money/2004/nov/25inter.htm (accessed May 29, 2018).

131 *"ever-decreasing resources":* Food and Drug Administration, "Drug Manufacturing Inspections Program (Foreign CGMP Pilot Protocol)," compliance program circular, October 1, 2006.

133 *becoming the bulk-drug capital of India:* Raksha Kumar, "Planned 'Pharma City' to Pump Out Cheap Indian Drugs Is Making Indian Villagers Sick with Anger," *South China Morning Post,* February 17, 2018, https://www.scmp.com/week-asia/business/article/2133347/planned-pharma-city-pump-out-cheap-indian-drugs-making-villagers (accessed September 21, 2018).

134 *a report by the U.S. Government Accountability Office:* U.S. Government Accountability Office, "Food and Drug Administration: Improvements Needed in the Foreign Drug Inspection Program," GAO/HEHS-98-21 (Washington, DC: GAO, March 17, 1988).

134 *In his summary of the inspection:* FDA, "Establishment Inspection Report," Ranbaxy Laboratories Ltd., Paonta Sahib, India, December 17–21, 2004.

135 *Gavini must have walked directly past:* Stein Mitchell & Muse LLP,

"Unregistered Use of 4°C Refrigerators to Conceal Drug Defects," February 21, 2011. This report, prepared by lawyers for Dinesh Thakur during the Ranbaxy case, notes: "In 2004, PSCWICO1, an 1800 liter ThermoLab stability refrigerator, was purchased in February 2004, was installed on May 1, 2004, and became operational on May 5, 2004."

CHAPTER 11: MAP OF THE WORLD

140 *His old boss, Barbhaiya, had been given:* Rajinder Kumar, "Food and Drug Administration Office of Criminal Investigations, Memorandum of Interview," April 10, 2007. In this interview with Dr. Kumar, FDA investigators noted: "Tempest told Kumar that Barbhaiya threatened that he would go public with information regarding fraudulent practices at Ranbaxy and received over one million dollars as a compensation package."

140 *Just eighteen months earlier, a project director:* V. K. Raghunathan, "Indian Engineer Killed for Exposing Graft," *Straits Times,* December 12, 2003.

141 *"Ranbaxy Laboratories in India is fooling you":* "Malvinder Singh," "PEPFAR & ARVs," email to Randall Tobias, Mark Dybul, and Adriaan J. Van Zyl, August 15, 2005.

142 *So he wrote again, this time more pointedly:* "Malvinder Singh," "Fwd: Pepfar & ARVs," email to Gary Buehler, Jane Axelrad, David Horowitz, Joseph Famulare, Steven Galson, Warren Rumble, and Robert West, August 17, 2005.

142 *In a forceful and urgent email:* "Malvinder Singh," "Re: Fwd: PEPFAR & ARVs," email to FDA commissioner Lester Crawford, August 29, 2005.

144 *In an email afterward:* "Malvinder Singh," "RE: PEPFAR & ARVs," email to Edwin Rivera-Martinez, September 9, 2005.

147 *The five-page assignment memo:* U.S. Department of Health and Human Services, Food and Drug Administration, Branch Chief, Investigations and Preapproval Compliance Branch, HFD-322, "Request for 'For Cause' Investigation FACTS #678634," memorandum sent to Director, Division of Field Investigations, HFC-130, October 7, 2005.

CHAPTER 12: THE PHARAOH OF PHARMA

149 *His father, Parvinder, had been austere:* Descriptions of Malvinder Mohan Singh's childhood, as well as his family's history, were drawn from Bhandari, *The Ranbaxy Story.*

150 *His management style was brash:* Depictions of Malvinder Mohan

Singh's management style, lifestyle, and personal tastes are drawn from numerous articles about him and his brother in the Indian media, including: "The Rediff Interview/Malvinder Singh, President, Ranbaxy"; Archna Shukla, "Ranbaxy Revs Up," *Business Today,* September 10, 2006, http://archives.digitaltoday.in/businesstoday/20060910/cover1.html (accessed June 8, 2018); Moinak Mitra and Bhanu Pande, "Ranbaxy's Singhs Ready to Build Empire," *Economic Times,* April 17, 2009, https://economictimes.indiatimes.com/magazines/corporate-dossier/ranbaxys-singhs-ready-to-build-empire/articleshow/4412356.cms (accessed June 8, 2018).

150 *dubbed him "the Pharaoh of Pharma":* Joe Mathew, "Newsmaker: Malvinder Mohan Singh: Pharaoh of Pharma," *Business Standard,* January 12, 2007, https://www.business-standard.com/article/beyond-business/newsmaker-malvinder-mohan-singh-107011201042_1.html (accessed June 6, 2018).

150 *to nineteenth in 2005:* Naazneen Karmali, "India's 40 Richest," *Forbes,* December 10, 2004, https://www.forbes.com/2004/12/08/04indialand.html#629040502bae (accessed June 15, 2018); Naazneen Karmali, "India's 40 Richest," *Forbes,* December 15, 2005, https://www.forbes.com/2005/12/15/india-richest-40_05india_land.html#5fa54b954faf (accessed June 15, 2018).

150 *"I want profit!":* Katherine Eban, "Dirty Medicine," *Fortune,* May 15, 2013.

150 *$100,000 champagne-colored Mercedes:* Archna Shukla, "Cars the Super Rich Drive," *Business Today,* October 22, 2006.

152 *in an interview for Duke University's business school alumni magazine:* John Manuel, "Singhing the Same Tune," *Exchange* (Summer 2001): 34–35.

152 *manufacturing plant in the northern state of Himachal Pradesh:* A full list of inspections conducted at Paonta Sahib can be found on the FDAzilla website, https://fdazilla.com/.

152 *the unregistered walk-in refrigerator, set to 4 degrees Celsius:* FDA, "Establishment Inspection Report," Ranbaxy Laboratories Ltd., Paonta Sahib, Simour District, India, February 20–25, 2006.

153 *"because we did not understand FDA to have requested it.":* "Re: Ranbaxy's Responses to Food and Drug Administration (FDA) Warning Letter of June 15, 2006." Alok Ghosh, Vice President, Global Quality, to Mr. Nicholas Buhay, Acting Director, Division of Manufacturing and Product Quality, August 29, 2006.

153 *"it remains unclear to us":* Nicholas Buhay, Acting Director, Division of Manufacturing and Product Quality, CDER, FDA, U.S. Department of

Health and Human Services, "Warning Letter" to Ramesh Parekh, Vice President, Manufacturing, Ranbaxy Laboratories Ltd., June 1, 2006, 4.

153 *Brown and Horan set out to inspect Dewas:* FDA, "Establishment Inspection Report," Ranbaxy Laboratories Ltd., Dewas, India, February 27–March 2, 2006.

154 *"after the fact or outside of the laboratory operation":* Ibid., 21.

CHAPTER 13: OUT OF THE SHADOWS

162 *"Since this is a more formal meeting":* "Malvinder Singh," "Re: Information Meeting," email to Debbie Robertson, September 19, 2006.

163 *On November 29, 2006, Singh led a delegation:* This scene was re-created, in part, through reference to the minutes of the Ranbaxy-FDA meeting, November 29, 2006.

166 *the assignment sheet for the inspection:* U.S. Department of Health and Human Services, FDA, Karen Takahashi, Consumer Safety Officer, HFD-325, "Request for 'For Cause' Assignment FACTS #792363, Firm: Ranbaxy Laboratories, Ltd., Paonta Sahib, Himachal Pradesh, India FEI: 3002807978," to Rebecca Hackett, Branch Chief, HFC-130, January 16, 2007.

166 *Hernandez arrived on January 26, 2007:* FDA, "Establishment Inspection Report," Ranbaxy Laboratories Ltd., Paonta Sahib, India, January 26–February 1, 2007.

CHAPTER 14: "DO NOT GIVE TO FDA"

170 *New Jersey to New Delhi, Ranbaxy issued a statement:* Patricia Van Arnum, "Ranbaxy Comments on Merck KGaA Generics Rumors, Confirms Federal Raid in NJ," *PharmTech,* February 15, 2007, http://www.pharmtech.com/ranbaxy-comments-merck-kgaa-generics-rumors-confirms-federal-raid-nj (accessed September 21, 2018).

170 *It was the company's own secret:* Ranbaxy, "Sotret—Investigation Report" (four-page internal document).

174 *the internal auditor at WorldCom:* Cynthia Cooper, *Extraordinary Circumstances: The Journey of a Corporate Whistleblower* (Hoboken, NJ: Wiley, 2009), 281.

174 *had represented Monica Lewinsky:* Saundra Torry, "Lewinsky Legal Team Brings Credibility," *Washington Post,* June 4, 1998.

176 *dated back to the Civil War:* Henry Scammell, *Giantkillers: The Team*

and the Law That Help Whistle-Blowers Recover America's Stolen Billions (New York: Atlantic Monthly Press, 2004), 36.

176 *the infamous $640 toilet seat:* Eric Wuestewald, "Timeline: The Long, Expensive History of Defense Rip-offs," *Mother Jones,* December 18, 2013, https://www.motherjones.com/politics/2013/12/defense-military-waste-cost-timeline/ (accessed September 21, 2018).

CHAPTER 15: "HOW BIG IS THE PROBLEM?"

180 *and they piled up in back offices:* The FDA's media relations office, in response to a written question about whether the agency routinely reviews annual reports, stated: "Reports are reviewed as appropriate."

183 *that required inspection by the FDA skyrocketed:* Pew Charitable Trust, "After Heparin: Protecting Consumers from the Risks of Substandard and Counterfeit Drugs," July 12, 2011.

192 *The FDA did not approve:* Seven months after Ranbaxy failed to address all the issues raised by investigators in the February 12, 2008, Form 483, the FDA issued a warning letter that stated, "Until all corrections have been completed and FDA can confirm your firm's compliance with CGMPs, this office will recommend disapproval of any new applications or supplements listing your firm as a manufacturing location of finished dosage forms and active pharmaceutical ingredients." Richard L. Friedman, Division of Manufacturing and Product Quality, Office of Compliance, Center for Drug Evaluation and Research, Silver Spring, MD, to Mr. Malvinder Singh, "Warning Letter 320-08-03," September 16, 2008.

192 *On December 12, 2007, USAID:* Jean C. Horton, Acting Director, Office of Acquisition and Assistance, USAID, "Re: Show Cause," letter to Venkat Krishnan, Vice President and Regional Director, Ranbaxy Laboratories Inc., December 12, 2007.

193 *By standing at the edge of the property:* FDA, "Establishment Inspection Report," Ranbaxy Laboratories Ltd., Paonta Sahib, Himachal Pradesh State, India, March 3–7, 2008.

194 *He became "agitated and desperate":* Ibid., 44.

194 *withdrew its tacrolimus application:* Dr. T. G. Chandrashekhar, "Re: Ranbaxy's Responses to Food and Drug Administration (FDA) Form 483 Observations of Batamandi during Inspection Conducted March 3–7, 2008," letter to John M. Dietrick, May 1, 2008.

CHAPTER 16: DIAMOND AND RUBY

195 *"I am an entrepreneur at heart":* Archna Shukla, "Like Father Like Son," *Business Today,* August 13, 2006.

195 *Ranbaxy's bottom line was "sagging":* "Corporate Profile—Finding a Cure for Ranbaxy's Ills," *AsiaMoney,* March 1, 2006.

196 *The surgeon arrived at work to:* Vidya Krishnan, "Private Practice: How Naresh Trehan Became One of India's Most Influential Doctor-Businessmen," *The Caravan—A Journal of Politics and Culture,* February 1, 2015.

197 *with manufacturing plants in eleven countries:* Daiichi Sankyo, "Ranbaxy to Bring in Daiichi Sankyo as Majority Partner; Strategic Combination Creates Innovator and Generic Pharma Powerhouse," news release, June 11, 2008, https://www.daiichisankyo.com/media_investors/media_relations/press_releases/detail/005635.html (accessed June 15, 2018).

197 *"India will be the trump card":* Eiichiro Shimoda, "Daiichi Sankyo Targets Generics," *Nikkei Weekly,* June 16, 2008.

202 *in the public domain:* In a written statement, Malvinder Singh categorically denied that he had deceived Daiichi Sankyo: "There was no misrepresentation or concealment of information from Daiichi Sankyo (Daiichi) since the fact of the US FDA & DOJ investigation was all in the public domain and was also specifically informed to Daiichi." He blamed the Japanese company, instead, for poor management of Ranbaxy. "All allegations of fraud/concealment raised by Daiichi after more than 3 years after acquiring control over Ranbaxy, pursuant to a due-diligence which lasted around 10 months, are false and a mere afterthought aimed at vindicating the losses that Ranbaxy (controlled by Daiichi) suffered due to the recall of Atorvastatin (Gx version of Lipitor) necessitated by the presence of glass particles in the drug, manufactured under Daiichi's watch which made the drugs unsafe for consumption (Interestingly the Atorvastatin recall happened in November 2012 the period when Daiichi also filed the arbitration proceedings)."

203 *called the sale an "emotional decision":* Eban, "Dirty Medicine."

204 *on July 3, 2008, the Maryland U.S. Attorney:* United States of America v. Ranbaxy Inc., and Parexel Consulting, Motion to Enforce Subpoenas and Points and Authorities (U.S. District Court for the District of Maryland, Southern Division, July 3, 2008).

205 *the twenty-seven others:* According to the FDA's *Orange Book,* a compendium of FDA-approved drug products, the agency approved twenty-seven unique Abbreviated New Drug Applications (ANDAs)

submitted by Ranbaxy from August 2005 to August 2008, which included different dosage forms of eleven different drugs.

CHAPTER 17: "YOU JUST DON'T GET IT"

207 *"pattern of systemic fraudulent conduct":* United States of America v. Ranbaxy Inc., and Parexel Consulting, Motion to Enforce Subpoenas and Points and Authorities (U.S. District Court for the District of Maryland, Southern Division, July 3, 2008).

207 *In a cagey call with reporters:* Eban, "Dirty Medicine."

209 *said Dr. Janet Woodcock, CDER's director:* "FDA Issues Warning Letters to Ranbaxy Laboratories Ltd., and an Import Alert for Drugs from Two Ranbaxy Plants in India," *FDA News,* news release, September 16, 2008.

209 *Deb Autor told reporters:* FTS-HHS FDA, "Transcript of Media Briefing on Ranbaxy Labs," news release, September 17, 2008.

210 *what leverage did the FDA have?:* In a written statement, Deb Autor responded to the characterization of her remarks provoking anger inside the agency: "No one at the Agency told me that they considered my remarks misleading or detrimental to the case FDA was trying to build against Ranbaxy. If they had done so, that would certainly have been thoroughly considered and addressed."

211 *The FDA announced that it would level the harshest punishment*: Saundra Young, "FDA Says India Plant Falsified Generic Drug Data," CNN, February 25, 2009, http://edition.cnn.com/2009/HEALTH/02/25/fda.india.generic.drugs/index.html (accessed June 11, 2018).

216 *should not mention the SAR:* In a written statement provided by a spokesperson, Lavesh Samtani said, "Although I cannot discuss anything covered by attorney-client privilege, I can tell you that all material I shared with Daiichi was done in a manner that protected such privilege. After resolving the FDA and DOJ issues on behalf of a Daiichi-controlled Ranbaxy, I chose to leave to start a new venture in 2014 and remain on good terms with Dr. Une."

CHAPTER 18: CONGRESS WAKES UP

222 *round-the-world airline trip and receipts:* "Guilty Plea in Drug Case," *New York Times,* May 26, 1989, https://www.nytimes.com/1989/05/26/business/guilty-plea-in-drug-case.html (accessed May 21, 2018).

222 *hired by Mylan, the respected generic drug company:* Milt Freudenheim, "Exposing the FDA," *New York Times,* September 10, 1989.

223 *a full-scale investigation and send the trash:* Edmund L. Andrews, "A Scandal Raises Serious Questions at the FDA," *New York Times,* August 13, 1989.

223 *Dingell's committee uncovered corruption:* Malcolm Gladwell and Paul Valentine, "FDA Battles for Authority amid Generic-Drug Scandal," *Washington Post,* August 16, 1989, https://www.washingtonpost.com/archive/politics/1989/08/16/fda-battles-for-authority-amid-generic-drug-scandal/54ef2d8b-4a9d-45b0-851a-4446d139137e/?noredirect=on&utm_term=.33b423bb01c7 (accessed July 31, 2018).

223 *"a swamp that must be drained":* William C. Cray and C. Joseph Stetler, *Patients in Peril? The Stunning Generic Drug Scandal* (n.p., 1991), 113.

223 *Congressional hearings in 1989:* See Cray and Stetler, *Patients in Peril?*

223 *"horrible world of overwhelming work":* Seaman and Landry, *Mylan,* 62.

224 *whose CEO had given Chang $23,000:* "Founder of Generic Drug Firm Fined, Gets Jail Term in Bribery," *Los Angeles Times,* September 15, 1989, http://articles.latimes.com/1989-09-15/news/mn-183_1_generic-drug (accessed May 21, 2018).

224 *fraud or corruption charges:* The prosecutorial dragnet finally caught Marvin Seife, the FDA's director of the Office of Generic Drugs. Though well-meaning and candid with investigators, he also had what the *Washington Post* dubbed a "lunch problem." He enjoyed regular lunches with industry executives, who lobbied him during the meals and then picked up the tab. He'd been reprimanded for the habit almost a decade earlier. At the scandal's height, he signed an affidavit claiming that the lunches had stopped, when they hadn't. In 1990, Seife was convicted on two counts of perjury and sentenced to ten months in a Texas prison. There, he was given ill-fitting shoes. Seife was a diabetic and prone to foot infections. By the time prison authorities got him to a hospital, one of his legs had to be amputated below the knee. Phil McCombs, "The Bungled Punishment of Prisoner Seife," *Washington Post,* April 3, 1992.

225 *confidence in the generic drug industry:* Joe Graedon and Teresa Graedon, "Generic Drugs Still a Good Buy," *Buffalo News,* September 13, 1989, http://buffalonews.com/1989/09/13/generic-drugs-still-a-good-buy/ (accessed May 21, 2018).

225 *90 million kilograms by 2008:* Pew Health Group, Pew Charitable Trusts, "After Heparin: Protecting Consumers from the Risks of Substandard and Counterfeit Drugs," white paper, March 201125.

225 *"a string of ticking time bombs":* Scenes depicting the tense relationship between the FDA and Congress were re-created based on transcripts of con-

gressional hearings, including: Janet Woodcock and Deborah Autor, "The Heparin Disaster: Chinese Counterfeits and American Failures," testimony before a hearing of U.S. House of Representatives Committee on Energy and Commerce, Subcommittee on Oversight and Investigations, April 29, 2008; William Hubbard, "FDA'S Foreign Drug Inspection Program: Weaknesses Place Americans at Risk," testimony before hearing of U.S. House of Representatives Committee on Energy and Commerce, Subcommittee on Oversight and Investigations, April 22, 2008.

225 *cheap active ingredients imported from China:* Cheryl A. Thompson, "FDA Admits to Lacking Control over Counterfeit Drug Imports," *Health-System Pharmacists News* (American Society of Health-System Pharmacists), June 9, 2000.

227 *a team worked around the clock:* Beth Miller, "Drama in the Dialysis Unit," *Outlook* (Office of Medical Public Affairs, Washington University in St. Louis) (Spring 2009), https://core.ac.uk/download/pdf/70380372.pdf (accessed May 28, 2018).

228 *with a similar-sounding name:* Marc Kaufman, "FDA Says It Approved the Wrong Drug Plant," *Washington Post,* February 19, 2008.

228 *finally traveled to Changzhou:* Richard L. Friedman, Public Health Service, FDA, "Warning Letter" to Dr. Van Wang, WL: 320-08-01, April 21, 2008.

228 *The ingredient mimicked heparin:* Amanda Gardner, "Researchers Identify Contaminant in Tainted Heparin," *Washington Post,* April 23, 2008.

230 *"continuing to aggressively investigate the situation":* FDA, "Postmarket Drug Safety Information for Patients and Providers—Information on Heparin," last updated November 1, 2018, https://www.fda.gov/Drugs/DrugSafety/PostmarketDrugSafetyInformationforPatientsandProviders/default.htm.

231 *"left or right and you'll see it":* Suketu Mehta, *Maximum City: Bombay Lost and Found* (New York: Random House, 2004), 192.

CHAPTER 19: SOLVING FOR X

233 *The story described how the United States was importing:* Richard Knox, "As Imports Increase, a Tense Dependence on China," NPR, *Morning Edition,* May 25, 2007.

235 *best defense for his patients was medication:* Bernard J. Gersh et al., "2011 ACCF/AHA Guideline for the Diagnosis and Treatment of Hypertrophic Cardiomyopathy," *Circulation* 124 (December 8, 2011): e783–831, http://circ.ahajournals.org/content/124/24/e783 (accessed May 29, 2018).

237 *"not related to product efficacy":* In 2013, the year after Karen Wilmering stopped taking Glenmark's pravastatin, Glenmark recalled multiple lots of three different drugs, including 246,528 bottles of pravastatin, after consumers complained that the medicines gave off a strong fishy odor. A Glenmark spokesperson noted that the company could not comment on Wilmering's specific complaint because her case was never reported directly to the company: "Patient safety is our first priority and we take all claims on adverse events or product quality complaints very seriously." See "Glenmark Recalls Three Drugs from U.S. Market," *Economic Times,* May 23, 2013.

237 *a trail of aggrieved patients:* Patient complaints made in 2013 about various Zydus drugs appear on the websites MedsChat.com and ConsumerAffairs.com.

238 *started marketing the first generic version:* Sarah Turner, "AstraZeneca to Launch Generic of Its Own Heart Drug," *MarketWatch,* November 22, 2006, https://www.marketwatch.com/story/astrazeneca-to-launch-generic-version-of-its-own-heart-drug (accessed May 29, 2018).

238 *Sandoz quietly recalled its drug:* Tom Lamb, "Sandoz Metoprolol Succinate ER Tablets Recall Has Been Done Rather Quietly," *Drug Injury Watch,* December 5, 2008, http://www.drug-injury.com/druginjurycom/2008/12/generic-drug-recall-metoprolol-er-tablets-by-sandoz--recall-metoprolol-er-tablets-by-sandozwwwipcrxcompharmacy-industry-n.html (accessed May 29, 2018).

238 *"Does not comply with the dissolution test of the USP monograph":* The Cleveland Clinic pharmacist wasn't the only person to question why Ethex would be permitted to sell a drug that did not comply with the USP monograph. The independent drug testing organization Consumer Lab also investigated the drug and package insert and published a report; see "Drug Investigation: Toprol XL vs. Generic Metoprolol Succinate Extended-Release (ER) Tablets," product review, ConsumerLab.com, December 31, 2008, http://coyo.ga/www.consumerlab.com/reviews/Toprol_vs_Generic_Metoprolol/Toprol/ (accessed May 29, 2018.

239 *Ethex announced a sweeping recall of more than sixty products:* Tom Lamb, "January 2009: ETHEX Corp. Issues Voluntary Recall of All Pills Due to Suspected Manufacturing Problems," *Drug Injury Watch,* February 2, 2009, http://www.drug-injury.com/druginjurycom/2009/02/ethex-corporation-issues-nationwide-voluntary-recall-of-products-press-release-includes-list-of-all-generic-drugs-by-ethex.html (accessed May 29, 2018).

239 *over $27 million:* Federal Bureau of Investigation, "Ethex Corporation, a Subsidiary of KV Pharmaceutical, Pleads Guilty to Two Felonies and Agrees to Pay United States $27,568,921 for Fine, Restitution, and For-

feiture," news release, March 2, 2010, https://archives.fbi.gov/archives/stlouis/press-releases/2010/sl030210.htm (accessed December 10, 2018).

242 *And Brown never did:* In a written statement, Dr. Reddy's asserted that its tacrolimus is safe and effective, and made under identical FDA standards as brand-name drugs. Since the drug's launch in 2010, the company says that it has manufactured more than 569 million capsules for the U.S. market and has received only twenty lack-of-effect complaints, the majority of them within the first two years of the drug's launch. "It is common behavior to get lack-of-effectiveness complaints at the beginning of a launch cycle as the patient population adjusts to the idea of generics replacing the innovator drugs," the company stated.

242 *"the Dr. Reddy's brand of Prograf":* According to the FDA's MedWatch database, the report from Loma Linda was filed on October 28, 2013, under the defect "Potency Questioned."

242 *which they published in 2017:* Rita R. Alloway, Alexander A. Vinks, Tsuyoshi Fukuda, Tomoyuki Mizuno, Eileen C. King, Yuanshu Zou, Wenlei Jiang, E. Steve Woodle, Simon Tremblay, Jelena Klawitter, Jost Klawitter, and Uwe Christians, "Bioequivalence between Innovator and Generic Tacrolimus in Liver and Kidney Transplant Recipients: A Randomized, Crossover Clinical Trial," *PLOS Medicine,* November 14, 2017, doi:10.1371/journal.pmed.1002428.

CHAPTER 20: A TEST OF ENDURANCE

249 *a claim that a Justice Department official would later deny:* In a written statement, in response to a number of questions, a Justice Department official stated, without being more specific, "The Department of Justice disagrees with these previously unvoiced allegations against its personnel. These allegations are false and could not have in any way affected the outcome of the investigation."

CHAPTER 21: A DEEP, DARK WELL

259 *For three decades, Joe and Terry Graedon:* The Graedons' views on generic drugs over the years, as well as consumer comments and complaints about specific drugs, can be found on their website, http://www.peoplespharmacy.com/.

260 *published these cases in a 1998 newspaper column:* Joe Graedon and Terry Graedon, "Are Generic Equivalents as Good as Brand Name Drugs?" part 2 of 3, *King Features Syndicate,* May 18, 1998.

260 *By 2002, Joe Graedon had contacted the FDA:* Joe Graedon and Teresa Graedon, "The Generic Drug Quandary: Questions about Quality," in *Best Choices from the People's Pharmacy: What You Need to Know before Your Next Visit to the Doctor or Drugstore* (New York: Rodale, 2006), 22.

261 *fall below 80 percent or rise above 125 percent: Statistical Procedures for Bioequivalence Studies Using a Standard Two-Treatment Crossover Design* (Washington, DC: FDA, CDER, 1992).

262 *patients taking it began writing in to the Graedons:* After both Toprol XL and Wellbutrin XL went generic, the Graedons saw an uptick in patient complaints, many of which can be found on their website (http://www.peoplespharmacy.com/).

263 *more than eighty-five reports:* Anna Edney, "Teva Pulls Version of Wellbutrin XL on Effectiveness," Bloomberg, October 4, 2012.

264 *ConsumerLab tested the 300-milligram dose:* "Generic Bupropion Is Not Always the Same as Brand-Name Wellbutrin," ConsumerLab.com, October 12, 2007, updated October 17, 2013, https://www.consumerlab.com/reviews/Wellbutrin_vs_Generic_Bupropion/Wellbutrin/ (accessed May 29, 2018).

265 *Graedon and Temple of the FDA were both asked to be guest:* Larry Mantle, producer, "Generic Drug Safety," KPCC, *AirTalk,* December 19, 2007.

266 *assured consumers that it had been right:* FDA, Division of Drug Information (DDI), "Drug Information Update—Review of Therapeutic Equivalence Generic Bupropion XL 300 Mg and Wellbutrin XL 300 Mg," news release, April 16, 2008.

267 *the FDA simply reviewed the bioequivalence data:* Although Teva marketed and sold generic Budeprion, the drug was manufactured and developed by Impax, which provided bioequivalence data in its 2003 application to the FDA. After the drug was pulled, Teva sued Impax for misrepresenting its bioequivalence. These events were reported in Roger Bate et al., "Generics Substitution, Bioequivalence Standards, and International Oversight: Complex Issues Facing the FDA," *Trends in Pharmacological Science* 37, no. 3 (December 2015), doi:10.1016/j.tips.2015.11.005; and Dan Packel, "Impax Must Pay for GSK Wellbutrin Settlement, Teva Says," *Law360*, August 31, 2017, https://www.law360.com/articles/959538/impax-must-pay-for-gsk-wellbutrin-settlement-teva-says (accessed May 29, 2018).

269 *Teva and Impax planned to conduct a bioequivalence study:* Andy Georgiades, "Teva Aims to Quell Concerns with Generic Wellbutrin Trial," *Wall Street Journal* (Toronto), December 2, 2009.

269 not *therapeutically equivalent:* After the FDA released its findings in 2012, Teva released a statement: "Upon receiving the communication

from the U.S. Food and Drug Administration, Teva has immediately ceased shipment of Impax's 300 mg Budeprion XL. This update to the FDA's guidance affects the bioequivalence rating of the product and does not reflect any safety issue. Teva's first priority is to our patients and providing them with quality medicines." Teva did not respond to repeated requests for comment for this book.

270 *vice president of global regulatory intelligence and policy—at Teva:* Pat Wechsler, *Bloomsberg News,* "Teva Hires Gary Buehler Away from FDA," *SFGate,* February 9, 2012, https://www.sfgate.com/business/article/Teva-hires-Gary-Buehler-away-from-FDA-3170563.php (accessed June 10, 2018).

270 *Horrified, she watched as the bug:* Carla Stouffer's complaint, though filed with the FDA, never reached Dr. Reddy's. The company stated: "From 2011 to date, Dr. Reddy's has not received any communication, complaint or intimation relating to insect in dosage forms of this product." Descriptions of adverse events were pulled from the FDA's Drug Quality Reporting System (DQRS) (MedWatch Reports) database. Carla Stouffer's complaint is logged as number 1603903 in the DQRS database. Each year the FDA receives millions of such reports from patients, caregivers, and the general public; see Lichy Han, Robert Ball, Carol A. Pamer, Russ B. Altman, and Scott Proestel, "Development of an Automated Assessment Tool for MedWatch Reports in the FDA Adverse Event Reporting System," *Journal of the American Medical Informatics Association* 24, no. 5 (September 1, 2017): 913–20, doi:10.1093/jamia/ocx022.

CHAPTER 22: THE $600 MILLION JACKET

273 *group of U.S. senators reminded the FDA commissioner in a March 2011 letter:* "U.S. Sen. Harkin and Others Urge FDA to Avoid Delays of Generic Drug Approvals," *Pharma Letter,* March 16, 2011, https://www.thepharmaletter.com/article/us-sen-harkin-and-others-urge-fda-to-avoid-delays-of-generic-drug-approvals (accessed December 11, 2018).

273 *owing to a settlement with Pfizer:* "Ranbaxy, Pfizer Sign Truce over Lipitor," *Economic Times,* June 19, 2008, https://economictimes.indiatimes.com/industry/healthcare/biotech/pharmaceuticals/ranbaxy-pfizer-sign-truce-over-lipitor/articleshow/3143801.cms (accessed December 11, 2018).

274 *"where this is going to come out":* Eban, "The War over Lipitor."

275 *company operated on razor-thin margins:* Ashish Gupta, "The Pills That Saved Ranbaxy," *Fortune India,* August 5, 2012, https://www.fortune

india.com/ideas/the-pills-that-saved-ranbaxy/100819 (accessed May 28, 2018).

275 *Since 2002, individual reviewers had flagged anomalies:* Depictions of anomalies in Ranbaxy's atorvastatin application, as well as back-and-forth correspondence between the FDA and Ranbaxy, were drawn from FDA, Center for Drug Evaluation and Research, *Approval Package for Application Number: ANDA 076477Orig1s000, Sponsor: Ranbaxy, Inc.,* November 30, 2011, https://www.accessdata.fda.gov/drugsatfda_docs/anda/2011/076477Orig1s000.pdf (accessed May 24, 2018).

276 *An inspection of Ranbaxy's research and development laboratory in Gurgaon:* FDA, "Establishment Inspection Report," Ranbaxy Laboratories, Gurgaon, India, April 27–May 12, 2009.

276 *In February 2009, it had leveled a rare Application Integrity Policy:* FDA, Center for Drug Evaluation and Research, "Enforcement Activities by FDA—Regulatory Action against Ranbaxy," updated May 15, 2017, https://www.fda.gov/Drugs/GuidanceComplianceRegulatoryInformation/EnforcementActivitiesbyFDA/ucm118411.htm (accessed May 28, 2018).

282 *a hard-hitting motion to dismiss Mylan's suit:* "District Court Dismisses Mylan's Complaint against FDA Concerning Generic Lipitor," *Orange Book Blog,* May 2, 2011, http://www.orangebookblog.com/2011/05/district-court-dismisses-mylans-complaint-against-fda-concerning-generic-lipitor.html (accessed May 28, 2018); Mylan Pharms. v. FDA and Ranbaxy Labs, 789 F.Supp.2d 1, Civil Action No. 11-566 (JEB) (U.S. District Court for the District of Columbia, 2011).

283 *Wall Street analysts had mapped out complex flow charts:* Eban, "The War over Lipitor."

283 *with a fateful recommendation:* Director, Office of Compliance, "Proposal to Review Ranbaxy's Atorvastatin ANDA," memo to Director, Center for Drug Evaluation and Research, Food and Drug Administration, May 11, 2011.

284 *once every two years:* At the time, the FDA endeavored to inspect each plant every two years. That changed in 2012 with the passage of a law called The FDA Safety and Innovation Act (FDASIA), which codified into law a risk-based model for assessing when to inspect plants. Jerry Chapman, "How FDA And MHRA Decide Which Drug Facilities To Inspect—And How Often," Pharmaceutical Online, July 13, 2018, https://www.pharmaceuticalonline.com/doc/how-fda-and-mhra-decide-which-drug-facilities-to-inspect-and-how-often-0001 (accessed February 7, 2019).

284 *FDA investigators arrived at Toansa:* FDA, "Establishment Inspection Report," Ranbaxy Laboratories, Toansa, India, November 21–25, 2011.

287 *At 8:12 p.m., the agency issued a press release:* "FDA Confirms Nod for Ranbaxy's Generic Lipitor," *Reuters,* December 01, 2011, https://www.reuters.com/article/us-ranbaxy/fda-confirms-nod-for-ranbaxys-generic-lipitor-idUSTRE7B007L20111201 (accessed June 11, 2018).

288 *Arun Sawhney, addressed his staff:* Vikas Dandekar, "Ranbaxy Launches AG Version of Caduet as CEO Likens Lipitor Deal with Teva to an Insurance Policy," *The Pink Sheet,* December 6, 2011.

288 *nearly $600 million within six months:* Ashish Gupta, "The Pills That Saved Ranbaxy," *Fortune India,* August 5, 2012.

289 *would prove dismayingly well founded:* "Ranbaxy Halts Generic Lipitor Production after Recall: FDA," *Reuters,* November 29, 2012, https://www.reuters.com/article/us-ranbaxy-lipitor-idUSBRE8AS1C620121129 (accessed May 24, 2018).

CHAPTER 23: THE LIGHT SWITCH

294 *The officials took them to the MP-11 plant:* FDA, Form 483, Inspectional Observations, Ranbaxy, Toansa, India, December 7–14, 2012.

295 *those who'd signed the logs had not actually:* Ibid., 2–3.

299 *to the United States soared:* According to the website FDAzilla, FDA investigator Muralidhara Gavini performed ninety facility inspections in India between 2001 and 2011. An analysis of these records showed that in at least forty-one of these inspections, the FDA had never visited the facility before and had to assess whether to approve the plant to export into the U.S. market. Gavini approved thirty-five of those forty-one facilities, a roughly 85 percent approval rate.

299 *On his fifth inspection in India, at RPG Life Sciences:* FDA, "Establishment Inspection Report," RPG Life Sciences Ltd., Ankleshwar, India, November 20–24, 2012.

299 *a manufacturing plant in Kalyani . . . owned by Fresenius Kabi:* FDA, "Form 483: Inspectional Observations," Fresenius Kabi Oncology Ltd., Nadia, India, January 14–18, 2013.

300 *he saw what looked like earlier tests:* Ibid., 1.

304 *On March 18, 2013, Peter Baker arrived at the main plant of Wockhardt Ltd.:* FDA, "Form 483: Inspectional Observations," Wockhardt Ltd., Aurangabad, India, March 18–22, 2013.

307 *"emergency plan in place prior to arrival":* FDA, Establishment Inspection Report, Wockhardt Ltd., Waluj, Aurangabad, March 18–22, 2013, 7. The threats the FDA team received at that inspection were documented in Pallavi Ail, "USFDA Says Team Threatened during Wockhardt Inspection," *Financial Express* (Mumbai), May 28, 2014.

CHAPTER 24: WE ARE THE CHAMPIONS

313 *$500 million to settle the case:* Ranbaxy Laboratories, "Ranbaxy Laboratories Sets Aside $500 Million to Settle U.S. Probe, Signs Consent Decree with FDA," news release, December 21, 2011.

314 *Beato responded angrily to a government lawyer:* Conversation with Andrew Beato, January 3, 2013, notes taken by Dinesh Thakur.

317 *Motz approved the settlement:* United States of America v. Ranbaxy USA, Inc., Ranbaxy Pharmaceuticals, Inc., Ranbaxy Laboratories, Inc., Ranbaxy, Inc., Ohm Laboratories, Inc., Ranbaxy Laboratories Ltd., filed by Dinesh S. Thakur, Settlement Agreement (U.S. District Court for the District of Maryland, Southern Division, May 10, 2013), PACER Case Locator Case 1:07-cv-00962-JFM.

318 *appeared on* Fortune *magazine's U.S. website:* Eban, "Dirty Medicine."

319 *Daiichi Sankyo issued a press release:* Daiichi Sankyo, Media Relations, "Ranbaxy Announces Improved Business Standards and Quality Assurance Initiatives," news release, May 22, 2013. https://www.daiichisankyo.com/media_investors/media_relations/press_releases/detail/005976.html (accessed December 16, 2018).

CHAPTER 25: CRASHING FILES

322 *with nine manufacturing plants worldwide:* Katie Thomas, "Mylan Buys Drug Maker of Generic Injectables," *New York Times,* February 27, 2013.

322 *largely credited to Bresch:* Gardiner Harris, "Deal in Place for Inspecting Foreign Drugs," *New York Times,* August 13, 2011.

323 *a sterile injectable plant in Bangalore, Karnataka:* FDA, "Form 483: Inspectional Observations," Mylan Laboratories Ltd., Bangalore, India, June 17–27, 2013.

323 *a dangerously sloppy plant:* Ibid.

324 *"disruptions of the unidirectional air flow":* Ibid., 3.

324 *the investigators found "crushed insects":* Michael Smedley, Acting Director, Office of Manufacturing and Product Quality, CDER, Office of Compliance, FDA, "Warning Letter" to Venkat Iyer, CEO, Agila Specialties Private Ltd., September 9, 2013, 2.

325 *he and his team exceeded targets:* Between 2011 and 2017, proxies that Mylan filed with the U.S. Securities and Exchange Commission show that the company exceeded its own targets for the number of applications submitted to global regulators. One such example includes a target of 140 global submissions in 2012, while the actual

number of applications submitted was 171. U.S. Securities and Exchange Commission, proxy statement, Mylan Inc., April 12, 2013, 26. As another example, in 2017, the target for applications was 135, while the actual number submitted was 184. U.S. Securities and Exchange Commission, proxy statement, Mylan N.V., May 30, 2018, 46. The targeted number of global regulatory submissions comprised one-quarter of Rajiv Malik's annual incentive compensation, which was one-third of his overall compensation, according to the proxies.

326 *It hiked the price:* Alex Nixon, "Firestorm Grows over Price Hikes on EpiPen," *Pittsburgh Tribune Review,* August 25, 2016.

326 *10 percent of the company's revenue:* Figures for Bresch's 2007 compensation are drawn from U.S. Securities and Exchange Commission, proxy statement, Mylan Inc., April 5, 2010, 26. Her 2015 compensation comes from U.S. Securities and Exchange Commission, proxy statement, Mylan N.V., May 30, 2018, 53. The 2014 compensation for Bresch and Malik comes from U.S. Securities and Exchange Commission, proxy statement, Mylan N.V., May 23, 2017, 62. The figure for EpiPen as providing 10 percent of the company's revenue comes from U.S. Congress, Full House Committee on Oversight and Government Reform, "Reviewing the Rising Price of EpiPens: Testimony of Heather Bresch, CEO of Mylan," 114th Cong., 2nd sess., September 21, 2016, https://oversight.house.gov/hearing/reviewing-rising-price-epipens-2/, 54.

326 *Overnight, Bresch became:* Andrew Buncombe, "Mylan CEO's Salary Soared by 671% as Firm Hiked EpiPen Prices," *Independent,* August 26, 2016.

326 *In a disastrous CNBC interview:* Dan Mangan, "Mylan CEO Bresch: 'No One's More Frustrated than Me' about EpiPen Price Furor," CNBC, *Squawk Box,* August 25, 2016.

327 *none of this made any sense:* On August 17, 2017, Mylan Inc. and Mylan Specialty LLP agreed to settle a False Claims Act lawsuit and pay $465 million to the U.S. Justice Department for knowingly misclassifying EpiPen as a generic drug, rather than as a brand drug, to avoid paying greater Medicaid rebates. U.S. Department of Justice, Office of Public Affairs, "Mylan Agrees to Pay $465 Million to Resolve False Claims Act Liability for Underpaying EpiPen Rebates," news release, August 17, 2017.

327 *the* Pittsburgh Post-Gazette *uncovered:* Patricia Sabatini and Len Boselovic, "MBA Mystery in Morgantown: Questions Raised over How WVU Granted Mylan Executive Her Degree," *Pittsburgh Post-Gazette,* December 21, 2007.

327 *allegations that Executive Chairman Robert Coury:* Tracy Staton, "Think EpiPen Is Mylan's First Scandal? Here's a Timeline of Jet Use, an Unearned MBA, and More," *FiercePharma,* September 2, 2016, https://www.fiercepharma.com/pharma/think-epipen-mylan-s-first-scandal-here-s-a-timeline-jet-use-resume-fakery-and-more (accessed June 13, 2018).

327 *Bresch found herself grilled under oath:* U.S. Congress, Full House Committee on Oversight and Government Reform, "Reviewing the Rising Price of EpiPens: Testimony of Heather Bresch, CEO of Mylan," 114th Cong., 2nd sess., September 21, 2016, https://oversight.house.gov/hearing/reviewing-rising-price-epipens-2/ (accessed June 19, 2018).

329 *Over the course of nine days:* FDA, "Form 483: Inspectional Observations," Mylan Laboratories Ltd., Sinnar, Nashik District, Maharastra, India, September 5–14, 2016.

330 *"instrument malfunction," "power loss":* Ibid., 7.

330 *Within two months, three investigators:* FDA, "Form 483: Inspectional Observations," Mylan Pharmaceuticals Inc., Morgantown, West Virginia, November 7–18, 2016.

332 *"the accuracy and integrity of data":* FDA, "Warning Letter 320-17-32" (re: Mylan Laboratories Ltd., Nashik, FDF), letter to Rajiv Malik, President, Mylan Pharmaceuticals Inc., April 3, 2017.

CHAPTER 26: THE ULTIMATE TESTING LABORATORY

337 *a very sick thirteen-year-old boy:* Jason W. Nickerson, Amir Attaran, Brian D. Westerberg, Sharon Curtis, Sean Overton, and Paul Mayer, "Fatal Bacterial Meningitis Possibly Associated with Substandard Ceftriaxone—Uganda, 2013," *Morbidity and Mortality Weekly Report* 64, nos. 50–51 (January 1, 2016), 1375–77, doi:10.15585/mmwr.mm6450a2.

337 *The throng of patients usually exceeded:* Chris Obore, "Time Bomb: The Inside Story of Mulago Hospital's Troubles," *Daily Monitor,* January 20, 2013, http://www.monitor.co.ug/News/National/Time-bomb-The-inside-story-of-Mulago-hospital-s-troubles/688334-1669688-akvcb7/index.html (accessed June 3, 2018).

340 *in the manufacture of drugs bound for the least-regulated markets:* Depictions of variations in drug quality found around the world were drawn from several scientific studies, including Roger Bate, Ginger Zhe Jin, Aparna Mathur, and Amir Attaran, "Poor Quality Drugs and Global Trade: A Pilot Study," Working Paper 20469 (Cambridge, MA: National Bureau for Economic Research, September 2014), doi:10.3386/w20469; and Richard Preston Mason, Robert F. Jacob, and Seth A. Gerard,

"Atorvastatin Generics Obtained from Multiple Sources Worldwide Contain a Methylated Impurity That Reduces Their HMG-CoA Reductase Inhibitory Effects," *Journal of Clinical Lipidology* 7, no. 3 (2013): 287, doi:10.1016/j.jacl.2013.03.096.

341 *He and Westerberg published a case report:* Nickerson et al., "Fatal Bacterial Meningitis Possibly Associated with Substandard Ceftriaxone—Uganda, 2013."

342 *every single sample failed quality evaluations:* Anita Nair, Stefanie Strauch, Jackson Lauwo, Richard W. O. Jähnke, and Jennifer Dressman, "Are Counterfeit or Substandard Anti-infective Products the Cause of Treatment Failure in Papua New Guinea?" *Journal of Pharmaceutical Sciences* 100, no. 11 (June 30, 2011): 5059–68, doi:10.1002/jps.22691.

342 *Peter Baker inspected the company's plant northeast of Beijing:* FDA, "Form 483: Inspectional Observations," CSPC Zhongnuo Pharm (shijiazhuang) Co. Ltd., China, March 23–27, 2015.

342 *less than 70 percent of the required active ingredient:* Elizabeth Pisani, "Losing the War on Bugs," *Prospect* (February 2016).

345 *They zeroed in on oxytocin and ergometrine:* Eric Karikari-Boateng and Kwasi Poku Boateng, *Post-Market Quality Surveillance Project: Maternal Healthcare Products (Oxytocin and Ergometrine) on the Ghanaian Market,* Ghana Food and Drugs Authority, Promoting the Quality of Medicines Program, USAID, February 2013.

346 *Shanghai Desano, whose factory in Binhai had been inspected:* World Health Organization, "Inspection Report," Shanghai Desano Chemical Pharmaceutical Co., China, March 18, 2011.

348 *Caudron and his colleagues published a landmark paper:* J.-M. Caudron, N. Ford, M. Henkens, C. Macé, R. Kiddle-Monroe, and J. Pinel, "Substandard Medicines in Resource-Poor Settings: A Problem That Can No Longer Be Ignored," *European Journal of Tropical Medicine and International Health* 13, no. 8 (August 13, 2008): 1062–72, doi:10.1111/j.1365-3156.2008.02106.x.

348 *the British government commissioned:* Jim O'Neill, "Antimicrobial Resistance: Tackling a Crisis for the Health and Wealth of Nations," *Review on Antimicrobial Resistance* (December 2014), https://amr-review.org/sites/default/files/AMR%20Review%20Paper%20-%20Tackling%20a%20crisis%20for%20the%20health%20and%20wealth%20of%20nations_1.pdf (accessed June 3, 2018).

349 *fingered substandard drugs as a culprit:* Ian Williams, "The Race to Contain Drug-Resistant Malaria," *NBCNews.com,* January 22, 2011, http://worldblog.nbcnews.com/_news/2011/01/22/5825008-the-race-to-contain-drug-resistant-malaria (accessed June 3, 2018).

349 *he coauthored an editorial:* Paul N. Newton, Céline Caillet, and Philippe J. Guerin, "A Link between Poor Quality Antimalarials and Malaria Drug Resistance?" *Expert Review of Anti-infective Therapy* 14, no. 6 (May 23, 2016): 531–33, doi:10.1080/14787210.2016.1187560.

350 *the first study to link:* Muhammad Zaman and Zohar B. Weinstein, "Evolution of Rifampicin Resistance Due to Substandard Drugs in E. Coli and M. Smegmatis," forthcoming in *Antimicrobial Agents and Chemotherapy,* posted online November 5, 2018.

350 *wrote in a 2015 report:* Elizabeth Pisani, "Antimicrobial Resistance and Medicine Quality," *AMR Review* (November 2015), https://amr-review.org/sites/default/files/ElizabethPisaniMedicinesQualitypaper.pdf (accessed November 30, 2018).

350 *a Nevada woman in her seventies returned home:* Lei Chen, "Notes from the Field: Pan-Resistant New Delhi Metallo-Beta-Lactamase-Producing Klebsiella Pneumoniae—Washoe County, Nevada, 2016," *Morbidity and Mortality Weekly Report* 66, no. 1 (January 13, 2017): 33, https://www.cdc.gov/mmwr/volumes/66/wr/mm6601a7.htm?s_cid=mm6601a7_w (accessed June 3, 2018).

350 *a "nightmare bacteria" with no known cure:* Sabrina Tavernise, "Infection Raises Specter of Superbugs Resistant to All Antibiotics," *New York Times,* May 27, 2016.

351 *Only one-tenth of African countries:* Margareth Ndomondo-Sigonda, Jacqueline Miot, Shan Naidoo, Alexander Dodoo, and Eliangiringa Kaale, "Medicines Regulation in Africa: Current State and Opportunities," *Pharmaceutical Medicine* 31 (November 3, 2017): 383–97, doi:10.1007/s40290-017-0210-x.

CHAPTER 27: FLIES TOO NUMEROUS TO COUNT

357 *In the blog post, Lal:* Altaf Ahmed Lal, "FDA in India: Going Global, Coming Home," *FDA Voice,* September 24, 2013.

359 *America's patients had:* Within a year of Peter Baker's inspection at Wockhardt, the company began a series of recalls of the drug. Eric Palmer, "Wockhardt Again Recalls Generic of AstraZeneca Drug after It Fails Testing," *FiercePharma* (blog), September 2, 2014.

359 *Harry Lever, the Cleveland Clinic cardiologist, had sent a detailed letter of concern:* Dr. Harry M. Lever to Dr. Janet Woodcock, Director, Center for Drug Evaluation and Research, FDA, December 12, 2012.

359 *Lever received a detailed response:* Lawrence Yu, "FW: Metoprolol Response," email to Harry Lever, MD, December 19, 2012.

360 *Dipesh Shah and Atul Agrawal, had arrived two days earlier:* FDA, "Establishment Inspection Report," Wockhardt Ltd., Aurangabad, Maharashtra, India, July 22–31, 2013.

360 *labeled "Default May 2013":* Ibid., 14.

361 *The man begged him:* Ibid., 21.

362 *import restrictions to the company:* Wockhardt issued a statement after the FDA's July 2013 inspection and import alert at its Chikalthana plant. The company made assurances that it had "already initiated several steps to address the observations made by the USFDA and shall put all efforts to resolve the matter at the earliest." "Wockhardt's Chikalthana Plant Hit by USFDA Import Restrictions," *Economic Times.* November 27, 2013, https://economictimes.indiatimes.com/industry/healthcare/biotech/pharmaceuticals/wockhardts-chikalthana-plant-hit-by-usfda-import-restrictions/articleshow/26466331.cms (accessed December 8, 2018).

365 *In the early hours of a Sunday morning:* FDA, "Establishment Inspection Report," Ranbaxy Laboratories Ltd., Toansa, Punjab, India, January 5–11, 2014.

366 *"flies TNTC" (too numerous to count):* Ibid., 33.

367 *as one FDA official later described it:* The FDA official, Tom Cosgrove, called the finding "shocking" at an industry conference. "International Pharmaceutical Quality: Inside the Global Regulatory Dialogue.: Lecture, 2015, https://www.ipqpubs.com/wp-content/uploads/2015/06/Cosgrove-box.pdf (accessed February 10, 2019).

367 *restricted from the U.S. market:* Barbara W. Unger, "Does an FDA Import Alert Automatically Equate to an Impending FDA Warning Letter?" *FDAzilla.com* (blog), April 30, 2016, https://blog.fdazilla.com/2016/04/does-an-fda-import-alert-automatically-equate-to-an-impending-fda-warning-letter/ (accessed December 7, 2018).

368 *Forty years of national reports:* Over the years, numerous reports by Indian drug experts have castigated the state of drug regulation in India. The most recent reports include: Government of India, Ministry of Health and Family Welfare, *Report of the Expert Committee on a Comprehensive Examination of Drug Regulatory Issues Including the Problem of Spurious Drugs,* November 2003; Rajya Sabha, Parliament of India, *Fifty-Ninth Report on the Functioning of the Central Drug Standard Control Organisation (CDSCO),* May 2012; *Report of the Prof. Ranjit Roy Chaudhury Expert Committee to Formulate Policy and Guidelines for Approval of New Drugs, Clinical Trials, and Banning of Drugs,* July 2013.

369 *"If I have to follow U.S. standards":* Sushmi Dey, "If I Follow U.S. Standards, I Will Have to Shut Almost All Drug Facilities: G. N. Singh

Interview with Drug Controller General of India," *Business Standard,* January 30, 2014, https://www.business-standard.com/article/economy-policy/if-i-follow-us-standards-i-will-have-to-shut-almost-all-drug-facilities-g-n-singh-114013000034_1.html (accessed June 18, 2018).

370 *superior to those of any other nation:* The FDA's internal battles over mutual recognition are documented in employee interviews in the FDA History Office's oral history program. For more details, see the oral histories of: Walter M. Batts, "History of the Food and Drug Administration," interviewed December 13 and 20, 2011; Stephanie Gray, "History of the Food and Drug Administration," interviewed April 11, 2000; Linda Horton, "History of the Food and Drug Administration," interviewed December 28, 2001; Gerald "Jerry" E. Vince, "History of the Food and Drug Administration," interviewed December 2, 1998; and Andrew Von Eschenbach, "History of the U.S. Food and Drug Administration," interviewed September 15, 2013.

370 *"Statement of Intent," a four-page document:* Dr. Margaret Hamburg, Commissioner of the U.S. Food and Drug Administration, and Keshav Desiraju, Secretary of India's Department of Health and Family Welfare, signatories to "Statement of Intent between the Food and Drug Administration of the United States of America and the Ministry of Health and Family Welfare of the Republic of India on Cooperation in the Field of Medical Products," New Delhi, India, February 10, 2014.

371 *The Ranbaxy executive took that moment to lobby:* Gardiner Harris, "Medicines Made in India Set Off Safety Worries," *New York Times,* February 14, 2014, https://www.nytimes.com/2014/02/15/world/asia/medicines-made-in-india-set-off-safety-worries.html (accessed June 18, 2018).

371 *G. N. Singh shot back:* Sumeet Chatterjee and Zeba Siddiqui, "UPDATE 1—U.S. Regulator on India Visit Calls for Greater Drug Safety Collaboration," *Reuters,* February 18, 2014, https://www.reuters.com/article/fda-hamburg-india/update-1-u-s-regulator-on-india-visit-calls-for-greater-drug-safety-collaboration-idUSL3N0LN38W20140218 (accessed June 18, 2018).

CHAPTER 28: STANDING

378 *had maligned him to journalists*: "Some Brands of Nationalism Can Be Injurious to Your Health!" *Governance Now,* March 8, 2016, https://www.governancenow.com/news/regular-story/some-brands-nationalism-can-be-injurious-your-health (accessed December 16, 2018).

378 *Thakur sent Vardhan:* Dinesh S. Thakur, Executive Chairman, Medassure, to Honorable Dr. Harsh Vardhan, Minister of Health and Family Welfare, Government of India, October 19, 2013.

378 *a "snake pit of vested interests":* Pritha Chatterjee, "MCI Corrupt, Clinical Trials Body a Snake Pit: Harsh Vardhan," *Indian Express,* July 18, 2014.

382 *caught their attention:* Richard Preston Mason, Robert F. Jacob, and Seth A. Gerard, "Atorvastatin Generics Obtained from Multiple Sources Worldwide Contain a Methylated Impurity That Reduces Their HMG-CoA Reductase Inhibitory Effects," *Journal of Clinical Lipidology* 7, no. 3 (2013).

383 *drew defensive attacks from the FDA:* Dr. Preston Mason's findings about low-quality generic Lipitor appeared in the May/June 2013 issue of the *Journal of Clinical Lipidology*—just six months after Ranbaxy's recall of its own generic Lipitor product due to the presence of tiny glass particles. The FDA reacted defensively to Mason's study. During an interview with a *Bloomberg* reporter, CDER director Janet Woodcock claimed that Mason's team didn't use proper testing methods and therefore contaminated its own samples. Woodcock later repeated this attack in a paper cowritten with Mansoor A. Khan, "FDA Analysis of Atorvastatin Products Refutes Report of Methyl Ester Impurities," *Therapeutic Innovation and Regulatory Science* 48, no. 5 (May 27, 2014): 554–56, doi:10.1177/2168479014536567. However, Mason had used the same testing method prescribed by the USP to test all his samples, and the methylated impurities that Woodcock claimed were proof of improper testing appeared only in some. Documents supporting this narrative include: Mason et al., "Atorvastatin Generics Obtained from Multiple Sources Worldwide Contain a Methylated Impurity That Reduces Their HMG-CoA Reductase Inhibitory Effects," 287; Anna Edney, "Disputing Study, U.S. FDA Says Generics from Abroad Safe," *Bloomberg,* March 25, 2014, http://www.bloomberg.com/news/articles/2014-03-25/disputing-study-u-s-fda-says-generics-from-abroad-safe (accessed July 13, 2018).

384 *"Zero Defect, Zero Effect":* Vishwa Mohan, "PM's Slogan: Zero Defect, Zero Effect," *Times of India,* August 16, 2014.

384 *suspending from the European market:* B. V. Mahalakshmi, "EU Bans 700 Generic Drugs for Manipulation of Trials by GVK," *Financial Express,* July 26, 2015.

385 *a former employee of GVK Biosciences:* On May 6, 2012, the whistleblower, Konduru Narayana Reddy, using the pseudonym "People Safety," wrote an email to drug regulators in France, Britain, the United States, Austria,

and the World Health Organization with the subject line "Regulatory Violations and Misconduct of Bioequivalence and Bioavailability Studies for the Past 5 Years by Head-Bio Analytical (V. Chandra Sekhar), GVK Biosciences Private Limited, CRO (India Based-Hyderabad)."

385 *LeBlaye laid out his findings:* ANSM (French Agency on Medicinal Products), Trials and Vigilance Inspection Department, "Final Inspection Report: Investigation of the Clinical Part of Bioequivalence Trials, with a Specific Focus on Electrocardiograms, May 19–23, 2014, GVK Biosciences," July 2, 2014.

386 *"bigger game being played out here":* Vidya Krishnan, "A Love Story That Cost GVK Its International Reputation," *The Hindu,* October 9, 2015.

386 *"Hold onto your hats! This is incredible":* Joe Graedon, "Hold onto Your Hats . . . This Is Incredible!," email to Harry Lever, Erin Fox, Roger Bate, Preston Mason, and Dinesh Thakur, August 12, 2015.

389 *They had crafted two lengthy petitions:* Dinesh S. Thakur v. Union of India; Central Drug Standards Control Organisation, Drugs Consultative Committee, Comptroller and Auditor General of India (January 24, 2016); Dinesh S. Thakur v. Union of India (January 28, 2016).

391 *had publicly attacked Thakur:* Zeba Siddiqui, "Pharma Crusader Dinesh Thakur Takes India's Drug Regulators to Court," *Reuters,* March 7, 2016, https://www.reuters.com/article/india-pharma-whistleblower/pharma-crusader-takes-indias-drug-regulators-to-court-idUSKCN0W90C8 (accessed June 20, 2018).

392 *he posted what he'd written, with the headline:* Dinesh Thakur, "A Sincere Attempt to Improve the Quality of Medicine for People around the World," *Dinesh Thakur* (blog), March 11, 2016, http://dineshthakur.com/?s=A sincere attempt to improve the quality of medicine for people around the world&x=0&y=0 (accessed June 20, 2018).

393 *"From what little I have read":* Dinesh Thakur, "FDC Ban," email to K. L. Sharma, March 23, 2016.

EPILOGUE

395 *Kobe Steel, a Japanese steel maker:* Jonathan Soble and Neal E. Boudette, "Kobe Steel's Falsified Data Is Another Blow to Japan's Reputation," *New York Times,* October 10, 2017, https://www.nytimes.com/2017/10/10/business/kobe-steel-japan.html (accessed June 9, 2018).

397 *column for an online publication,* The Wire: Dinesh Thakur, "Lessons from Ranbaxy: Suffocating Silence Prevented Us from Questioning

the Rot in the System," *The Wire,* February 19, 2018, https://thewire.in/business/ranbaxy-suffocating-silence-prevented-us-questioning-rot-system (accessed June 9, 2018).

397 *G. N. Singh, was removed from his job:* Zachary Brennan, "India's Drug Regulator Sees Top-Level Shakeup," *Regulatory Affairs Professionals Society,* February 21, 2018, https://www.raps.org/news-and-articles/news-articles/2018/2/india's-drug-regulator-sees-top-level-shakeup (accessed June 9, 2018).

397 *the Ranbaxy company no longer exists:* Though Ranbaxy Laboratories ceased to exist when it was acquired by Sun Pharma on March 25, 2015, Sun Pharma has continued to sell Ranbaxy-branded generics in markets where they were popular. Documents describing these events include: Sun Pharma, "Sun Pharma Announces Closure of Merger Deal with Ranbaxy," news release, March 25, 2015, https://www.sunpharma.com/sites/default/files/docs/Press%20Release%20-%20Closure%20of%20Sun%20Pharma%20&%20Ranbaxy%20merger.pdf; Malvika Joshi and C. H. Unnikrishnan, "Sun Pharma to Retain Ranbaxy Brand Wherever It's Strong," *LiveMint,* April 10, 2014, https://www.livemint.com/Companies/rSdzvCSLvJesbEaSzgawVJ/Sun-Pharma-to-retain-Ranbaxy-brand-wherever-its-strong.html (accessed July 27, 2018); "Ranbaxy's Journey as a Company to End after Merger with Sun," *Hindu BusinessLine,* April 20, 2014, https://www.thehindubusinessline.com/companies/ranbaxys-journey-as-a-company-to-end-after-merger-with-sun/article20756422.ece# (accessed July 27, 2018); Sun Pharmaceutical Industries Ltd., "Annual Report of Subsidiary Companies," 2017–2018, http://www.sunpharma.com/investors/annual-report-of-subsidiary-companies (accessed July 27, 2018).

398 *sold the tarnished company cheaply:* Chang-Ran Kim and Zeba Siddequi, "India's Sun Pharma to Buy Struggling Ranbaxy for $3.2 Billion," *Reuters,* April 7, 2014, https://www.reuters.com/article/us-daiichi-sankyo-ranbaxy-sunpharma/indias-sun-pharma-to-buy-struggling-ranbaxy-for-3-2-billion-as-daiichi-sankyo-retreats-idUSBREA3600L20140407 (accessed June 9, 2018).

398 *Daiichi Sankyo emerged victorious from its arbitration in Singapore:* Prabha Raghavan, "Delhi High Court Upholds Daiichi's Rs 3,500-Crore Arbitral Award against Singh Brothers," *Economic Times,* February 2, 2018, https://economictimes.indiatimes.com/industry/healthcare/biotech/pharmaceuticals/delhi-high-court-upholds-daiichis-rs-3500-crore-arbitral-award-against-singh-brothers/articleshow/62723186.cms (accessed June 9, 2018).

399 *new allegations that they'd siphoned $78 million:* Ari Altstedter, George

Smith Alexander, and P. R. Sanjai, "Indian Tycoons Took $78 Million Out of Hospital Firm Fortis," *Bloomberg*, February 9, 2018.

399 *They also faced a similar allegation:* Ari Altstedter, "Billionaire Singh Brothers Accused by New York Investor of Siphoning Cash," *Bloomberg*, January 28, 2018, https://www.bloomberg.com/news/articles/2018-01-28/billionaire-singh-brothers-accused-in-lawsuit-of-siphoning-money (accessed June 9, 2018).

399 *"unsustainable debt trap":* Arun Kumar, "Fortis Founder Shivinder Singh Drags Elder Brother Malvinder Singh to NCLT," *Economic Times*, September 5, 2018.

399 *that her sons enter mediation:* ET Bureau, "Malvinder Singh and Shivinder Singh Ready for Mediation," *Economic Times*, September 15, 2018.

400 *faced a new and serious allegation:* Generic Pharmaceuticals Pricing Antitrust Litigation, Plaintiff States' (Proposed) Consolidated Amended Complaint (Eastern District of Pennsylvania, October 21, 2017).

400 *and possible cross-contamination between drugs:* FDA, Establishment Inspection Report, Mylan Laboratories Ltd., Morgantown, West Virginia, March 19–April 12, 2018.

401 *a response to the warning letter:* Mylan N.V., "Mylan Statement in Response to FDA Warning Letter Relating to Morgantown Plant," news release, November 20, 2018.

402 *an extensive "multidisciplinary investigation":* Dr. John Peters, Director, Division of Clinical Review, Office of Generic Drugs, FDA, to Dr. Harry Lever, Medical Director, Hypertrophic Cardiomyopathy Clinic, March 31, 2014.

402 *recalled their metoprolol succinate:* Zeba Siddiqui, "Dr Reddy's Recalls over 13,000 Bottles of Hypertension Drug—FDA," *Reuters*, July 19, 2014.

403 *had approved only a handful of visas:* Adam Minter, "Is China Blocking FDA Inspectors?" *Bloomberg*, February 28, 2014.

404 *the man "began running and fled the laboratory premises":* FDA, "Form 483: Inspectional Observations," Zhejiang Hisun Pharmaceutical Co., Taizhou, China, March 2–7, 2015, 7.

404 *Fifteen minutes later, a manager returned:* FDA, "Warning Letter" to Zhejiang Hisun Pharmaceutical Co., Taizhou, China, December 31, 2015.

405 *ended its partnership with Zhejiang Hisun:* On November 10, 2017, roughly two and a half years after Baker's inspection, Pfizer sold its 49 percent equity share in Hisun-Pfizer Pharmaceuticals, though it retained rights to manufacture, sell, and distribute the current and pipeline products that had resulted from the joint venture. Pfizer, "Pfizer

Sells Its 49% Equity Share in Hisun-Pfizer Pharmaceuticals," news release, November 10, 2017.

405 *Baker went to Dalian in the Liaodong Peninsula:* FDA, "Form 483: Inspectional Observations," Pfizer Pharmaceuticals Ltd., Dalian, China, April 13–17, 2015.

405 *There, too, he found:* In response to the events at Dalian, a Pfizer spokesperson said, "Pfizer has responded and addressed the issues raised during the pre-approval inspection of our manufacturing site at Dalian. The issues cited in the FDA Form-483 do not indicate any quality or safety concerns and do not have any impact on products currently on the market manufactured at the Dalian site."

405 *where fraud was endemic:* In 2016, an investigation by China's own State Food and Drug Administration (SFDA) found that 80 percent of clinical trial data submitted by Chinese companies to regulators to gain approval for new drugs was fabricated. Fiona Macdonald, "80% of Data in Chinese Clinical Trials Have Been Fabricated," *Science Alert,* October 1, 2016, https://www.sciencealert.com/80-of-the-data-in-chinese-clinical-trial-is-fabricated (accessed September 30, 2018).

407 *That, in turn, jeopardized FDA funding:* In 1992, the Prescription Drug User Fee Act (PDUFA) permitted the FDA to collect fees from companies seeking approval for new drugs. Subsequent laws expanded the so-called "user fee" system to encompass the generic drug and medical device industries. Collectively, user fees now comprise about 40 percent of the FDA's total budget. The system is not without critics, with some arguing that the revenue generated from user fees might limit the FDA's ability to make impartial regulatory decisions. An FDA spokesperson said prescription drug user fees have helped the FDA to provide "timely review" of drug applications. J. Carroll, "PDUFA Faces Rough Reauthorization," *Biotechnology Healthcare* (July 2007); see also Tara O'Neill Hayes and Anna Catalanotto, "Primer: FDA User Fees," *American Action Forum,* August 22, 2017.

407 *"shade the facts . . . increasingly uncomfortable":* Michael Mezher, "FDA Official Highlights Foreign Supply Chain Challenges," *Regulatory Affairs Professionals Society,* May 5, 2017, https://www.raps.org/regulatory-focus™/news-articles/2017/5/fda-official-highlights-foreign-supply-chain-challenges (accessed June 9, 2018).

407 *to make the final classifications less severe:* In Zhejiang province, China, Bangli Medical Products held an FDA investigator hostage in its conference room, trapping her there for hours and demanding that she destroy the photographs she'd taken. It seemed clear to the FDA's staff in China that the company had refused a for-cause inspection and its drugs needed to be blocked. An FDA supervisor wrote back to officials in Maryland:

"Needless to say, they first refused the inspections and refused to recognize our investigator's authority to inspect the premises. We need to immediately put this firm on import alert." But an official at FDA headquarters quickly sounded a note of caution about "declaring that we have 'authority' in the foreign arena." Another official weighed in, stating that it didn't appear the plant manager who'd imprisoned the FDA's investigator "was making a specified refusal." The incident is documented in: Kelli Giannattasio, "Re: For Cause Inspection of Bangli Medical Products," email to Susan F. Laska and Sherry Bous, July 27, 2016.

407 *made by bureaucrats in Maryland:* In a written statement, an FDA spokesperson explained, "The FDA can and does change assessments of a plant's compliance. After the initial data gathered by the investigator is reviewed by both the Office of Regulatory Affairs and the Center for Drug Evaluation, additional information can be taken into account. Oftentimes, a firm is not able to provide paperwork at the time of an inspection but can produce documents later on that provide more insight into the matter. Assessments can also change based on how willing a firm is to cooperate and fix issues that are found."

408 *the agency lifted the restriction on about half of the drugs:* E. J. Lane, "U.S. FDA Ingredient Exceptions from Banned Zhejiang Hisun Plant Draw Scrutiny," *FiercePharma,* July 25, 2016, https://www.fiercepharma.com/pharma-asia/u-s-fda-ingredient-exceptions-from-banned-zhejiang-hisun-plant-draw-scrutiny (accessed June 9, 2018).

409 *but the agency downgraded that to VAI:* Tamara Felton Clark, Branch Chief, Global Compliance Branch 4, "Reclassification of Surveillance Inspection: VAI as Inspection Classification," CMS File—Work Activity 161861, Zheijiang Huahai Pharmaceutical Co., Ltd. (FEI 3003885745), September 7, 2017.

410 *the FDA would notify India's companies in advance:* Mathew Thomas, Dean Rugnetta, Solomon Yimam, Daniel Roberts, and Shiva Prasad, "Office of International Programs, U.S. FDA India Office (INO) Meeting Minutes," proceedings of FDA, IPA, CDSCO meeting, India International Centre, New Delhi, November 3, 2016. Internally, FDA officials had abruptly stopped the India pilot program, and the short- and no-notice inspections in July 2015 but notified Indian companies sixteen months later. In 2018, when asked by a journalist why the FDA stopped the program, an agency spokesperson responded in a written statement, "After evaluation of the pilot a decision was made to discontinue the pilot."

INDEX